The Evolution of Human Hunting

The Evolution of Human Hunting

Edited by

Matthew H. Nitecki

and

Doris V. Nitecki

Field Museum of Natural History
Chicago, Illinois

PLENUM PRESS • NEW YORK AND LONDON

Library of Congress Cataloging in Publication Data

Spring Systematics Symposium (9th: 1986: Chicago, Ill.)
The evolution of human hunting / edited by Matthew H. Nitecki and Doris V. Nitecki.
p. cm.
"Proceedings of the Field Museum of Natural History Ninth Annual Spring Systematics Symposium on the Evolution of Human Hunting, held May 10, 1986, in Chicago, Illinois."
Bibliography: p.
Includes index.
ISBN 0-306-42821-0
1. Hunting. Primitive—Congresses. I. Nitecki, Matthew H. II. Nitecki, Doris V. III. Title.
GT5820.S97 1986
639′.1′0901—dc19 87-34302
CIP

Proceedings of the Field Museum of Natural History Ninth Annual Spring Systematics Symposium on the Evolution of Human Hunting, held May 10, 1986, in Chicago, Illinois

A Division of Plenum Publishing Corporation
233 Spring Street, New York, N.Y. 10013

Printed in the United States of America

Preface

The successful early adaptations of man involve a complex interplay of biological and cultural factors. There is a rapidly growing number of paleontologists and paleoanthropologists who are concerned with hominid foraging and the evolution of hunting. New techniques of paleoanthropology and taphonomy, and new information on human remains are added to the traditional approaches to the study of past human hunting and other foraging behavior. There is also a resurgence of interest in the early peopling of the New World.

The present book is the result of the Ninth Annual Spring Systematics Symposium, on the Evolution of Human Hunting, held on May 10, 1986, in the Field Museum of Natural History in Chicago.

We are grateful to the NSF (grant no. BNS 8519960) for partial financial support in arranging the symposium. In preparation of this volume we have received assistance from many people, particularly the reviewers of individual chapters; it is impossible to name them all. We must however single out Drs. Richard G. Klein and Glen H. Cole for their encouragement at various stages of preparation of the symposium and this volume, and for being a help to the anthropological knowledge. Zbigniew Jastrzebski assisted with the figures and Paul K. Johnson diligently typed the camera-ready copy, and patiently coordinated the endless book-making chores.

Matthew H. Nitecki

Doris V. Nitecki

Contents

The Idea of Human Hunting

Matthew H. Nitecki

Department of Geology
Field Museum of Natural History
Chicago, Illinois

Paleontology in the broadest sense is the study of the history of life, and paleoanthropology is the part of this history that deals with man. Like all historical scientists, paleontologists and paleoanthropologists assume that this history consists of individual events joined together into the Great Chain of Events leading to the present. They believe that history is thus directional, continuous and composed of identifiable minor events eventually transformable into the major events such as mass extinctions, biotic crises, origins of hunting and origins of agriculture. These significant events are interpreted and become powerful concepts in the development of scientific thought.

The essential feature of all historical sciences is, of course, time. Time is a measure of events that can be arranged in sequence and that can be interpreted segmentally or as discrete phenomena. The frequency of identical events is measured by the relative distances between these events on the time scale. The faster the occurrence of events, the shorter is the interval of time, and fewer events imply a longer time span. Thus the continuous or discontinuous appearance of events is decoded. The sequence of events is read from the stratigraphic superposition of fossil data. Human hunting is one such sequence of human behavior that we desire to place in a time series. However, we know little about this sequence. Did it start suddenly at an identifiable time or are there no measurable "hunting events" on that time scale? Once hunting started did it soon become universal and involve mass hunts, and hence may not be easily recognized in the fossil record? Or was the development of hunting gradual and thus with obscure beginnings?

Anthropological events are grouped into the relatively short, agriculturo-technological period, and into the immeasurably (in the anthropological sense of lack of events) long and excavatable part of the pre- or nonagricultural period. It is in this latter, longer era that hunter-gatherer, forager and scavenger periods began.

It is perhaps the inevitable nature of the fossil record that the oldest events are rare and farther apart, while the most recent are abundant and therefore close together. This always gives us the impression that events accelerate and are characterized by geometric progression, or that they grow exponentially. All such exponential curves suggest the sudden explosion of events. What is also always associated with events that occur geometrically in time is that they not only increase geometrically in numbers, but also in rate. Thus we see the evidence of late Pleistocene hunting almost as an explosion in the pattern of human behavior.

We are here concerned with hunting as procurement of animals only. We are not concerned whether there may have been other functions to hunting, and we are assuming that hunting was a reflection of the need for securing meat, skin, sinew, bone, and marrow.

When hunting started depends on what is really meant by hunting. If all people prior to the domestication of plants and animals are considered hunters (and gatherers) then, of course, man has been a hunter since the beginning. If, on the other hand, hunting is defined as the sole means of securing proteins, then the claim that the pre-agricultural humans were hunters must become a matter of inference. The fossil evidence is poor. If man were a solitary hunter, little evidence of his activity would have been preserved. The monkey shot from the tree, or the arrow with which it was shot, are unlikely subjects for fossilization, or for taphonomic analysis of hunting. If man gathered shellfish, eggs, or creepers and crawlers, he was not a hunter. It is only the highly organized, communal enterprise of stalking with specialized tools, and the concomitant accumulation of large amounts of bone that may provide good evidence for the evolution of hunting.

Contributions by Klein and by Binford ask for sound proof that *Homo erectus*, 400,000 B.P., was a hunter. Both authors are concerned with the archeological data, and show that there is no solid evidence to support the existence of hunting at the important site at Torralba, and that therefore the hunting practices of *H. erectus,* if demonstrable at all, were very feeble. How many such sites were in reality water laid and not accumulated by man? Would not entire carcasses float down and accumulate?

Richard Klein is an eminent anthropologist and biologist at the University of Chicago, and an author of numerous books and articles on early man and how he gained subsistence. Here he argues that human ancestors more than four million years ago were probably not carnivorous, but that the much later historic hunting-gathering societies depended heavily on meat procurement. Klein questions what was in between these still unresolved events. It is possible that the improvement in technology of meat procurement was *gradual* and progressively more efficient with the evolution of human behavior. It is also possible that between 100,000 and 40,000 B.P. big game hunting occurred *suddenly* with the emergence of modern man. It is also possible that *scavenging* lasted a long time; but likewise it is also possible that hunting has a long history. However, there is good evidence of human involvement with bone accumulations in the late Pleistocene, and at about 40,000 B.P. there is a major advancement in human hunting.

Binford uses statistical methods to show that the romantic view, that pre-*H. sapiens sapiens* were mighty hunters, is an exaggeration. Lewis Binford, a distinguished professor at the University of New Mexico represents a subfield of anthropology, which he, more than most specialists, has done his best, with marked success, to change. In the past, it had been generally accepted that man before *H. erectus* was a marginal scavenger. Binford, as others in this volume, now argues that this is also true for *H. erectus*. Hominids were not successful hunters as late as 400,000 years ago.

To us, nonanthropologists, it seems unlikely that early man was herbivorous. It is possible, however, that the ancestors of man were herbivorous, insectivorous, or omnivorous. For the herbivore to become a meat eater the biological innovation of meat-digesting enzymes is required as is the development of an appropriate dentition and the behavioral adjustment to hunting or scavenging. The animal source of meat must be available, and there must also be selection, to prefer meat over vegetable nutrition. No evidence exists to suggest that such major changes occurred within *H. sapiens* or *H. erectus*; therefore, they must have already been present in these lines. In any case some closely related hominids (i.e., chimps) occasionally eat meat, as do certain other primates. We can assume that the ancestors of *H. sapiens* and *H. erectus* had the biological potential to have been meat eaters.

It is also reasonable that hunting, or at least the organized big hunts, belong only to modern man. This does not seem surprising since we define hunting in modern terms, and hence it can only be a property of *H. sapiens sapiens*. We very strongly believe, however, that the difference between *H. sapiens and all other animals* is so fundamental that the present classification of man based only on morphology makes little sense. Modern man, and modern man alone, has evolved the ability to create not only new environments, but also a new universe made up of what Karl Popper calls objective contents of man's thoughts. Animals may build nests and webs, but only man can produce and record thoughts as independent objects, and pass them on to future generations. Therefore it may be futile to look to the human anatomical characters for the *explanata* of human behavior. The advent of high technology, industrial revolution, agriculture or the science of anthropology *does not require* any morphological changes in our anatomies. The same must apply to the origin of hunting. It is important to notice that the high speed of transmission of new patterns is also a sole human character. Once discovered, the advantageous behavioral pattern also proceeds with geometrical progression. Thus human behavior itself may produce punctuated events against gradualism. Those are unique human traits!

Nevertheless, we do not, even for a moment, deny the existence of anatomical changes in man, or their significance for the evolution of human behavior. The enlargement of brain size, the changes in jaw, nose, pelvis and the reduction of body hair are all new features. But hunting may have also resulted only from the cultural activities of man, from the objective development of his thoughts; the invention of his tools, shelter, fire and from changes in acquiring, processing and consuming his meat.

The human anatomical adaptations of large brain, bipedal locomotion and manipulative faculties have been observed to distinguish man from nonhuman primates. Trinkaus argues that these adaptations are results of selective pressures due to the early human predatory way of life. Erik Trinkaus is a professor of anthropology at the University of New Mexico, and for many years has been a central figure in the physical anthropological studies of modern man. Through the detailed study of functional anatomy, Trinkaus now shows that australopithecines had terrestrial and arboreal habits, and hence must have used both "food resources," but they did very little hunting or none at all. He also demonstrates that *H. habilis, H. erectus* and archaic *H. sapiens* (including the neanderthals) were probably also not efficient predators; only the anatomically modern humans have hands that can manipulate with precision, and only they have hafted tools that could have been used in efficient hunting. Trinkaus's morphological studies show that the anatomical basis for predation appears with the genus *Homo*, but that efficient hunting is a characteristic of modern man.

We assume that man was first a solitary hunter and gatherer, and that human evolution progressed toward the later-developed high degree of social cohesion and cooperation that was a necessary precondition of massive purposeful hunting. The lack of evidence of very early hunting may mean that early humans did not hunt, or it may indicate a lone hunter, who, however mighty, could have not accumulated massive bone beds. The assumption is also made that mass hunting was not instinctive (the way some carnivores hunt instinctively), but resulted from the higher level of mental development characteristic of man.

Straus, Frison, Fisher and Todd discuss these efficient, organized, specialized and mass modes of hunting of *H. sapiens sapiens*. Straus examines Western Europe, and Frison, Fisher and Todd the New World; Morlan shows the possible path of the Asian hunter to North America. There is excellent evidence for Late Upper Pleistocene (from 35,000 to 30,000 B.P.) hunting particularly in the periglacial regions. Magdalenian people of Western Europe, and the much younger Paleoindians of the New World were undoubtedly hunters of big game -- the bison, reindeer and the proboscideans. There are very good hunting tools preserved, butchered bones are relatively plentiful, and there is cave art to prove the hunting behavior. However, these hunters are anatomically, and we suspect in all other ways, already modern man.

Lawrence Straus describes the "full-fledged" Solutrean and Magdalenian mass hunters of Western Europe at about the time of the last pleniglacial maximum. Straus is on the faculty of the University of New Mexico, and is an anthropologist with wide knowledge of the Pleistocene hunters of the Old World. From studies of the newest excavations he describes the ecology, flora and fauna and their ungulate "biomass," and the human diet that in addition to mammals consisted of molluscs, birds, fish and plant foods. The main objects of the hunt were reindeer, red deer, bear, ibex and bison. Straus studies region by region, and analyzes which animals dominated Germany, France and Spain. Straus's efficient big game hunter specialized with a great array of weapons and hunting techniques. He had a battue, a drive line and a surround; he used a sophisticated understanding of topography and of animal behavior. These imply a high level of cultural sophistication, and find supporting evidence in the art work of European caves.

Surely, we can assume that such cultural and transmittable phenomena as tools and communication brought changes in man's fate and in his better manipulation of the environment. This is a distinct arrow of progress that we see. Once these cultural elements developed, they conferred advantages on those who possessed them, became operative by natural selection and increased their chances of further transmittal and improvement. The same must have also been true for cooperative behavior above the family level, such as mass hunting.

George Frison's contribution deals with the technique of hunting. Frison is an anthropologist at the University of Wyoming, and has for many years been concerned with present day and Pleistocene hunting. The first true North American hunters were the Clovis cultural complex, between 12,000 and 11,000 years B.P. In many ways, Frison is a comparative behaviorist. Himself an experienced elephant hunter, he is able to draw comparisons between the behavior of the Paleoindian hunter and that of the modern African hunter. He presents a fine model of the actual practice of hunting. He understands the physiography of the land and depicts well the topography of the old hunts. He even finds evidence of religious structures. Frison is concerned with mammoth, mountain sheep and pronghorn, but his main interest lies with the communal hunts of bison on the High Plains. He sees the Paleoindian hunter as basically a modern hunter. Paul Martin's claim that the faunal extinction at the end of the Pleistocene was a dramatic hunting blitzkrieg is well known. Martin may, or may not, be right that the disappearance of large mammals is related to the possible prehistoric overkill. Frison argues against Martin's overkill hypothesis, but the arguments presented in this book do not actually disprove the proposition that human hunting is the reason for the extinction of the mammoth, mastodont, sabre-toothed cat, and glyptodonts.

Lawrence Todd in his very detailed taphonomic analysis of the New World kill-butchery sites demonstrates that the Paleoindian hunting pattern is much different from that previously held. Todd is on the faculty of the University of Denver, and as can be seen from his contribution to this volume is an accomplished student of Paleoindian adaptations. He demonstrates that Paleoindians were responsible for relatively few of the extensive bone beds, claimed for a long time to have been man made, and that these bone beds do not show any intensive processing of meat carcasses. He concludes that the Paleoindian goal was not the storage of large quantities of food, but rather procurement and processing of meat for a relatively short time.

We suppose that if the rate of accumulation of these bare bones could be measured, and if these bones were indeed the result of hunting, then in theory the size of human groups necessary for such accumulations could be calculated. At the present time our knowledge is insufficient to answer such questions.

Richard Morlan presents a very detailed analysis and interpretation of the earliest human occupations of Beringia and the associated evidence of hunting and its spread in that region. Morlan is on the curatorial staff of the National Museum of Man and is best known for his work on the archeology of Beringia. He points out that while there may be disagreement on the exact date of man's entry to North America, there is a consensus that it was relatively late and that the route from North-Eastern Asia was through the Beringian land bridge. This route was not only the path of man, but also of many mammals that crossed the Pleistocene low land bridge in both directions. Mammoth and bison came from Asia, and it is of interest to speculate whether man did not follow the migrating herds. It seems indisputable that humans were present in western Beringia at about 35,000 B.P., and there also seems to be some evidence for human association with bone accumulations. This is not surprising in view of the existence of the even older associations in Western Europe and the Ukraine. It appears that man entered eastern Beringia not later than 12,000 B.P. Thus it is of no surprise that the Paleoindian hunters appear to be descended from Siberian stock which brought the Old World culture and hunting pattern with them. Morlan analyzes in detail the published Soviet evidence of Siberian hunting and the Paleoindian hunting technology derived from Asia.

Daniel Fisher, in a most remarkable analytical study, discusses how attributes such as the sex, age, and season of death of butchered animals may be used to evaluate competing interpretations of the means of carcass procurement. Fisher, a professor at the University of Michigan, is a paleobiologist who has lately concentrated much of his attention on the taphonomy of the Late Pleistocene proboscideans and subsistence activities of the Paleoindians. He is not dealing with the mass hunts, but with evidence of butchery and with distinguishing hunting from scavenging, using data from Michigan late-Pleistocene sites dating from about 12,000 to 10,000 B.P. A particularly interesting facet of his work is the determination (from examination of the incremental lamination of tusks and molar teeth) of patterns of seasonal mortality in mastodonts. By comparing the seasons of death (and other attributes) of butchered and nonbutchered mastodonts, Fisher is able to argue that the butchered animals were hunted rather than scavenged. It will be interesting to extend his studies back in time to gain further evidence on the date of entry of man into North America, and on whether or not hunting was indeed responsible for the late Pleistocene mass extinctions.

Anna Behrensmeyer examines the fossil bone assemblages and the taphonomic evidence for carnivory. Behrensmeyer, a paleoecologist on the curatorial staff of the National Museum of Natural History, concentrates on the general problems of vertebrate taphonomy and paleoecology. She has extensive field experience in Africa, particularly in Amboseli Park. She points out that it is generally difficult to differentiate hunting from scavenging, although there are numerous and detailed arguments used in considering dense concentrations of bones: marks on bones; patterns of selectivity and frequency of various parts of skeletons; the taxonomic selectivity of bone accumulations, or the particular sizes of carcasses; age distributions among carcasses; and seasonality of death. Behrensmeyer concludes that there are no unequivocal methods of distinguishing hunting from scavenging, and that in cases where cultural evidence is absent, anthropologists should work on the problems of the evolution of meat procurement and carcass utilization, rather than on the evolution of hunting itself. In this manner, while Fisher suggests ways to recognize hunting, Behrensmeyer asks anthropologists to be cautious and to use taphonometric techniques in interpreting the archeological sites.

Thus our volume demonstrates that there is still much interest in the problems associated with the evolution of human hunting, and that these problems are far from solved. Nevertheless, by doubting many previously accepted assumptions, the book further questions the relationship of hunting and early human evolution. It also introduces paleontological arguments from taphonomy which strongly criticize some archeological techniques of interpreting bone assemblages.

Reconstructing How Early People Exploited Animals: Problems and Prospects

Richard G. Klein

Department of Anthropology
University of Chicago, Chicago, Illinois

If we can judge by the behavior of our closest living relatives among the apes, our remotest ancestors, who evolved in Africa more than four million years ago, probably ate little or no meat. In contrast, meat was a significant dietary staple in most historic hunter-gatherer societies. It has generally been assumed that meat-eating became progressively more important as people evolved, but it is also possible that the shift occurred in relatively short bursts coinciding with major evolutionary events, such as the emergence of modern humans sometime between 100,000 and 40,000 years ago. It is also possible that there was a stage, perhaps a very long one, in which meat was acquired mainly by scavenging, or alternatively that hunting was important even early on, perhaps becoming more important and more successful as time passed.

Determining how early people obtained meat is not easy, because the available archeological data, particularly animal bones from ancient sites, are often ambiguous in their implications. Often it is even difficult to determine whether people were the principal bone collectors at a site. This is especially true for early and middle Pleistocene sites (before 130,000 B. P.), including such well-known ones as Elandsfontein in southern Africa (Singer and Wymer 1968; Klein 1978), Olduvai Gorge, Olorgesailie, Koobi Fora, Melka Kuntouré, and others in East Africa (Bunn 1982; Bunn et al. 1980; Chavaillon 1982;

Geraads 1985; Isaac 1977, 1984; Leakey 1971; Potts 1984), Ternifine in North Africa (Balout et al. 1967; Geraads 1981), 'Ubeidiya in southwest Asia (Bar-Yosef 1975, 1980), Torralba/Ambrona, Torre in Pietra, Terra Amata, Hoxne, Bilzingsleben, Vértesszöllös, and other sites in Europe (Binford 1985; Freeman and Butzer 1966; Howell 1966; de Lumley 1975; Malatesta et al. 1978; Mania 1976; Roe 1981; Stuart 1982; Vértes 1975, Villa 1983), and Zhoukoudian in East Asia (Jia 1975; Wu and Lin 1983; Liu 1985; Binford and Ho 1985). Excepting Zhoukoudian and some very late mid-Pleistocene sites in Europe, the crucial early and middle Pleistocene sites are mainly spring-, stream-, or lake-side localities. In these sites, it is difficult to separate bones that may represent natural deaths from ones reflecting carnivore or human kills, perhaps followed by carnivore or human scavenging.

Interpreting the animal bones from late Pleistocene and Holocene sites (after 130,000 B. P.) is often much less problematic, since the sites include numerous caves and some open-air localities with good evidence that people were the principal bone collectors. Equally important, the late Pleistocene and Holocene sites include ones occupied under similar environmental conditions and with similar postdepositional histories, so that attention may be focused on collector behavior as the principal cause of differences among bone assemblages. From such sites, it is possible to determine if human ability to obtain animal resources evolved through time.

My purpose here is first to illustrate the interpretive problems of early and middle Pleistocene sites, from the Torralba and Ambrona sites, and then to demonstrate the interpretive potential of late Pleistocene and Holocene sites, from a set of caves at the southern tip of Africa.

HISTORY OF INVESTIGATION AT TORRALBA AND AMBRONA

The Torralba and Ambrona sites are located near villages by the same names approximately 150 km northeast of Madrid (Fig. 1). Bones were first found at Torralba in 1888 during railway construction, and excavations by the Marqués de Cerralbo in 1909-1911 demonstrated the stratigraphic association of the bones with handaxes and other Acheulean artifacts. Cerralbo also

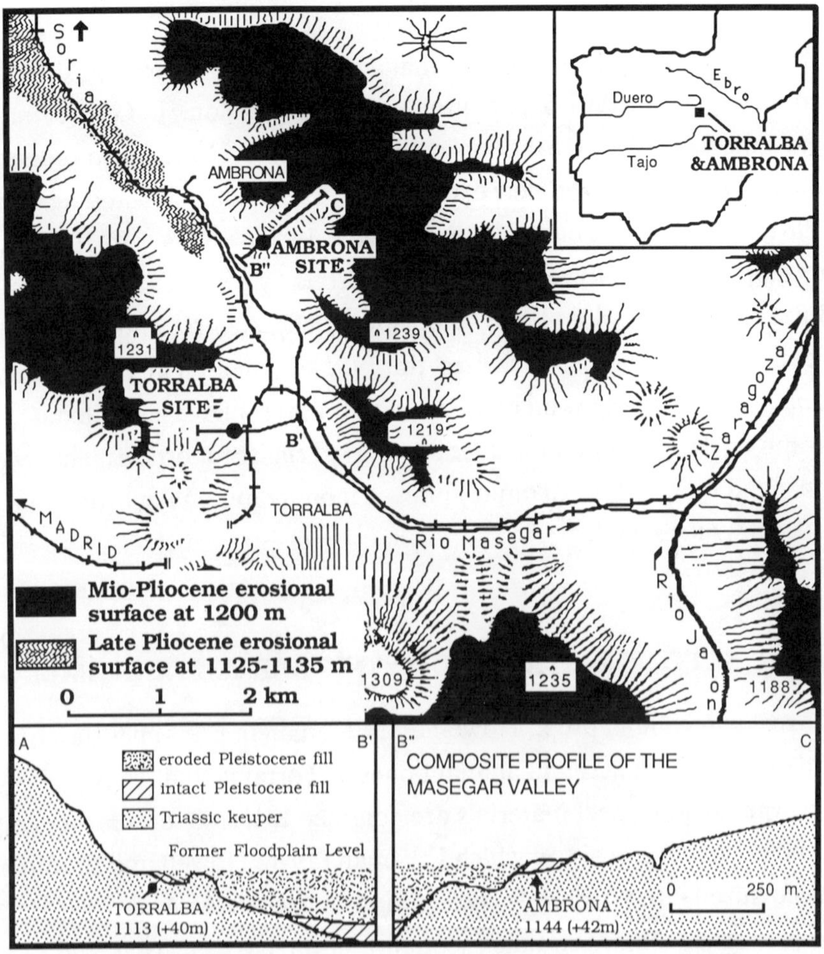

Figure 1. The approximate locations of the Torralba and Ambrona sites (modified after Butzer 1965).

discovered and tested the very similar Ambrona site, approximately 2 km away, but he never published full descriptions of his findings at either Torralba or Ambrona.

In 1960, F. C. Howell established that a portion of the Torralba site survived intact, and in 1961-63, he directed a re-excavation. L. G. Freeman supervised the Torralba excavations in 1962 and 1963. Howell also rediscovered Ambrona, where he directed new excavations in 1962 and 1963 and again in 1980, 1981, and 1984. Freeman co-directed the 1980-81 work. In 1973, E. Aguirre supplemented Howell's work at Ambrona by a small salvage excavation.

Butzer (1965), Howell (1966), Biberson (1968), Aguirre and Fuentes (1969), and Freeman (1975) have summarized the 1961-63 research at Torralba and Ambrona. In addition, Howell and Freeman (in press) have compiled a monograph on Torralba that includes several specialist reports, especially on the fauna, and Howell (pers. comm.) plans a monograph on Ambrona that will cover all years of research.

GEOLOGIC SETTING AND ANTIQUITY OF TORRALBA AND AMBRONA

Butzer (1965 and in prep.; Howell and Freeman in press) has described the geologic setting of artifacts and animal bones at Torralba and Ambrona. At Torralba, the archeological materials are confined to a suite of interbedded fluvial and colluvial deposits in which Freeman (1978; Howell and Freeman in press) recognized 9 to 10 discrete "occupation surfaces." At Ambrona, artifacts and bones occur in a more complex sequence of fluvial, lacustrine, and colluvial sediments. No discrete occupation surfaces have been identified at Ambrona, but stratigraphically, it is possible to divide the materials between a "Lower Occupation," probably broadly coeval with the occupation surfaces at Torralba, and an "Upper Occupation" that is somewhat younger.

Unfortunately, neither Torralba nor Ambrona contain materials suitable for radiometric dating, and the geologic age of the sites is not precisely known. Judging from the fauna and the geologic setting, the bones and artifacts are probably 500,000 ± 200,000 years old, that is, broadly middle Pleistocene.

SPECIES REPRESENTATION AT TORRALBA AND AMBRONA

Figure 2 shows the abundance of each mammalian species in the Howell/Freeman excavations at Torralba and Ambrona, based on the percentage of identifiable bones (NISP) assigned to each species. The counts follow identifications I made with help from Kathryn Cruz-Uribe. The placement of Torralba below Ambrona is heuristic and does not imply greater antiquity. The figure does not include birds, represented at both sites by a handful of bones (Howell and Freeman in press), and tortoises, represented by a small number of carapace and plastron fragments.

The numbers on which Figure 2 is based are only approximate, because some bones from the 1961-63 excavations could not be located during my study (1980, 1981, and 1984) and others had lost their provenience designations. Most important perhaps, I was unable to locate the very small number of macaque bones from the "Lower Occupation" at Ambrona.

The figure documents the well-known prominence of elephant and the rarity of most other species, especially carnivores and other non-ungulates. Horse, however, is also very abundant and in fact dominates the "Upper Occupation" at Ambrona. Although the relative abundance of species in Figure 2 need not closely reflect their live abundance near the sites, the prominence of grazers (straight-tusked elephant, horse, and aurochs) implies that grasses were common locally. Only the fallow deer -- historically a mixed browser/ grazer with a preference for forest-edge habitats or for relatively dense woodland with clearings -- indicates the existence of forests or thickets. Overall, the fauna implies relatively open vegetation, in keeping with the pollen from both sites (Freeman 1975) which suggests open woodland or parkland. Thickets and forests were probably restricted to the edges of watercourses or to the margins of floodplain ponds and marshes. The significant increase in horse relative to elephant in the "Upper Occupation" at Ambrona probably reflects environmental change, as suggested by the sediments and the more sparse pollen record. Whereas the "Lower Occupation" occurred under relatively warm, moist conditions, the "Upper" occurred under cold to very cold, semiarid, and more strongly seasonal ones (Butzer pers. comm.).

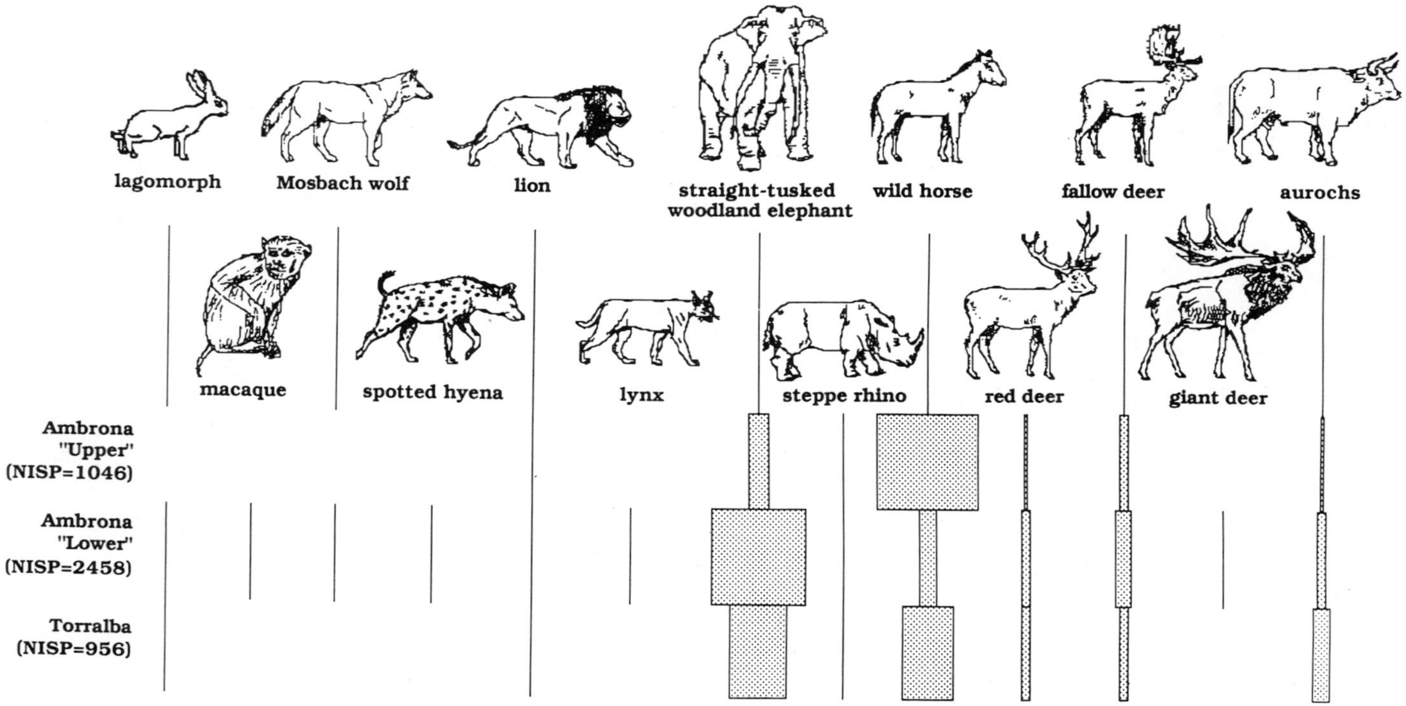

Figure 2. Taxonomic abundance at Torralba and Ambrona, based on the number of identifiable specimens (NISP) assigned to each species. Note that Torralba is placed below Ambrona for display purposes only. The fossiliferous deposits at Torralba are roughly the same age as those of Ambrona "Lower".

THE SOURCE OF THE BONES AT TORRALBA AND AMBRONA

Until very recently, it was assumed that the presence of artifacts at a site indicated that people had killed and butchered the animals represented by bones. Thus, the Torralba and Ambrona sites were often cited as proof that middle Pleistocene people were highly successful big-game hunters. It was only in the 1970's, when specialist studies of animal bones from archeological sites burgeoned and paleoanthropology began to draw heavily on taphonomy and paleoecology, that the assumed functional link between artifacts and animal bones at sites like Torralba and Ambrona was seriously questioned. Binford (1981) publicized the uncertainties widely. Today, we realize that nothing can be assumed about the relationship between artifacts and bones at any archeological site. Instead we must investigate the relationship by detailed analyses of several classes of evidence, such those that I illustrate here with data from Torralba and Ambrona.

Species Abundance

In general, the largest available ungulates, such as elephants and rhinoceroses, are rare or absent even in the largest faunal samples from stone age cave sites where contextual evidence indicates that people were the main bone accumulators. Carnivore bones, particularly from large species like lion and hyenas, are also rare at such sites. Lion, for example, is completely absent in the very large southern African cave samples discussed in the second part of this paper, and hyenas are extraordinarily rare. In contrast, at open-air sites very large ungulates are often common, sometimes dominating the fauna as at Torralba and Ambrona, and large carnivores, particularly lion, hyenas or both, are usually represented.

Limited human ability to transport heavy items may partly explain why large ungulate bones are rare or absent in stone age caves, but this could not explain the rarity or absence of large carnivore bones. In addition, the contrast in large ungulate abundance between open-air sites and caves is often too great to be explained simply by transportation problems. More likely, the occupants of the caves simply did not have frequent access to useable carcasses of very large ungulates. It follows that many ungulates and also carnivores at open-air sites like Torralba and Ambrona may have died without human notice.

Context and Associations

In most cases, it is reasonable to assume that ungulate bones found in caves were carried there by people or carnivores, since free-ranging ungulates rarely enter caves. In contrast, ungulate bones found on the margins of ancient streams or ponds at sites like Torralba and Ambrona could derive from natural deaths. At the same time, in either caves or open-air sites, items associated with bones can obviously provide clues as to how the bones accumulated. Numerous carnivore coprolites (fossilized feces) obviously suggest that carnivores killed or scavenged some of the animals. In contrast, numerous artifacts certainly imply a human role, which is indicated even more strongly when the artifacts and bones are tightly concentrated or are accompanied by structural ruins or fireplaces. Dense clustering of artifacts and bones together has been used to argue forcefully that people accumulated bones at some early Pleistocene sites in East Africa (Potts 1983, 1984; Isaac 1984).

It is difficult to draw firm conclusions from the context and associations of the bones at Ambrona and Torralba. Both artifacts and hyena coprolites are present. Although artifacts are more numerous than coprolites, many coprolites may not have survived the intensive leaching that occurred at both sites and others may not have been recognized in the excavation. In addition, both bones and artifacts tend to be dispersed (Howell and Freeman in press and in prep.), and artifacts are extremely rare in parts of the Ambrona site. Structural ruins and fireplaces are completely absent, and charcoal scattered through the deposits could have resulted from brush fires. In sum, the artifacts and coprolites at Torralba and Ambrona indicate that both people and hyenas were present, but do not prove that either people or hyenas were directly responsible for the bones at either site. From context and associations alone, we could even imagine that the artifacts, coprolites, and bones became associated by chance near ancient water sources that repeatedly attracted people and animals.

Condition of the Bones

Shipman (1981), Brain (1981), and others have discussed how damage from human butchering and food preparation, rodent gnawing, carnivore chewing, trampling, fluvial transport, and so forth can elucidate the origins of a bone assemblage. From bone modification, a careful and experienced analyst can demonstrate human butchering even in the absence of stone artifacts (Fisher 1984). East African animal bones damaged by stone tools, together with stone tools polished from meat cutting (Keeley and Toth 1981), constitute the oldest secure evidence for human use of animal resources, roughly 1.5 to 2 million years ago (Bunn 1981, 1983; Potts and Shipman 1981). With respect to early Pleistocene animal bones from Olduvai Gorge, Shipman (1983, 1986) has argued that the failure of presumed cutmarks to concentrate near joints implies that the people were mainly scavengers who were not engaged in primary disarticulation of carcasses. Occasional scavenging is also implied by presumed cutmarks overlying carnivore tooth marks, though the opposite also occurs.

Unfortunately, like many other fossil assemblages, the Torralba and Ambrona ones are not ideal for a study of surficial marks because intense post-depositional leaching has corroded bone surfaces. Leaching was especially intense in the "Upper Occupation" at Ambrona, resulting in the destruction of nearly all bones except teeth, and even these were often reduced to enamel fragments. Because the composition of the Ambrona "Upper" assemblage was modified so heavily by post-depositional destructive processes, it is not very useful for investigating assemblage origins and will therefore be ignored.

The alterations introduced by leaching and corrosion are compounded by the massive fragmentation that occurred during and after burial at both Torralba and Ambrona. Even most "identifiable" bones are only small portions of whole skeletal elements. Still, it is notable that one-third of the 1779 bones from Torralba and one-sixth of the 4326 bones from Ambrona "Lower" exhibit edge-rounding that Butzer (pers. comm.) suggests occurred during limited fluvial transport on seasonally activated valley margins or during net transport of sandy alluvium that partly embedded the bones. Many bones that are not conspicuously rounded show a distinctive polish or luster and probably would exhibit abrasion or edge rounding under magnification. Morlan (1980) found that virtually all bones from the Old Crow Flats (Yukon, Canada) exhibited signs of rounding, when microscopically examined. Using a hand-held glass on a sample of lustrous Torralba and Ambrona bones, Butzer (pers. comm.) in fact found parallel microstriations from abrasion by sand-sized particles on every one. Similarly, Shipman and Rose (1983) found edge-rounding on nearly every Torralba and Ambrona specimen they examined with a scanning electron microscope.

At both sites, conspicuous rounding is proportionately much more common on "unidentifiable" than on "identifiable" bones, suggesting that abrasion was an important factor in making many bones "unidentifiable." Abrasion perhaps also removed other significant surficial markings, particularly ones made by stone tools or carnivore teeth. Most important, however, the high frequency of rounding indicates that many Torralba and Ambrona bones have been transported. The problem is particularly acute for Torralba, where the high incidence of rounding is correlated with its extreme development, and many bones are in essence pebbles or cobbles. It is less serious at Ambrona, where the frequency of rounding is smaller and where it tends to be restricted to parts of bones. Such bones were embedded and in part imbricated in alluvial lags, and abrasion can best be attributed to other large particles moving over them.

Howell and Freeman (1983) suggest that 36 remarkably similar Ambrona and Torralba tusk tips were artificially shaped by fracture and polishing so that they resemble the stone artifacts called "shouldered points." However, the Ambrona and Torralba pieces closely resemble naturally fractured and polished tusk tips collected by Conybeare and Haynes (1984) in Hwange National Park, Zimbabwe. I have examined some of the fossil and recent specimens, and I think that they may have a similar natural origin.

Excepting abrasion and corrosion, Cruz-Uribe and I found little other damage on the Torralba and Ambrona bones. Carnivore chewing was apparent on only fourteen from Torralba and three from Ambrona "Lower", while possible stone tool cut marks were present on twenty-two Torralba bones and none from Ambrona "Lower". Shipman and Rose (1983) also found few stone tool marks on the Torralba and Ambrona bones, and it is possible that the few marks they or we identified resulted from trampling before burial, from scratching by sharp particles in a stream, or from profile compaction after burial. Andrews and Cook (1985) and Behrensmeyer et al. (1986) have shown that trampling of bones by large animals can produce marks that closely mimic cut marks. This finding complicates the interpretation of presumed cut marks even where they are more abundant, as at Olduvai Gorge.

It is possible of course that abrasion and corrosion removed many cut and chew marks from the at Torralba and Ambrona bones. There were also approximately 200 Ambrona bones, mostly large ones, whose surfaces, at the time of our examination, were wholly or partly masked by plaster or plastic casts. From field observations, Freeman (pers. comm.) reports that some of these may bear the most obvious stone tool marks in the entire assemblage. Nonetheless, by itself, surficial bone damage could clearly not be used to argue that either people or carnivores played an important role in forming the Torralba and Ambrona assemblages.

Skeletal Part Representation

Figures 3 to 5 illustrate skeletal part representation for the most abundant species in the bone assemblages from Torralba and Ambrona "Lower," based on the minimum numbers of individuals (MNI) necessary to account for each skeletal part. The numbers are small, and chance could account for the disproportionate abundance of some parts within each species or for some

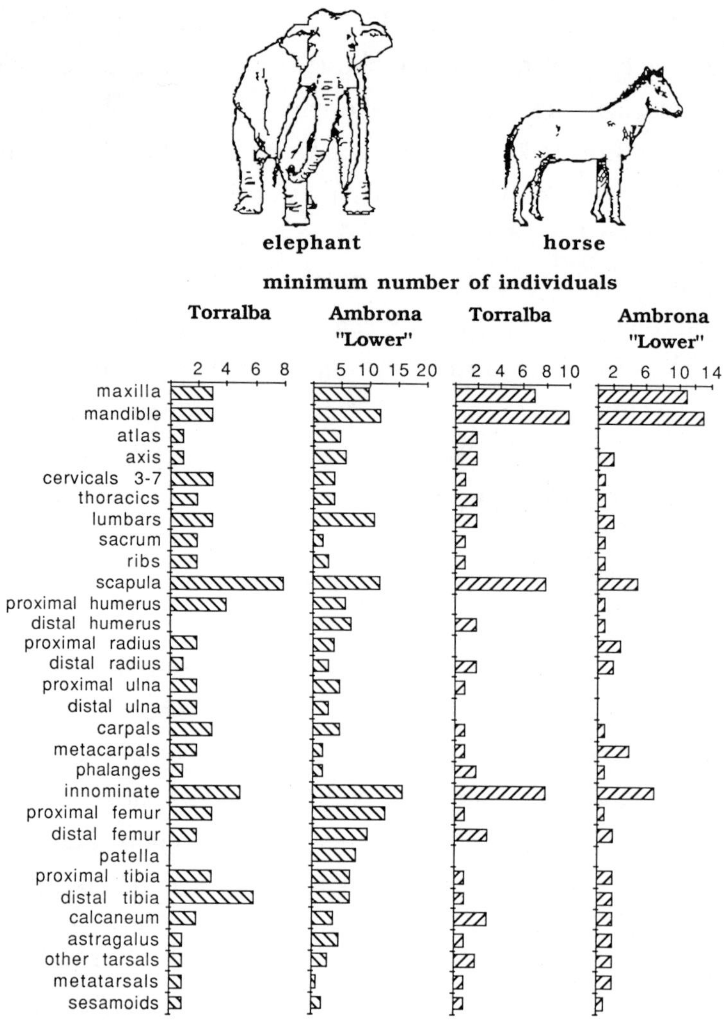

Figure 3. The minimum numbers of elephants and horses represented by various skeletal parts at Torralba and Ambrona "Lower".

differences in skeletal part representation among species. Additionally, smaller skeletal parts may be underrepresented because they were missed during the

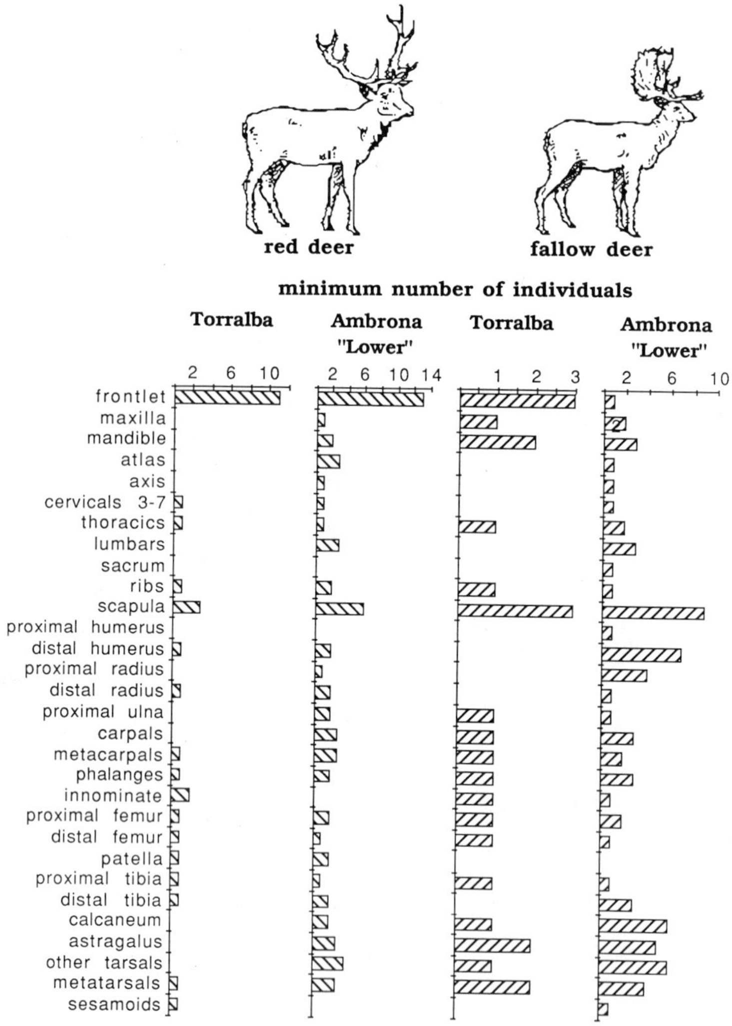

Figure 4. The minimum numbers of red deer and fallow deer represented by various skeletal parts at Torralba and Ambrona "Lower".

excavations, where there was no fine screening. Nonetheless, to some extent, the patterns in the illustrations almost certainly reflect the pre- and

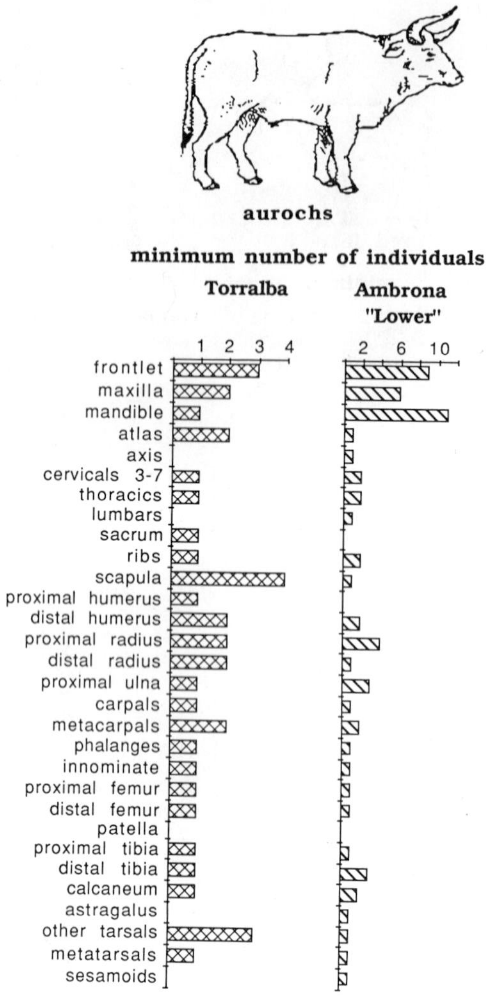

Figure 5. The minimum number of aurochs represented by various skeletal parts at Torralba and Ambrona "Lower".

Thus, greater durability probably explains the tendency for teeth to be among the most abundant skeletal parts in all species and for those long bone epiphyses that are densest and fuse first (the distal humerus, proximal radius, proximal femur, and distal tibia) to be more common than their less dense, later fusing counterparts. Brain (1967, 1969) has stressed the crucial importance of durability in explaining skeletal part abundance. Combined with greater resistance to fragmentation and displacement before and during burial, greater durability probably also explains why, compared to other animals, elephant is much better represented by post-cranial bones (vs. teeth).

Given the sedimentary context and the high incidence of abrasion on the Torralba and Ambrona "Lower" bones, it is likely that sorting during water transport accounts for some skeletal part discrepancies in Figures 3 to 5. The factors controlling bones as sedimentary particles in a stream include bone size, density, and shape, as well as water velocity, fluid density, and bed gradient and configuration (Behrensmeyer 1975). The relative importance of these factors at ancient sites is generally unknown (and perhaps unknowable), making it difficult to "predict" the precise effects of fluvial sorting on a particular fossil assemblage. Still, in the case of the Torralba and Ambrona "Lower" assemblages, it seems likely that fluvial transport or winnowing is at least partly responsible for the remarkably high incidence of relatively flat bones. These include scapulae, innominates, and basal antlers (frontlets), as well as elephant rib and longbone wall fragments whose abundance is not reflected in Figures 3 to 5 because the pieces involved are too fragmented for meaningful MNI calculation.

The relative abundance of innominates may also partly reflect the open-air context of the sites, where, after death, it was the animals' appendicular bones rather than their axial skeletons that were most likely to be removed by carnivores or people. [For more detailed discussion, see Shipman and Phillips-Conroy (1976), Hill (1983), and Behrensmeyer (1983) on what bones hyenas leave behind; compare Wheat (1972) or Frison (1974) on bison bones at prehistoric kill sites in western North America with White (1953, 1954a,1954b) on bison bones in prehistoric habitation sites.] The tendency for the axial skeleton to remain at the death site probably also explains why vertebrae appear better represented at Torralba and Ambrona than at stone age cave sites (like those discussed below), though the Torralba/Ambrona numbers are too small for statistical demonstration.

A final significant point regarding skeletal part representation at Torralba and Ambrona "Lower" is the abundance of basal antlers (frontlets) relative to other deer bones, particularly from red deer. A partial explanation is that approximately half the antlers were naturally shed and thus entered the system from animals who need not have contributed any other bones. More importantly, the shed antlers are a potent reminder of the complex mix of factors that probably shaped the composition of the Ambrona and Torralba bone assemblages.

More can probably be learned from skeletal part representation at Torralba and Ambrona, particularly from comparisons to other sites with similar bone preservation, from study of skeletal part distribution over the surface of the sites, and from analysis of the extent to which different parts are abraded. Whatever the outcome of such studies, however, the available data already indicate that the principal factors shaping skeletal part representation were probably fluvial sorting and winnowing, followed by relatively intense leaching. Except for the tendency of axial elements to be abundant relative to appendicular ones, there is no evidence of a human role in forming the bone assemblages, and the abundance of axial bones could as well be due to carnivores. Finally, the shed antlers indicate that the assemblages may include many bones that have nothing to do with either people or carnivores.

Mortality Profiles

Studies by Kurtén (1953), van Valen (1964), and Voorhies (1969) in paleontology, and by Frison (1978) and his colleagues in archeology (Reher 1978, Wilson 1980, with further references), demonstrate that establishing the ages of animals at time of death can help explain how the animals died and how their bones accumulated at a site. The individual ages for each species in a sample are most frequently grouped together as a mortality (or age) profile. The profiles are then commonly interpreted with respect to two theoretically expectable models (Figure 6, top).

In the first model, successively older age classes contain progressively fewer individuals. This is the typical age profile of a living population of large mammals that is essentially stable in size and structure, a condition to which natural populations tend. This kind of age profile could be fossilized if an entire living population were suddenly wiped out by a flash flood, a volcanic eruption, or an epidemic disease. As a result, it is often known in paleobiology as a "catastrophic" age profile.

The second kind of theoretically expectable age profile is one in which prime-age (reproductively active) adults are underrepresented relative to their live abundance, while very young and old individuals are overrepresented. Such a profile will contain those individuals who die of starvation, accidents, predation, endemic disease, and other routine, attritional mortality factors that affect the very young and the old most heavily. In paleobiology, this profile is commonly referred to as "attritional." Depending upon the actual rate of mortality in different age classes, an "attritional" profile may be either U-shaped, with a large peak in the youngest class and a smaller one in a class beyond 40-50% of potential lifespan, or L-shaped, with no obvious peaks beyond the one in the youngest age class (Klein 1982).

Teeth are the most dependable elements for estimating individual age in most mammals, including those at Torralba and Ambrona. Unfortunately, only the elephant sample from Ambrona "Lower" contains enough intact teeth to construct a useful age profile. Figure 7 presents this profile in a format suggested by Toepfer (1963). The age determinations in the figure assume that the timing of dental eruption and wear in the straight-tusked elephant was the same as in the living African elephant, as presented by Laws (1966) and Hanks (1979). Since the straight-tusked elephant was larger than the African elephant, it is possible that the ages shown are all slight underestimates, which, however, would not affect the overall pattern.

It is important to realize that very young individuals (under 5-10 years) are probably seriously underrepresented in Figure 7, because young teeth would be especially vulnerable to the intense pre- and post-depositional destruction that affected the Ambrona bones. In any case, young individuals do not aid in

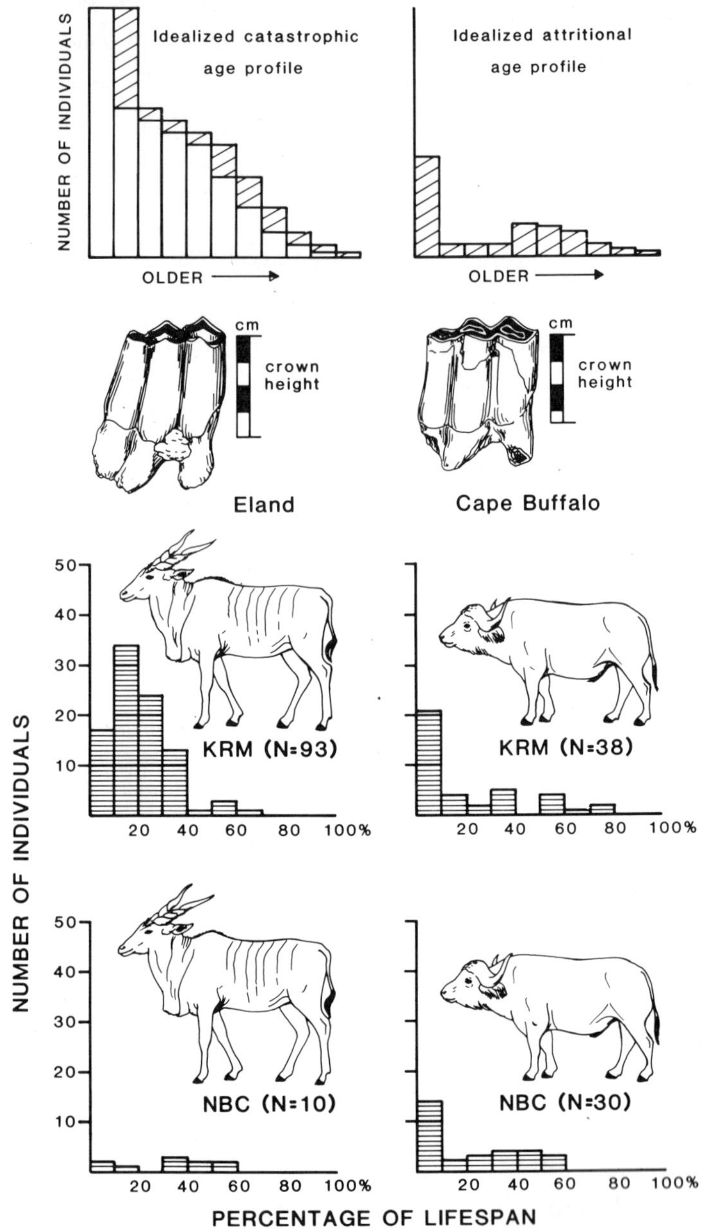

Figure 6. Top left: the numbers of individuals in successive age classes in an idealized, schematic catastrophic age profile (blank bars), and the number of individuals who die between

distinguishing between catastrophic and attritional mortality profiles, since they tend to be equally abundant in both. Excluding the young individuals then, Figure 7 (below) suggests that successively older age classes contain subequal or even increasing numbers of individuals, as in an attritional profile, and not progressively smaller numbers of individuals, as in a catastrophic one. Although the Torralba elephant age profile is based on a much smaller number of individuals, it is in fact very similar (Howell and Freeman in press), and if the Torralba and Ambrona "Lower" samples are combined, the case for attritional mortality is especially strong.

Among elephants, where adults are essentially immune from predation, an attritional mortality pattern suggests death mainly from accidents, disease, and old age. It clearly could not reflect human destruction of entire elephant groups with spears and large rocks after the animals had been driven into a marsh and mired, as hypothesized by Howell et al. (1965). This would produce a catastrophic mortality profile. Attritional mortality does not, however, rule out the possibility that people scavenged dead elephants, which in fact they almost certainly did, at least occasionally.

Figure 6 (*continued*) successive age classes (hatched bars). Top right: an idealized, schematic attritional age profile, comprising a separate display of the hatched bars in the profile immediately to the left. Although the basic shapes of catastrophic and attritional profiles will be the same for all ungulate populations that are approximately stable in size and age structure, the precise shapes will vary from population to population, depending upon species biology and upon mortality factors that affect the populations.

Center: Lower third molars (M3s) of eland and Cape buffalo showing the crown height dimension used to estimate individual age in the mortality profiles below.

Bottom: Mortality profiles of eland and Cape Buffalo from the Middle Stone Age layers of Klasies River Mouth Cave 1 (KRM) and the Later Stone Age layers of Nelson Bay Cave (NBC). It is probable that post-depositional leaching, profile compaction, and other destructive factors have selectively destroyed teeth of very young eland and buffalo at both sites. Keeping this in mind, the eland profile from Klasies is clearly catastrophic, while the buffalo profiles from both sites are attritional. The eland profile from Nelson Bay contains too few individuals for assignment to profile types. (After Klein 1983).

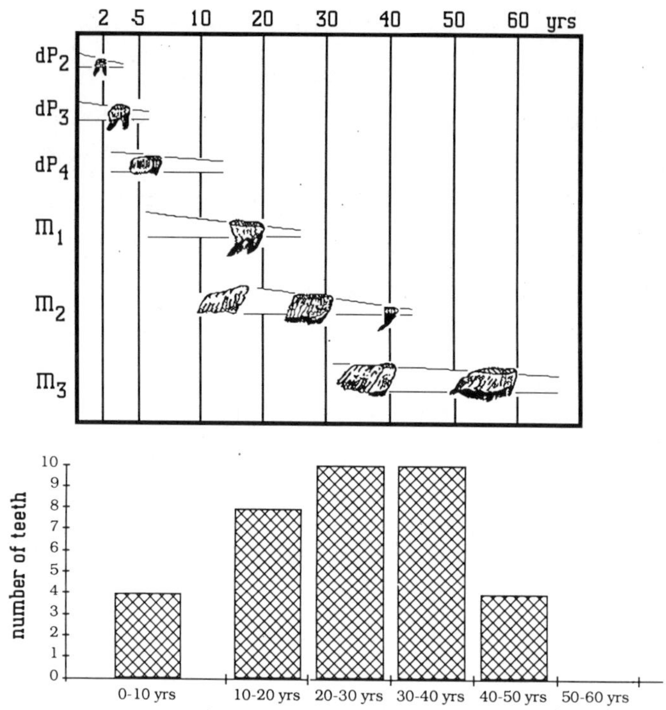

Figure 7. The age (mortality) profile of elephants in the sample from Ambrona "Lower", based on eruption and wear of the lower dentition. Analysis of the upper dentition provides a very similar profile. The format for the diagram was borrowed from Toepfer (1963). Eruption and wear determinations by K. Cruz-Uribe and R. Klein.

Comparison of the Torralba/Ambrona elephant mortality profile to the mortality profiles of recent elephants or of elephants from other fossil sites is hindered by differences in the criteria used to estimate individual age, by the small size of some fossil samples, and by the unstable demographic structure of the recent elephant populations. The conventional paleobiologic concepts of "attritional" and "catastrophic" mortality do not apply well to populations rapidly changing in size. This is a potential problem for interpreting the Torralba and Ambrona data, but the geologic setting suggests the elephant bones accumulated over centuries, if not millennia, during which fluctuations in population size and structure were probably time-averaged toward the stable norm.

If problems of comparability were resolved, it would be particularly interesting to compare the Torralba/Ambrona mortality profiles with those of fossil mammoths from Paleoindian (end-Pleistocene) sites in the western United States. Saunders (1980), with admittedly small samples, concluded that the mammoth age profiles from some key Paleoindian sites reflected destruction of whole groups by determined and sophisticated human hunters. In contrast, Haynes (1985), from observations of mortality in recent elephants from Hwange National Park, Zimbabwe, suggests that the same mammoth mortality profiles may reflect attritional death in declining populations stressed by severe drought. In this case, Paleoindian exploitation of elephants need not have been any more sophisticated than that represented hundreds of thousands of years earlier at Torralba and Ambrona.

Statistical Associations

Freeman (1978) inferred patterned, culturally meaningful associations between various artifact and bone classes from a multivariate (principal components) analysis of the Torralba artifacts and bones. In hindsight, these associations should be regarded as very tentative, since many bones had been incorrectly identified. Additionally, it is doubtful that much culturally significant patterning could have survived the small-scale but cumulative transport of particles and often also of bones and artifacts across most of the occupation surfaces at Torralba. Still, Freeman's basic approach is clearly worth pursuing, especially with the now much larger samples available from Ambrona. It should be particularly useful to look for statistical or spatial associations among artifacts and bones in the vicinity of the occasional clusters of semi-articulated elephant bones found in low-energy deposits, probably very near their original positions.

WHAT DO TORRALBA AND AMBRONA REPRESENT?

Circumstantially, it is probable that people used some of the artifacts found at Torralba and Ambrona to kill or butcher some of the animals. However, even if the bones indicated this directly, it probably would still be impossible to determine how many animals the people killed or scavenged, how many were killed or scavenged by carnivores, or how many died and decomposed without human or carnivore intervention. Present evidence suggests that the ancient setting at Torralba and Ambrona probably resembled the margins of a historic African stream or waterhole, where the events that produce and disarticulate carcasses can be complex and need not involve people.

Clearly, if we wish to determine how ancient people obtained meat and how successful they were at it, we need sites where it is reasonably certain that people introduced most, if not all of the animal bones. In this regard, caves are better than open-air sites, but unfortunately, nearly all fossiliferous stone age sites antedating the late middle Pleistocene are open-air sites, with problems of interpretation broadly similar, if not identical to those at Torralba and Ambrona. The principal exception is the "Peking Man" cave of Zhoukoudian, which unhappily was not excavated with sufficient care to distinguish levels in which bones may have been introduced mainly by people from ones where they were probably introduced mainly by hyenas. To illustrate the potential of cave sites to inform on prehistoric human-animal relationships, we must turn to late Pleistocene and Holocene localities, such as the southern African ones I discuss now.

SOUTHERN AFRICAN CAVES

The southern African evidence for late Pleistocene and Holocene exploitation of animals by stone age people comes mainly from large faunal assemblages excavated in the Middle Stone Age layers of the Klasies River Mouth Caves and Die Kelders Cave 1, and in the Later Stone Age layers of

Nelson Bay Cave, Byneskranskop Cave 1, and other caves in the southern and southwestern Cape Province of South Africa (Figure 8.) Volman (1984) and J. Deacon (1984a, 1984b) have published comprehensive accounts of the Middle Stone Age (MSA) and Later Stone Age (LSA) respectively, while specific information on the sites that provided the bone assemblages may be found in Singer and Wymer (1982) on Klasies; Schweitzer (1976) and Tankard and Schweitzer (1976) on Die Kelders; Klein (1972a, 1972b) and J. Deacon (1978) on Nelson Bay; and Schweitzer and Wilson (1982) on Byneskranskop. The present discussion is abbreviated, since most of the evidence and interpretations have been published already (Klein 1983 and references therein.)

On stratigraphic evidence, the Middle Stone Age (MSA) layers at Klasies probably formed mainly during the last interglaciation *sensu lato*, that is, between ca. 130,000 and 75,000 years B. P. (Butzer 1978; H. J. Deacon et al. 1986). The latest MSA at Klasies and the entire MSA at Die Kelders probably date from the early part of the last glaciation, shortly after 75,000 B. P. The Later Stone Age (LSA) layers at all the sites are younger than 20,000 years B. P. and most are Holocene (or present interglaciation), after 12-10,000 B. P. The time when the LSA replaced the MSA is not well-fixed, but based on evidence from sites other than those providing the faunal remains considered here, was certainly before 30,000 years ago. The MSA/LSA interface (?transition) may prove very difficult to date, partly because it lies at or beyond the limits of conventional radiocarbon dating and partly because most southern African sites appear to have been abandoned for tens of thousands of years between the latest MSA and the earliest LSA occupations. The abandonment probably reflects a sharp decline in human and animal populations due to widespread hyperaridity during most of the last glaciation.

EVOLUTION IN SOUTHERN AFRICAN STONE AGE FORAGING PROFICIENCY

It is essential to emphasize at the outset that stone age people probably introduced the overwhelming majority of large bones to all the caves. This is based on the presence of superabundant artifacts, fossil fireplaces (sometimes

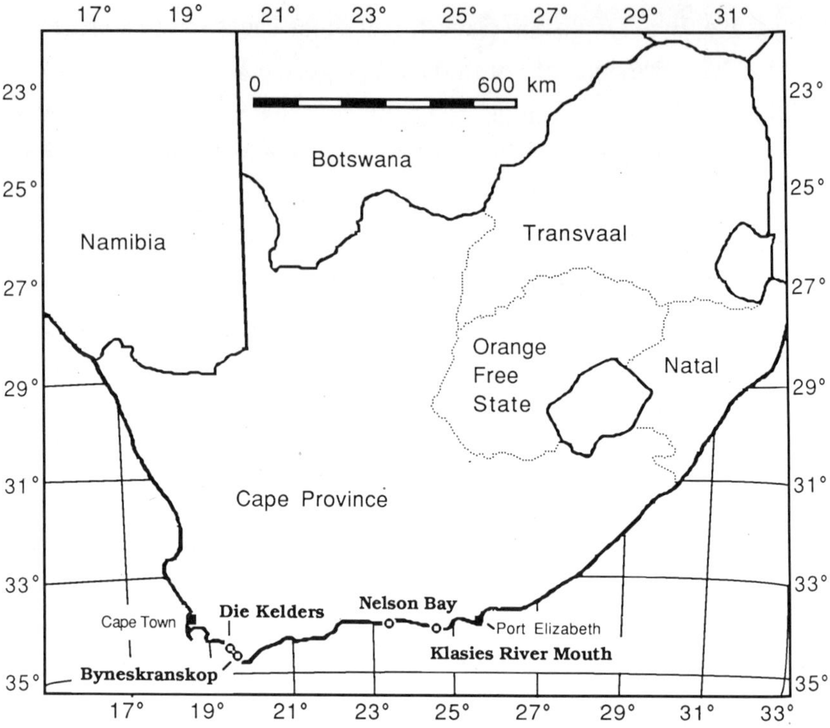

Figure 8. Locations of the southern African sites mentioned in the text.

accompanied by other evidence for structural modification of cave floors), and relatively extensive damage to bones by stone artifacts, combined with the rarity of carnivore coprolites and limited damage to bones from carnivore or rodent teeth. Carnivores and porcupines were probably responsible for only a small minority of bones.

The most interesting information comes from comparisons of bones accumulated by MSA people during the last interglaciation and by LSA people during the Holocene. Both circumstantial and direct evidence indicate that environmental conditions were very similar in these two periods. Therefore, environmental change can be dismissed as a cause of differences between bone assemblages. Similarly, the context of the assemblages and the bones themselves indicate that the assemblages do not differ because of different post-depositional histories.

As might be expected, shellfish are abundant at both MSA and LSA sites that were located near the coast. Both types of sites also contain remains of marine vertebrates, and the MSA layers at Klasies in fact provide some of the oldest evidence in the world for human exploitation of coastal resources. However, there is also an important difference in the composition of the marine organisms between MSA and LSA sites. The bones from MSA layers are mainly from seals and penguins; bones of flying seabirds (cormorants, gulls, etc.) are relatively rare, and bones of fish are nearly absent. In contrast, in LSA sites with comparable amounts of seal and penguin bones, bones of fish and flying seabirds are abundant to superabundant. The conclusion that only LSA people actively fished and fowled (for flying birds) is supported by the fact that artifacts reasonably interpreted as fishing and fowling gear ("gorges" and sinkers) are known only in LSA assemblages.

Striking differences have also been found in the relative abundance of land animals between MSA and LSA sites occupied under similar environmental conditions. In the MSA sites, eland is the most common large ungulate, though historic observations locally and elsewhere in Africa suggest it was probably very uncommon on the ground. Wild pigs (warthog and bushpig) are barely represented, although one or both was probably common near the sites. In comparable LSA sites, eland is far less frequent and pigs far more so, and their relative numbers probably correspond much more closely to their live numbers near the sites.

In their response to attackers eland were probably among the least dangerous available large prey, while pigs were among the most dangerous, suggesting that the differences between MSA and LSA faunas reflect the enhanced ability of LSA hunters to obtain dangerous game. This was probably a result of technological innovation -- LSA development of projectiles or of snares and traps -- that allowed attack from a relatively safe distance. In this regard, it is pertinent that, on ethnographic analogy, some of the bone and stone artifacts found in LSA sites served as parts of composite arrows. There is no evidence for the bow and arrow in MSA sites. More generally, greater LSA cultural sophistication is clearly indicated by the presence of numerous formal bone tools and art objects, especially beads and pendants, unknown in MSA contexts, and by an increase in artifactual variability through time and space, implying greater LSA capacity for innovation or invention in the face of environmental differences or change.

The age profiles of the large bovids in MSA faunas provide further evidence that MSA people were relatively ineffective hunters. In most of the bovid species present, very young and relatively old individuals predominate, and prime-age adults are rare (see the buffalo profile in Figure 6). This pattern is attritional, and it suggests that, like lions and other large predators, MSA hunters could take only the most vulnerable (very young and old) individuals. The age data for most of the species could even imply that the MSA people were scavenging carcasses of animals that died naturally or that were killed by other predators. However, active hunting is suggested by the abundance of very young individuals in the bone assemblages. In Africa today, the carcasses of very young prey are usually completely destroyed by hyenas, lions, and other scavengers/carnivores with whom MSA people had to compete. The people could have obtained so many young carcasses by scavenging only if they were able to locate carcasses before other predators did, which seems unlikely.

In contrast to the other ungulates, the eland in MSA sites include a large proportion of prime-age adults, and the age profile has a catastrophic shape (Fig. 6). The most likely explanation is that MSA people had learned that, unlike most other large African bovids, eland can be easily driven, without much personal risk. An eland herd caught in the right position could be forced over a cliff or into a trap, places where differences in individual vulnerability due to age would have little meaning. There are eminently suitable cliffs very near the Klasies and Die Kelders sites.

However, MSA people could not have driven eland herds to their death very often or the species would have become extinct, since its reproductive vitality would have been sapped by the continuing loss of a large proportion of the available prime adults. Not only did the eland survive, but there is no evidence that it became less numerous during the long MSA time span. The reason must surely be that MSA eland, like historic eland, lived in sparsely distributed, wide-ranging herds that were difficult to locate in a position suitable for driving to a trap. MSA people were thus able to obtain only a small portion of the available herds.

Thus, MSA people were probably not very successful at hunting eland, and this makes it especially interesting that eland is the most abundant ungulate in the MSA faunas. The clear implication is that MSA people must have been even less successful at hunting other species that are less common in the sites but were more common in the environment. In short, MSA impact on the large mammal fauna was negligible. By extension, it may be argued that LSA peoples, in whose sites eland and other species are represented more in proportion to their live abundance, probably took a higher proportion of game overall. In short, LSA people were almost certainly more proficient hunters.

So far, the technological or other cultural advances that allowed LSA people to obtain more animals are not reflected in mortality profiles. The profiles of most large ungulates remain attritional. Thus, while LSA people probably obtained more buffalo than their predecessors, they were still unable to acquire prime-age adults in proportion to their live abundance. The technology to accomplish this was probably unavailable until the appearance of Europeans just a few hundred years ago.

Patterns of skeletal part representation are also very similar between MSA and LSA sites. Most notably, in both, small ungulates tend to be well-represented by a wide range of skeletal parts, while larger ungulates are represented mainly by skull and foot bones. With respect to Klasies, Binford (1984) interprets this to mean that MSA people hunted smaller animals, but only scavenged larger ones.. His reasoning is based on the observation that the skull and foot bones by which large ungulates are best represented have relatively little nutritional or other value and would thus often occur at carcasses on which nonhuman predators had already fed.

Binford's hypothesis is intriguing, and it is certainly possible that the people scavenged carcasses of those large ungulates characterized by attritional mortality profiles. However, it is difficult to imagine how scavenging could create a catastrophic mortality profile like that for eland. Another problem with Binford's suggestion is that the contrast in skeletal part representation between smaller and larger ungulates is extremely widespread. It not only characterizes Klasies, Nelson Bay Cave, Die Kelders, Byneskranskop, and other Stone Age sites throughout the world, but also later sites, like African Iron Age sites, where the smaller and the larger ungulates were domesticated species (Voigt 1983). Like Voigt (1983), I think the pattern partly reflects greater fragmentation of large bones during butchering and food preparation, combined with post-depositional leaching, profile compaction and other destructive processes that tend to remove fragmented bones first, or that make them unidentifiable. It may also partly reflect the phenomenon that Perkins and Daly (1968) called the "schlepp effect", in which people were more likely to transport large carcasses in sections, causing greater spatial separation of different large animal skeletal elements.

Besides taxonomic abundance, mortality profiles, and skeletal part representation, there are other classes of data that illuminate MSA and LSA foraging behavior. Perhaps most significant is the contrast between MSA and LSA sites in the average size of limpets and tortoises. Both limpets and tortoises tend to be larger in MSA sites than in LSA ones occupied under broadly comparable environmental conditions, or at least under conditions where environmental differences could not account for the size contrast. The most plausible explanation for the smaller mean size of tortoises and limpets in LSA sites is that people were pressing harder on the species, probably because of higher human population densities. These in turn were probably a result of the kinds of cultural advances reflected in the faunal and artifactual data discussed previously.

SUMMARY AND CONCLUSION

It is difficult to determine how successful early and middle Pleistocene people were at obtaining animals, because most known sites were open-air localities near ancient water sources, where a functional relationship between artifacts and bones cannot be simply assumed. Circumstantially, the artifacts suggest that people killed or butchered at least some of the animals represented by bones, and at some sites this is strongly supported by bones damaged by stone tools. In some cases, human activity may also best explain dense, localized concentrations of artifacts and bones. Finally, there are abstract observations that may imply advances in early human ability to obtain meat. These include an early Pleistocene reduction in the number of large African carnivore taxa, broadly coincident with the emergence of *Homo erectus* and the Acheulean Cultural Tradition (Walker 1984), and the first human penetration of north temperate latitudes, probably at or shortly before the beginning of the middle Pleistocene. The biological advance of *Homo erectus* over preceeding *H. habilis* and the cultural advance of the Acheulean over the preceding Oldowan almost surely were accompanied by enhanced human ability to extract energy from the environment.

However, there are still no early or middle Pleistocene sites where the true nature or extent of human exploitation of animals can be demonstrated. For the moment, perhaps the most secure basis for assessing the ability of early and middle Pleistocene people to obtain meat is by extrapolation from their immediate late Pleistocene successors as discussed here. If it is fair to project backwards from these late Pleistocene (Middle Stone Age) people, yet earlier people were probably very ineffective hunters.

Animal bones from Upper Pleistocene and Holocene sites provide a much firmer basis for gauging human foraging abilities, largely because these sites include many localities where it is reasonably certain that people introduced most of the large bones. Analysis of late Pleistocene and Holocene bone assemblages from the southern African caves, discussed above, suggests that there was a quantum advance in human ability to extract animal protein from nature, probably coincident with the advent of the Later Stone Age, 30-40,000 years ago or before. This in turn perhaps coincided with the emergence of totally modern people (*Homo sapiens sapiens*). Further checking for a possible relationship between artifactual and biological change on the one hand and an increased ability to obtain animals on the other depends less on the development of new ideas or techniques than on the carcful excavation and analysis of large faunal and artifactual samples from as many sites as possible.

ACKNOWLEDGEMENTS

I thank the National Science Foundation for financial support and the Museo Arqueológico Nacional (Madrid) and the South African Museum (Cape Town) for study facilities. Karl Butzer, Kathryn Cruz-Uribe, Matthew Nitecki, Pat Shipman, and Cornelia Wolf kindly commented on a draft of the manuscript.

APPENDIX

Technical names of the mammalian taxa mentioned in the text and figures.

VERNACULAR NAME	TECHNICAL NAME
aurochs	*Bos primigenius*
bushpig	*Potamochoerus porcus*
Cape buffalo	*Syncerus caffer*
eland	*Taurotragus oryx*
fallow deer	*Dama dama*
giant deer	*Megaceros* sp.
lion	*Panthera leo*
lynx	*Felis lynx*
macaque (monkey)	*Macaca* sp.
Mosbach wolf	*Canis lupus mosbachensis*
red deer	*Cervus elaphus*
spotted hyena	*Crocuta crocuta*
steppe rhinoceros	*Dicerorhinus hemitoechus*
straight-tusked elephant	*Elephas antiquus*
warthog	*Phacochoerus aethiopicus*

REFERENCES

Aguirre, E., and C. Fuentes 1969. Los vertebrados fosiles de Torralba y Ambrona. *Etudes sur le Quaternaire dans le Monde*:433-437. Paris: Centre National de la Recherche Scientifique.

Andrews, P. A., and J. Cook. 1985. Natural modifications to bones in a temperate setting. *Man* 20:675-691.

Balout, L., P. Biberson, and J. Tixier. 1967. L'Acheuléen de Ternifine (Algérie), gisement de l'Atlanthrope. *L'Anthropologie* 71:217-237.

Bar-Yosef, O. 1975. Archaeological occurrences in the Middle Pleistocene of Israel. In: *After the Australopithecines*, ed. K. W. Butzer and G. L. Isaac, pp. 571-604. The Hague: Mouton.

Bar-Yosef, O. 1980. Prehistory of the Levant. *Annual Review of Anthropology* 9:101-133.

Behrensmeyer, A. K. 1975. The taphonomy and paleoecology of the Plio-Pleistocene vertebrate assemblages east of Lake Rudolf, Kenya. *Bulletin of the Museum of Comparative Zoology* 146:473-578.

Behrensmeyer, A. K. 1983. Patterns of natural bone distribution on recent land surfaces: implications for archaeological site formation. *British Archaeological Reports* 163:93-104.

Behrensmeyer, A. K., K. D. Gordon, and G. T. Yanagi. 1986. Trampling as a cause of bone surface damage and pseudo-cutmarks. *Nature* 319:768-771.

Biberson, P. 1968. Les gisements Acheuléens de Torralba et Ambrona (Espagne). Nouvelles précisions. *L'Anthropologie* 72:241-278.

Binford, L. R. 1981. *Bones: ancient men and modern myths*. New York: Academic Press.

Binford, L. R. 1984. *Faunal Remains from Klasies River Mouth*. Orlando, Florida: Academic Press.

Binford, L. R. 1985. Human ancestors: changing views of their behavior. *Journal of Archaeological Anthropology* 4:292-327.

Binford, L. R., and C. K. Ho. 1985. Taphonomy at a distance: Zhoukoudian, "The Cave Home of Beijing Man?" *Current Anthropology* 26:413-442.

Brain, C. K. 1967. Hottentot food remains and their meaning in the interpretation of fossil bone assemblages. *Scientific Papers of the Namib Desert Research Station* 32:1-11.

Brain, C. K. 1969. The contribution of Namib Desert Hottentots to an understanding of Australopithecine bone accumulations. *Scientific Papers of the Namib Desert Research Station* 39:13-22.

Brain, C. K. 1981. *The hunters of the hunted? An introduction to African cave taphonomy*. Chicago: University of Chicago Press.

Bunn, H. T. 1981. Archaeological evidence for meat-eating by Plio-Pleistocene hominids from Koobi Fora and Olduvai Gorge. *Nature* 291:574-577.

Bunn, H. T. 1982. *Meat-Eating and Human Evolution: Studies on the Diet and Subsistence Patterns of Plio-Pleistocene Hominids in East Africa*. Unpublished Ph.D. Thesis, University of California-Berkeley.

Bunn, H. T. 1983. Evidence on the diet and subsistence patterns of Plio-Pleistocene hominids at Koobi Fora, Kenya and Olduvai Gorge, Tanzania. *British Archaeological Reports International Series* 163:21-30.

Bunn, H. T., J. W. K. Harris, Z. Kaufulu, E. Kroll, K. Schick, N. Toth, N., and A. K. Behrensmeyer. 1980. FxJj50: an early Pleistocene site in northern Kenya. *World Archaeology* 12:109-136.

Butzer, K. W. 1965. Acheulean occupation sites at Torralba and Ambrona. Their geology. *Science* 150:1718-1722.

Butzer, K. W. 1978. Sediment stratigraphy of the Middle Stone Age sequence at Klasies River Mouth, Tzitzikamma coast, South Africa. *South African Archaeological Bulletin* 33:141-151.

Chavaillon, J. 1982. Position chronologique des hominidés fossiles d'Ethiopie. *Congrès International de Paléontologie Humaine* 1:766-797.

Conybeare, A., and G. Haynes. 1984. Observations on elephant mortality and bones in water holes. *Quaternary Research* 22:189-200.

Deacon, H. J., V. B. Geleijnse, A. I. Thackeray, J. F. Thackeray, M. L. Tusenius, and J. C. Vogel. 1986. Late Pleistocene cave deposits in the southern Cape: current research at Klasies River. *Palaeoecology of Africa* 17:31-37.

Deacon, J. 1978. Changing patterns in the late Pleistocene/early Holocene prehistory of southern Africa as seen from the Nelson Bay stone artifact sequence. *Quaternary Research* 10:84-111.

Deacon, J. 1984a. Later Stone Age people and their descendants in southern Africa. In: *Southern African Prehistory and Paleoenvironments*, ed. R. G. Klein, pp. 221-328. Rotterdam: A. A. Balkema.

Deacon, J. 1984b. The Later Stone Age of Southernmost Africa. *British Archaeological Reports International Series* 213.

de Lumley, H. 1975. Cultural evolution in France in its paleoecological setting during the Middle Pleistocene. In: *After the Australopithecines*, ed. K. W. Butzer and G. L. Isaac, pp. 745-808. The Hague: Mouton.

Fisher, D. C. 1984. Taphonomic analysis of late Pleistocene mastodon occurrences: evidence of butchery by North American Paleo-Indians. *Paleobiology* 10:338-357.

Freeman, L. G. 1975. Acheulian sites and stratigraphy in Iberia and the Maghreb. In: *After the Australopithecines*, ed. K. W. Butzer and G. L. Isaac, pp. 661-743. The Hague: Mouton.

Freeman, L. G. 1978. The analysis of some occupation floor distributions from Earlier and Middle Paleolithic sites in Spain. In: *Views of the Past: Essays in Old World Prehistory and Paleoanthropology*, ed. L. G. Freeman, pp. 57-93. The Hague: Mouton.

Freeman, L. G., and Butzer, K. W. 1966. The Acheulean station of Torralba (Spain). A progress report. *Quaternaria* 8:9-21.

Frison, G. C. 1974. *The Casper Site*. New York: Academic Press.

Frison, G. C. 1978. Animal population studies and cultural inference. Plains Anthropologist Memoir 14:44-52.

Geraads, D. 1981. Bovidae et Giraffidae (Artiodactyla, Mammalia) du Pléistocène de Ternifine (Algérie). *Bulletin du Museum national d'Histoire naturelle* 4e sér., 3:47-86.

Geraads, D. 1985. La fauna des gisements de Melka-Kunturé (Ethiopie). *L'Environnement des hominidés au Plio-Pléistocène*:165-174. Paris: Masson.

Hanks, J. 1979. *A struggle for survival. The elephant problem*. Cape Town: C. Struik.

Haynes, G. 1985. Age profiles in elephant and mammoth bone assemblages. *Quaternary Research* 24:333-345.

Hill, A. 1983. Hyaenas and early hominids. *British Archaeological Reports International Series* 163:87-92.

Howell, F. C. 1966. Observations on the earlier phases of the European Lower Paleolithic. *American Anthropologist* 68(2/2):88-201.

Howell, F. C., and the Editors of Time-Life Books. 1965. *Early Man*. New York: Time-Life Books.

Howell, F. C., and L. G. Freeman. 1983. Ivory points from the earlier Acheulean of the Spanish Meseta. *Homenaje al Prof. Martin Almagro Bosch* I:41-59. Madrid: Ministerio de Cultura.

Howell, F. C., and L. G. Freeman. In press. *Torralba: An Acheulean butchering station on the Spanish Meseta*. Chicago: University of Chicago Press.

Isaac, G. L. 1977. *Olorgesailie: Archaeological studies of a Middle Pleistocene lake basin in Kenya*. Chicago: University of Chicago Press.

Isaac, G. L. 1984. The archaeology of human origins: studies of the Lower Pleistocene in East Africa 1971-1981. *Advances in World Archaeology* 3:1-87.

Jia, L. P. 1975. *The Cave Home of Peking Man*. Peking: Foreign Languages Press.

Keeley, L. H., and N. Toth. 1981. Microwear polishes on early stone tools from Koobi Fora, Kenya. *Nature* 293:464-465.

Klein, R. G. 1972a. Preliminary report on the July through September 1970 excavations at Nelson Bay Cave, Plettenberg Bay (Cape Province, South Africa). *Paleoecology of Africa* 6:177-208.

Klein, R. G. 1972b. The late Quaternary mammalian fauna of Nelson Bay Cave (Cape Province, South Africa): its implications for megafaunal extinctions and environmental and cultural change. *Quaternary Research* 2:135-142.

Klein, R. G. 1978. The fauna and overall interpretation of the "Cutting 10" Acheulean site at Elandsfontein (Hopefield), southwestern Cape Province, South Africa. *Quaternary Research* 10:69-83.

Klein, R. G. 1982. Patterns of ungulate mortality and ungulate mortality profiles from Langebaanweg (early Pliocene) and Elandsfontein (middle Pleistocene), south-western Cape Province, South Africa. *Annals of the South African Museum* 90:49-94.

Klein, R. G. 1983. The Stone Age prehistory of southern Africa. *Annual Review of Anthropology* 12:25-48.

Kurtén, B. 1953. On the variation and population dynamics of fossil and recent mammal populations. *Acta Zoologica Fennica* 76:1-122.

Laws, R. M. 1966. Age criteria for the African elephant, *Loxodonta africana. East African Wildlife Journal* 4:1-37.

Leakey, M. D. 1971. *Olduvai Gorge*. Vol. 3. *Excavations in Beds I and II, 1960-1963*. Cambridge: Cambridge University Press.

Liu, Z. 1985. Sequence of sediments at locality 1 in Zhoukoudina and correction with loess stratigraphy in Northern China and with the chronology of deep-sea cores. *Quaternery Research* 23:139-153.

Malatesta, A., A. Jaccbacci, G. Nappi, V. Conato, V. Molinari Paganelli, A. Van der Werff, S. Durante, F. Settepassi, L. Caloi, M. R. Palombo, P. Cassoli, M. Piperno, and I. Biddittu. 1978. Torre in Pietra, Roma. *Quaternaria* 20:205-577.

Mania, D. 1976. Altpaläolithische Rastplatz mit Hominidresten aus dem Mittelpleistozän Travertin-Komplex von Bilzingsleben (DDR.) *IXe Congrès UISPP, Nice, 1976, Colloque IX*:35-45.

Morlan, R. E. 1980. Taphonomy and archaeology in the Upper Pleistocene of the Northern Yukon Territory: a glimpse of the peopling of the New World. *Archaeological Survey of Canada Paper* 94:1-380.

Perkins, D., and P. Daly. 1968. A hunters' village in Neolithic Turkey. *Scientific American* 219(11):97-106.

Potts, R. 1983. Home bases and early hominids. *American Scientist* 72:338-347.

Potts, R. 1984. Hominid hunters? Problems in identifying the earliest hunter-gatherers. In: *Hominid Evolution and Community Ecology*, ed. R. Foley, pp. 129-166. New York: Academic Press.

Potts, R., and Shipman, P. 1981. Cutmarks made by stone tools on bones from Olduvai Gorge, Tanzania. *Nature* 291:577-580.

Reher, C. A. 1978. Buffalo population and other deterministic factors in a model of adaptive process on the shortgrass plains. *Plains Anthropologist Memoir* 14:23-39.

Roe, D. A. 1981. *The Lower and Middle Palaeolithic Periods in Britain*. London: Routledge and Kegan-Paul.

Saunders, J. J. 1980. A model for man-mammoth relationships in late Pleistocene North America. *Canadian Journal of Anthropology* 1(1):87-98.

Schweitzer, F. R. 1976. Excavations at Die Kelders, Cape Province, South Africa. *Annals of the South African Museum* 78:101-233.

Schweitzer, F. R., and M. L. Wilson. 1982. Byneskranskop 1: a late Quaternary living site in the southern Cape Province, South Africa. *Annals of the South African Museum* 88:1-203.

Shipman, P. 1981. *Life History of a Fossil*. Cambridge: Harvard University press.

Shipman, P. 1983. Early hominid lifestyle: hunting and gathering or foraging and scavenging? *British Archaeological Reports International Series* 163:31-49.

Shipman, P. 1986. Scavenging or hunting in early hominids: theoretical framework and tests. *American Anthropologist* 88:27-44.

Shipman, P., and J. Phillips-Conroy. 1976. Scavenging by hominids and other carnivores. Current Anthropology 17:170-192.

Shipman, P., and J. Rose. 1983. Evidence of butchery and hominid activities at Torralba and Ambrona: an evaluation using microscopic techniques. *Journal of Archaeological Science* 10:465-474.

Singer, R., and J. Wymer. 1968. Archaeological investigations at the Saldanha skull site in South Africa. *South African Archaeological Bulletin* 25:63-74.

Singer, R., and J. Wymer. 1982. *The Middle Stone Age at Klasies River Mouth in South Africa*. Chicago: University of Chicago Press.

Stuart, A. J. 1982. *Pleistocene Vertebrates of the British Isles*. London: Longman.

Tankard, A. J., and F. R. Schweitzer. 1976. Textural analysis of cave sediments: Die Kelders, Cape Province, South Africa. In: *Geoarchaeology*, ed. D. A. Davidson and M. L. Shackley, pp. 289-316. London: Duckworth.

Toepfer, V. 1963. *Tierwelt des Eiszeitalters*. Leipzig: Akademische Verlagsgesellschaft Geest and Portig K.-G.

Van Valen, L. 1964. Age in two fossil horse populations. *Acta Zoologica* 45:93-106.

Vértes, L. 1975. The lower Palaeolithic site of Vértesszöllös, Hungary. In *Recent Archaeological Excavations in Europe,* ed. R. Bruce-Mittford, pp. 287-301. London: Routledge and Kegan Paul.

Villa, P. 1983. Terra Amata and the Middle Pleistocene archaeological record of southern France. *University of California Publications in Anthropology* 13:1-303.

Voigt, E. A. 1983. Mapungubwe: an archaeozoological interpretation of an Iron Age community. *Transvaal Museum Monograph* 1:1-203.

Volman, T. P. 1984. Early prehistory of southern Africa. In: *Southern African Prehistory and Paleo-environments*, ed. R. G. Klein, pp. 169-221. Rotterdam: A. A. Balkema.

Voorhies, M. R. 1969. Taphonomy and population dynamics of an early Pliocene vertebrate fauna, Knox County, Nebraska. *University of Wyoming Special Contributions to Geology Paper* 1:1-69.

Walker, A. 1984. Extinction in hominid evolution. In: *Extinctions*, ed. M. H. Nitecki, pp. 153-190. Chicago: University of Chicago Press.

Wheat, J. B. 1972. The Olsen-Chubbock site: a Paleo-Indian bison kill. *Memoirs of the Society for American Archaeology* 26.

White, T. E. 1953. Observations on the butchering technique of some aboriginal peoples No. 2. *American Antiquity* 19:160-164.

White, T. E. 1954a. Observations on the butchering technique of some aboriginal peoples No. 4. *American Antiquity* 19:257-259.

White, T. E. 1954b. Observations on the butchering technique of some aboriginal peoples No. 5. *American Antiquity* 19:259-262.

Wilson, M. 1980. Population dynamics of the Garnsey Site Bison. *Museum of Anthropology (University of Michigan) Technical Report* 12:88-129.

Wu Ru Kang, and S. L. Lin 1983. Peking Man. *Scientific American* 248(6):86-94.

Were There Elephant Hunters at Torralba?

Lewis R. Binford

Department of Anthropology
University of New Mexico, Albuquerque, New Mexico

In recent years there has been growing skepticism among some students of the pre-*Sapiens sapiens* hominids that the earlier romantic views, which pictured early man as a mighty hunter, are an accurate construction of the past. In fact, the trend in much recent work has been to modify this view and to see as unwarranted much of the evidence previously cited in support of the "mighty hunter" view of the past. Some have begun the serious investigation of the distinct possibility that early man was more commonly a scavenger of animal carcasses than a successful predator. This view, while seriously discussed for the pre-*Homo erectus* hominids, has not been popularly adopted for the investigation of *Homo erectus* himself. In fact, many theorists consider *Homo erectus* to be the author of what is referred to as the "hunting way of life" and believe that this shift may in fact stand behind the species' successful radiation into new environmental zones (Shipman 1984).

Recent analysis (Binford 1985; Binford and Stone 1986) of the archaeological remains of Europe and Asia seeking to document the successful penetration of *Homo erectus*, or at least "*erectus*-grade" hominids, into the temperate zone yields ambiguous results suggesting that hominids were not successful hunters even as late at 200,000 years ago! Each time the results of the new investigations are discussed the conversation inevitably shifts to the site of Torralba (see Klein this volume, Fig. 1, p. 13), where the excavators and interpreters of the archaeological remains have certainly made the most known and generally accepted case for big-game hunting as basic to the early way of life for an *erectus*-grade hominid anywhere in the world.

Clearly, if any changed view of hominid life during the evolutionary era dominated by *erectus*-grade hominids is to be sustained, the data from Torralba must be faced. That has now been done, using only the published data. This analysis will make possible a new discussion of the meaning of Torralba in the assessment of the role of hunting in hominid evolution.

LOOKING BACK AT THE CLAIMS

During the early 1960s, when it was *assumed* by nearly all prehistorians that Lower Paleolithic hominids were hunters, the interpretation of the data from Torralba was presented in support of this assumption. Certainly the well-published pictures of *in situ* elephant bones and associated stone tools together with well-executed reconstructive drawings of what life was like at Torralba (Howell 1965) made this site the most visible evidence for Lower Paleolithic big-game hunting in the English language literature. Subsequent interpretative literature (Freeman 1975a, 1978) regarding this site has reinforced and enlarged this picture of the past.

I am an archaeological methodologist. My concern is with how we as archaeologists warrant or justify our constructions of a past. The past is gone, and our constructions of the past are all based on inferences. The crucial task of inference justification is the linchpin of archaeological investigation. My concern in the following discussion is focused on this "linchpin." Can the argument for Lower Paleolithic big-game hunting survive a detailed evaluation from a methodological perspective?

The Early Days of "Self-Evident" Interpretation

During the early years of the excavations at Torralba I was a colleague of F. Clark Howell at the University of Chicago. I had the privilege of seeing the wonderful maps of bones and archaeological specimens as they were brought from the field to our busy laboratories at Chicago. I was as excited as most others by archaeological data collected in a detailed and well-documented manner from what in those days were called "living floors." Like most of my fellow researchers, I believed the traditional wisdom of our field, which claimed that early hominids were hunters. I also viewed the field maps and other data as self-evident testimony to the validity of our belief in early man as a hunter. Howell (1965) summarized the evidence in his extremely influential book, *Early Man:*

> What made the fossil sites at Torralba and Ambrona interesting was the enormous number of elephant bones that they contained....These were far and away too numerous to be explained away as having gotten there by accident. Furthermore, their condition and their position in the ground were extremely unusual....
>
> Mixed in with the bones were many signs of ancient human presence. Stone tools, of a type associated with *Homo erectus* in Africa, were abundant. There were also bone tools and even pieces of wood, pointed or vaguely spatulate at one end.
>
> There was also a quantity of material that shows different degrees of burning. These materials were not so concentrated in any one place as to suggest the presence of continuous fires over a long period of time. Rather they were thinly and very widely scattered. Whoever had been lighting these fires was apparently burning grass and brush over large areas. This evidence, plus that of the elephant bones concentrated in what was once a bog, suggests that the setting of those fires had been purposeful -- to drive the unwieldy elephants into the mud....
>
> In general, *Homo erectus* may be labeled as a hunter, and an extremely successful one at that. (Howell 1965:84)

At the time, at least to judge from the number who cite the Torralba data in textbooks as evidence for big-game hunting (e.g., Bordes 1968; Gowlett 1984; Jolly and Plog 1976; Pfeiffer 1978), most archaeologists accepted the suggested construction of a past, a view created by the archaeologists for *Homo erectus*. It must be recalled that the early sixties was an intellectual era when two trends were present in archaeology. The first traditional trend was made consistent by a strict empiricist bias on the part of archaeologists who generally believed that if we looked harder and increased our skills for seeing through the use of computers, better excavation strategies, and better recording and recovery methods, we would surely succeed in "seeing" an accurate past. Linked with this approach was the belief that our knowledge grew primarily as a result of seeing new and previously unseen things. This "progress-through-discovery" approach is perhaps nowhere better illustrated than in the many news conferences held by the Leakeys and in the regular news accounts, which claimed that some new discovery completely changed our view of human evolution or of what the past was like. It was commonplace for adherents to this view to seek their professional fortune by finding new and previously unseen things. This attitude is at least partially responsible for the "instant" interpretations of the Torralba data and the emphasis given to alleged "previously unknown" bone and wooden tools recovered from the site.

At the same time another trend was developing that focused on the logic of inferences made from archaeological observations. It was suggested that the assumptions archaeologists made about the nature of evidence and its implication for past events and conditions were, in the long run, a greater obstacle to the achievement of accurate inferences as to the nature of the past than were our poor observational techniques. From this perspective it was argued that increasing our observational skills was in fact an idle endeavor if pursued in ignorance as to what constituted evidence indicative of past events and conditions! Even more important, claims for discovery of previously unseen things and conditions were generally considered to be "grandstand" behavior. If they were new and previously unseen, how would we immediately know what they mean or imply in terms of the past? This is the epistemologically oriented perspective that I have consistently presented and used over the course of my career. In turn, the "success through discovery" and the "truth through looking harder" approaches have commonly been the target of some critical essays.

Analytical Approaches to Seeing

Leslie Freeman, my student at the University of Chicago during the early 1960s, along with many others was engaged with me in exploring new ways of "seeing" or recognizing patterning in the archaeological record. As we explored pattern recognition techniques we were frequently made aware of incompatibilities between what we saw, using new techniques, and what earlier archaeologists had inferred from the same facts, as "seen" or summarized by them. One of our major research interests was the investigation of archaeological assemblages as potential compounds built up by the manufacture and use of different types of tools in different activity contexts (Binford and Binford 1966). We were fond of citing evidence from relatively modern hunter-gatherers to demonstrate that activities were spatially differentiated on sites. We argued that at least some variability in assemblage content not only reflected different activity regimens at different places but also was the consequence of an excavational strategy that sampled various activity-differentiated segments within a site instead of uncovering complete sites.

It is not suprising that having invested so much, both during excavation and later in descriptive analysis, Leslie Freeman should try to apply to the Torralba data some of the pattern recognition strategies with which we were experimenting during his student days at Chicago. Results of Freeman's endeavors in this direction are summarized in his later paper (Freeman 1978), and some of his attempts to expand the kinds of inference archaeologists could make from archaeological facts (something else we were trying to do in those Chicago days) are seen in his earlier paper (1975a; also see Freeman 1975b, 1981). What new or expanded views of the past did these efforts facilitate?

In Freeman's (1975a) initial synthesis he accepts the arguments and assumptions made during those early years of Torralba interpretation:

> The Torralba/Ambrona site complex is a set of special purpose hunting and preliminary butchering stations along a valley favored by game animals....
>
> Each utilization of the Torralba site seems to have been of extremely short duration -- one might almost say ephemeral. Each occupation could very well represent the results of a single hunt, to judge from the fact that parts of several individuals (sometimes from all the individuals represented) were subjected to simultaneous processing in the secondary butchering areas. (Freeman 1975a:679-680)

Here we see a restatement of the original assumptions, specifically that the remains at Torralba were the result of hominid activity and that the activity indicated is the killing of game facilitated by game drives. A new element has been introduced, however: the assumption of "activity" areas and the citation of their heterogeneous content as evidence for the "duration" of site use by man. Freeman concludes that we see at Torralba the results of essentially single-episode phenomena in each of the several levels at Torralba. The important role of the original interpretation, used here as a basic assumption, is clear. If multiple animals were killed by hominids in one hunting episode, then the presence of multiple individuals in a separately recognized spatial localization within a single level proves that there was only one episode of hunting!

Another assumption hidden in Freeman's interpretations is that spatial differentiation in the dispersion and content of localizations on a site directly indicates consistently localized activities differentiated within a single occupation. Clearly, if one believes these things about the archaeological record, then the conclusion is self-evident that given the demonstration of localized tool and bone concentrations, and the observation that multiple individual animals occurred in such areas, single-episode occupations are demonstrated! This conclusion is of course a tautology since with most deductive reasoning strategies one cannot reason to a position that contradicts the original assumptions. Importantly, the assumptions used here were historical assumptions about what the past was like and not theoretical assumptions about the way the world works and hence about what causes patterning in the archaeological record.

How could we arrive at Freeman's interpretations if (a) the original assumption about game drives was wrong and (b) there were other formation contexts that could result in heterogeneous localizations? We could not. Freeman's interpretative arguments summarized thus far are dependent upon the truth of the original interpretations of the site. In short, no middle-range research (Binford 1981) or reasoning stands behind the Torralba interpretations.

Let's continue with other interpretations.

> In some levels, there are several spatially isolated concentrations of statistically indistinguishable materials, representing multiple synchronous performances of the same set of game processing activities. Often the same body parts of several individuals of different species will be recovered from each such concentration. This situation strongly suggests a relatively egalitarian sharing of the product of the hunt among a series of probably similarly constituted social units -- perhaps different teams or individuals who cooperated in the drive. (Freeman 1975a:681)

Once again the assumptions of episodal game drives and synchronous activity areas are made. Given these assumptions, the conclusion is drawn that a recurrent pattern of body-part associations in spatially discrete locations means an internally differentiated work force. Once again we must ask (a) what if the original interpretation of game drives is wrong, and (b) what if there are other formation contexts for organizationally variable localizations? If either were suspected, then the interpretations would collapse. Nevertheless, Freeman continues:

> Since most of the meatier body parts were carried away from the site, presumably to some nearby living site, or base camp, it is possible, even likely, that those active in meat processing at the Torralba butchering

site did not include the total personnel of a coresident group. (Freeman 1975a:681-682; earlier Freeman had estimated the number of workers in the "team"-based work force engaged in butchering meat at between 10 to 35 individuals.)

Here we see inferences drawn from inferences; each is increasingly dependent upon the accuracy of the initial inferences instead of being justified in terms of arguments about the nature of causal processes operative in the past that produced the patterning in the archaeological record. The propositions used as assumptions in the interpretative arguments are consistently unwarranted inferences as to what the past was like. For instance, how does Freeman justify the assumption that meat was removed from the site of Torralba? Simple. He assumes that all the parts were originally present because he assumes that the animals were hunted by hominids and killed in groups on the site. Given that such assumptions are unwarranted, it is clearly reasonable to consider the possibility that the parts present at Torralba were introduced to the location and that the missing parts had never been there! This possibility is never considered simply because the interpreters of Torralba assumed as factual both the hominid role in the accumulation of the deposit and the particulars of the character of that role -- big-game hunting using game drives in which multiple prey individuals were killed on the spot in single episodes.

Clearly, the crucial arguments to address are the original interpretations. If they were unwarranted, all derivative arguments would fall like a house of cards. Before taking on this important task, it is important to discuss from a methodological perspective the observations and arguments thus far presented about "activity areas" since they will be important in my assessment of the original interpretation that hominids were big-game hunters using game drives.

In Freeman's (1978) second paper he makes his case for activity areas. He devotes some time to the history of activity area research and presents what he believes to be the assumptions underlying activity area recognition strategies. He states these as follows:

> Assumption 1: Prehistoric men performed markedly different activities in different places whenever space permitted.
> Assumption 2: The implements used together in a specific activity or its products and by-products (or both) were usually abandoned together in a restricted area (whether or not abandonment was intentional).
> Assumption 3: Materials exclusively related to different activities were abandoned in different places.
> Assumption 4: The place of abandonment of materials related to an activity tends generally to coincide with the place of activity performance. (Freeman 1978:67)

Freeman notes that these assumptions have not been "proven valid," but he then concludes that "we must proceed as though they might be true" (Freeman 1978:67). Here we have a classic example of the point that Schiffer (1976) has made in the critical literature, namely, questioning whether a given model of archaeological formation processes is *relevant* (see Binford 1983:157-167) to the data to which it is applied for interpretive purposes. Even more important in the case of Freeman's use of a model is the question of the relevance of alternative models that might account for localizations on Paleolithic sites.

Freeman generates a list of classes to accommodate the contents at the Torralba site and seeks to defend his use of nonnormalized data (raw counts) in a multivariate analysis of the samples from Torralba. (The use of raw counts completely violates the assumptions of the statistical techniques used by Freeman -- namely, that one is working with normally distributed data.) Freeman presents his results as if (a) they were valid pattern-recognition results, and (b) they were exclusively understandable in terms of the activity area "assumptions" cited above. He summarizes his pattern recognition endeavors as shown in Table 1. After presenting these results, Freeman restates his assumptions as follows:

> All evidence from the Torralba site suggests that it was the locus of a range of activities involving the killing of game animals, dressing out

Table 1. Summary of Freeman's Pattern Recognition Results.

FACTOR CONTENT Tools	Animal Bones	Associated (non-exclusively determined) Variables
	FACTOR I	
(1) perforators (2) becs	(1) elephant skull fragments	(notches)
	FACTOR II	
(1) utilized flakes (2) flakes	(1) unworked equid limbs (2) bovid foot bones	
	FACTOR III	
(1) denticulates	(1) bovid skull fragments (2) cervid limb fragments	
	FACTOR IV	
(1) endscrapers (2) sidescrapers	(1) elephant teeth (2) unworked tusks (3) worked tusks (4) elephant scapulae (5) elephant pelves (6) elephant vertebrae (7) equid skull fragments (8) equid scapulae (9) equid vertebrae (10) cervid scapulae	(1) cervid antlers (2) worked elephant ribs (3) equid teeth
	FACTOR V	
		(1) equid pelves
	FACTOR VI	
	(1) cervid foot bones	
	FACTOR VII	
(1) burins (2) paraburins		

meat from the carcasses, some secondary stages in meat preparation, and probably, limited consumption on the spot. The clusters (factors) seem to reflect these activities. (Freeman 1978:87)

Freeman then interprets the results of his factor analysis in these terms. The association among becs, perforators, and elephant skulls as well as denticulates (Factor III) and bovid skull fragments are interpreted as follows:

> The skulls of both wild oxen and elephants have features that are necessary to sustain long, heavy horns or much heavier tusks. Extraction of the edible material from these creatures' heads requires some means of smashing the structure apart and some device for getting at edible fragments in nooks and crannies. The perforator-becs at Torralba would be suitable sharp-edged pointed probes for the latter process, and denticulates unite a series of small-pointed projections (and a series of hollow notches) in a working edge, which might have served a similar purpose. The factor I cluster contains nothing like a battering tool, but the cervid limb bones (including the metapodials) in the Factor III cluster would serve the purpose well enough. (Freeman 1978:87-88)

This is a classic example of a post hoc accommodative argument: that is, the a priori assumption is invoked to give meaning to the recognized patterns. Further post hoc accommodative arguments are offered to tailor the patterns to fit the original assumptions. I do not need to go through all of Freeman's post hoc accommodations; needless to say, they all fit the data.

What Is Done with the Recognized Patterns?

Freeman correctly recognizes that his patterns were generated by virtue of interassemblage comparisons. That is, the summed data from different levels at Torralba were the units among which patterning was sought. Given his commitment to the position that such assemblages should be variable exclusively by virtue of spatial differentiations among activities *within single, short-term occupations of essentially the same type* (game drives and subsequent processing of big game for meat) he proceeds to "strengthen" his original assumptions by demonstrating that the defining variables for the factors isolated in his pattern recognition endeavors do in fact differentially distribute when single excavated levels are studied spatially.

Freeman chooses for his demonstrations Occupations 7 and 8 from Torralba (Figs. 1 and 2). He reports the results of a chi-square test to strengthen his argument that the defining variables are differentially distributed within single excavation units. I doubt that anyone viewing these distributions would question that they are distributed differentially, but Freeman does not discuss the patterning itself. Certain questions might reasonably be asked about the illustrated patterns. For example, in Occupation 7 there are clearly two forms of patterning: (a) a linear pattern running diagonally from square 9-F to square 24-J and (b) a circular,.clustered distribution above the "L" line (see Freeman 1978:Fig. 1). What is in each? In the linear distribution are primarily diagnostics of Factor II. Above the L line are primarily diagnostics of Factor IV. And what are these diagnostics? Factor II is defined by small items, utilized flakes, unretouched flakes, unworked equid limbs, and bovid foot bones. These are lineally arranged with apparently a greater concentration in the square 9-F area. What is diagnostic of Factor IV? Clearly the most abundant diagnostics are large or heavy items: elephant teeth, tusks, scapulae, pelves, and vertebrae, along with scapulae of both cervids and equids, and equid vertebrae. Is it not possible that what we are seeing is the structured result of differential erosion? Freeman never considered this possibility since he already assumed the hominid behavioral cause for his structured results.

A remarkable situation is apparent in the distributions summarized in Freeman's illustrations. Almost all of the remains in Occupation 8 form a linear distribution in exactly the same place as the linear distribution in Occupation 7 (Figs. 1 and 2). Now, however, there is a greater concentration of diagnostics in the area of squares 9-F and E. Examination reveals that this "lower" distribution (which extends into squares 9-C through E) is similar to that found in Occupation 7, in that the small items of Factor II are more concentrated while the large items diagnostic of Factor IV are more common in the "upper" part of the distribution. In addition we note that the small item diagnostics of Factor I occur in the lower end of the linear distribution,

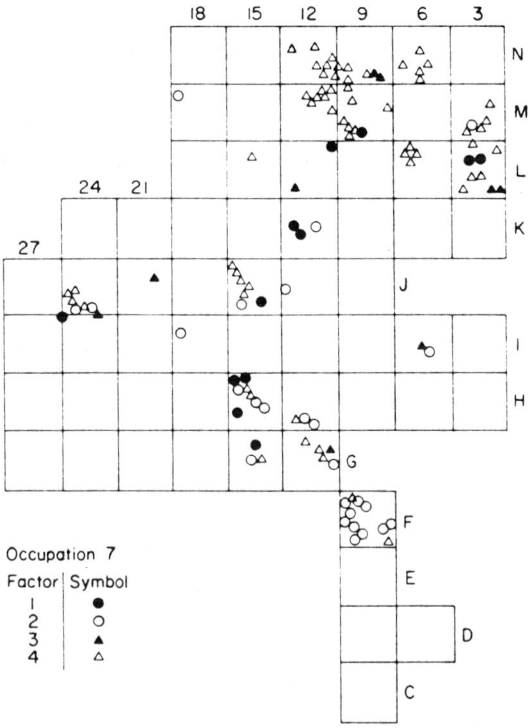

Figure 1. The distribution of factor-specific items in Occupation 7 at Torralba. Distance between grid lines equals three meters. From Freeman 1978, p.90.

which is not as evident in the Occupation 7 distribution.

What we see is a remarkable set of facts. If we are to be convinced by Freeman's arguments, we must imagine two groups of big-game hunters conducting independent big-game drives, separated in time by some unknown number of years, and placing analogous activities in exactly the same spot on the land surface. These separate activity areas yield analogous and similarly structured internal patterns -- small items occur at the lower end of a linear pattern and large items are located at the upper end of this distribution. In addition, abruptly retouched pieces, items sometimes considered indicative of cryoturbation or analogous geological processes, are heavily loaded in Factor II

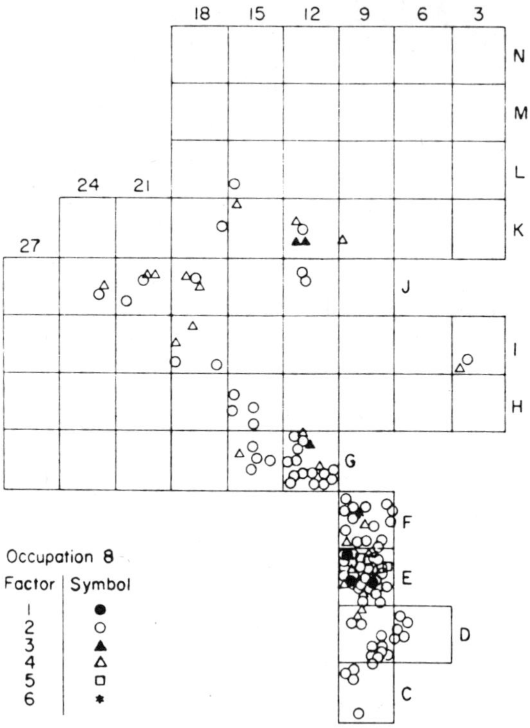

Figure 2. The distribution of factor-specific items in Occupation 8 at Torralba. Distance between grid lines equals three meters. From Freeman 1978, p.91.

in Freeman's results (Freeman 1978:88). I see no evidence for "activity areas" in Freeman's results. I see suggestive evidence for the operation of natural processes as the major cause for the differential spatial distributions illustrated by Freeman.

I must conclude that all discussion, interpretation, and analysis thus far conducted has been carried out under the assumption that the site represents the structured remains of the activities of big-game hunters primarily taking game in game drives with an emphasis on elephants. This important interpretation was originally presented in the logical form of an argument from lack of evident alternatives. As Howell (1965) has described the archaeological evidence,

> What made the fossil sites of Torralba and Ambrona interesting was the enormous number of elephant bones that they contained. These belonged to a straight-tusked species now extinct and somewhat larger than the African elephant of today. They were far and away too numerous to be explained away as having gotten there by accident. Furthermore, their condition and their positions in the ground were extremely unusual.... Mixed in with the bones were many signs of ancient human presence. Stone tools...bone tools...and even pieces of wood. (Howell 1965:84)

Two alternative explanations were considered to account for the prevalence of elephant bones and their provocative distributions: (a) man was responsible for the archaeological patterning by virtue of the evidence of his presence -- stone tools -- or (b) the bone distributions were *accidental*. The latter alternative appeared to be improbable; therefore, the romantic view of "man the elephant hunter" was claimed. No alternative possibilities, such as natural deaths or geologically structured deposits, were investigated. The role of man appeared to be self-evident. In this reasoning process one quite literally assumes the truth of what should be the object of investigation.

It should be clear by now that the basic flaw in all the interpretative work thus far done at Torralba rests with the assumption that the remains are exclusively referable to the organized actions of hominids, and that these organized activities were multiple game drives. Given these initial interpretations, which served as the basis for all further interpretative arguments, we have an internally consistent construction of past conditions of life for the hominids who are assumed to have structured the remains.

Pattern Recognition: The Basis of our Knowledge of the Past

The first principle of successful inference as to the nature of the past must be the explanation of patterning made explicit by the interaction of our units of observation (our taxonomic summary of entities recovered and observed) viewed within a frame of reference. For the majority of practicing archaeologists, this frame of reference is most often formal, chronological, and/or spatial.

In the case of Torralba the chronological frame of reference is approximated by the stratigraphic sequence of levels or "occupation units" identified by the excavators. The spatial frame of reference is provided by the horizontal grid system in terms of which the site was excavated as well as the point-provenience maps generated during excavation. The formal frame of reference is provided by our strategies for observing the patterns of statistical interaction among our basic units of observation, our types or classes of phenomena. Patterning results when we use our observations to discover the statistical interaction between our classificatory units and our frames of reference. Patterning recognizable at this juncture is not imposed on the archaeological record by our conventions for observation. It is only this patterning, seen in terms of our frames of reference, that can directly inform us (in our own terms, of course) about the past. The task of archaeologists is to explain the resulting patterning, and insofar as we are successful, we will have constructed an accurate past. We cannot assume such explanation, nor can we use unjustified inferences as the assumptive basis for still further inference, as I have demonstrated has been done at Torralba.

Obviously the archaeologist can construct false frames of reference, date units of observation inaccurately, and lay out the grid incorrectly or record inaccurate provenience data for specific items. For our purposes, the most common source of error in many attempts to observe formal patterning (structure) in our data resides in using inappropriate analytical procedures when seeking the recognition of patterning. I have suggested that Freeman has committed such an error.

A NEW MULTIVARIATE ANALYSIS OF FREEMAN'S DATA

When one uses multivariate analytical techniques one seeks to discover patterning among the variables, which for archaeologists are commonly summary frequency counts for different recognized categories of things -- e.g., tool "types" or anatomical and/or species classes of faunal remains. One is attempting to assess the common or shared variance among pair-wise comparisons for the variables being studied. The common variance of interest is not that which arises from the fact that some samples are large, having many items, while others are small, having few items. This variance may arise from numerous conditions that do not interest archaeologists at this stage of pattern recognition study. Some units may be small in terms of size of excavation, or variable in size as a result of differential placement of past occupations relative to the particular "grid squares" excavated through a stratified deposit. Such variance may well be a function of excavation strategy, and we do not want to confuse this patterning with patterning that might result from organized dynamics in the past. We want to see the latter clearly and unambiguously. For this reason, multivariate analysis *must* be carried out on normalized data, data that has been transformed in such a way as to eliminate variance arising from sample size differences.

In the early days of experimentation with multivariate techniques, several different normalizing conventions were used. Some of these include the calculation and study of percentages rather than raw counts, the calculation and study of ratio values rather than raw counts, and normalization with Z-scores. The problem of normalization was essentially solved by using a chi-square calculation. That is, chi-square values were calculated for each raw cell count in a matrix. These values were then used as the data for analysis using multivariate techniques. This is a very elegant procedure, since what one looks at are the patterned and interactive suites of variables that consistently deviate from the summed character of all the cases. If all cases were identical there would be no result, even though they all shared an identical pattern. We are quite literally looking at the dimensionally summarized sets of variables that regularly deviate from the normative pattern described by the matrix as a whole. This approach allows us to see patterning that is structured not by our viewing techniques but by dynamic conditions in the past that structured our archaeological samples. Once data are normalized the matrix is processed using a singular value decomposition (Harpending and Rogers 1985) to produce a type of principal component multivariate result. I will consistently use this approach in the following analyses.

Contrasting my approach to those used by earlier analysts, I do not assume I know what the past was like and use such assumptions to guide my analysis. On the contrary, I assume that it is the *patterning* that is the source of information about the past, and the knowledge available to me for understanding patterning is the ultimate justification of any subsequent interpretations. In short, it is the middle-range knowledge of process and pattern that makes interpretation possible.

Analysis Step One: Looking at Classes of Remains

If we adopt Freeman's last three middle-range assumptions, which stipulate that implements used together are usually abandoned together, that different activities were generally conducted in different places, and therefore that the place of abandoned materials generally coincides with the place of activity performance, what conclusions must be drawn from the following analysis?

I will begin the study of Torralba from a gross data perspective and then move to finer and finer resolution of observations. The first matrix was constructed using the ten stratigraphic levels as the cases (rows) and the seven classes of material remains as the variables (columns). Summed frequencies were calculated by level for each class of remains as reported by Freeman (1978:72-73); these classes were (1) all tools, (2) all elephant anatomical units, (3) all equid remains, (4) all bovid remains, (5) all cervid remains, (6) all unworked unidentified bone, and (7) all worked unidentified bone. These data are presented in Table 2.

The previously outlined multidimensional analysis resulted in a three-component solution with the following eigenvalues: Dimension 1, 12.55 (38 percent); Dimension 2, 6.50 (20 percent); and Dimension 3, 5.08 (16 percent). This three-component solution (Table 3), which accounts for 74 percent of the

Table 2. Classes of remains by occupation.

Occupation	Tools	Elephant	Equid	Bovid	Cervid	Unworked Unidentified Bones	Worked Unidentified Bones
1	97	182	80	15	28	85	64
2	30	3	6	5	2	10	2
3	56	42	32	2	13	21	26
4	107	99	54	24	19	66	60
5	45	31	33	9	18	26	10
6	29	13	8	5	5	25	9
7	30	17	12	5	8	10	11
8	19	12	9	2	2	7	3
9	88	140	44	6	8	31	25
10	124	25	57	9	17	42	24

variance in the data, exhibits some very provocative patterning. The most robust pattern occurs in Dimension 1, where tools are associated with all species *except* elephants. One can think of this situation as one in which tools and the non-elephant species are more common than in the matrix norm, while elephant remains and "worked" unidentified bone are less common. On the other hand, when elephant remains and worked bone are common, tools and the other species are less common than usual. This pattern documents the fact that the tools, when considered in a single group, covary with all the non-elephant material. In short, if we accept Freeman's assumptions, we would have to conclude that most of the events at Torralba indicated by faunal remains and tools were unrelated to events involving elephants!

Turning to Dimension 2 we see that all unidentified bone fragments tend to associate with bovids and cervids, while tools now covary with the greatly emphasized elephant bone, together with equids. Since Dimension 1 accounts for 38 percent of the total variance, clearly the deposition of tools was *greater* with the bovid and equid remains at the site than with the much discussed elephant remains. We must keep in mind that the two dimensions record independently varying patterns, and thus the episodes in which tools were structured in association with the bovid, cervid, and equid remains appear to be independent of the conditions that structured the association of tools with elephant remains. Based on these results alone we would have to conclude that the deposits at Torralba are palimpsests of independent depositional events. The most common situations are those in which bovid, cervid, and equid remains were deposited in regular ways *with* tools, while elephant remains were deposited independently and in a manner *unrelated* to tools.

The next most common situation is one in which elephant bones were deposited in association with tools independently of conditions that tended to result in breakage and modification of bones beyond recognition. In addition, bovid and cervid remains formed a separate suite that was unassociated with tools. The presence of these two suites suggest another set of depositional conditions, one in which bones get broken independently of hominid involvement (i.e., independent of tools). This is in marked contrast to the

Table 3. Analysis Step One: three component solution,faunal remains and lithic artifacts by class.

Dimension 1		Dimension 2		Dimension 3	
Tools	6.56	Unidentified unworked bone	3.30	Bovid remains	1.94
Bovid remains	2.23	Unidentified worked bone	2.87	Tools	1.42
Cervid remains	1.91	Bovid remains	2.33	Unidentified unworked bone	1.31
Unidentified unworked bone	1.57	Cervid remains	1.49	Elephant remains	0.68
Equid remains	1.27				
- - -	- - -	- - -	- - -	- - -	- - -
Unidentified worked bone	-1.24	Equid remains	-1.07	Unidentified worked bone	-0.48
Elephant remains	-10.01	Elephant remains	-1.45	Equid remains	-2.85
		Tools	-3.48	Cervid remains	-3.06

depositional situation in which elephant remains covary with tools. This pattern is consistent with some published observations (e.g., Freeman 1978) that record the presence of elephant bones in "low" areas suggestive of ponds or marshes while other faunal remains are found "farther from the bog" on higher ground. In the latter context, bones could be exposed to more breakage, and the association of both "worked" and "unworked" bone in a context unrelated to tools suggests that what have been judged as worked bone are more likely to be remains exhibiting normal breakage unrelated to human activities.

Turning now to Dimension 3, which accounts for 16 percent of the variance, we see that tools again appear to be associated with bovids and so-called unworked unidentifiable bone fragments. These items vary independently of cervid and equid bone as well as "worked" bone. This dimension reinforces the earlier observation that most of the patterns of regular covariation between tools and bones are clearly with non-elephant species.

When we view all of the dimensions comparatively we note some interesting things. First, all species appear in covariant associations independent of or inversely related to tools (negative Dimension 1, positive Dimension 2, and negative Dimension 3). This can be seen as documenting the fact that all the species represented at the site were on occasion distributed independently of human activities (i.e., independent of tools). The most economical way of viewing this phenomenon is that the deposit is a palimpsest of both natural depositional events and hominid tool-using events. The vast majority of the latter involved bovids and cervids, with elephants being least important with regard to associated hominid activities. Second, elephant, bovid, and equid bones were regularly deposited independently of tools, while cervids tended to be deposited with both bovids and equids but not with elephants. So-called worked bone was exclusively associated with all these cases (negative Dimension 1, positive Dimension 2, and negative Dimension 3). On the other hand, in all cases of tool-associated suites of variables, only "unworked" unidentified bone was covariantly related. This highly consistent pattern strongly suggests that the properties of broken bone identified by the earlier analysts as being diagnostic of "worked" pieces were predominantly unrelated to hominid actions. This view is supported by my own observations of bones photographed by Aguirre (part of the unpublished Torralba site report) that are cited as evidence for human modification. These specimens appear to be hydrologically rolled, and the details of original modification are frequently obscured. In those cases where the modifications are preserved to some extent, they exhibit many properties of animal gnawing or natural abrasion and breakage. This view is further supported by a recent detailed study of alleged cutmarks on bones from Torralba (Shipman and Rose 1983). These researchers found few marks that could be identified as cutmarks and concluded that

> An obvious explanation for the paucity of cutmarks on these bones is that Acheulian people had little to do with the bones comprising these assemblages. An alternative explanation is that the ubiquitous sedimentary abrasion on the Torralba and Ambrona bones has obscured the diagnostic features of nearly all of the cutmarks that were once present on these bones. (Shipman and Rose 1983:472)

Even more important to the analysis being undertaken here was the recognition by Shipman and Rose of the nearly ubiquitous evidence for sedimentary abrasion:

> An important result of the entire study involves the extent of sedimentary abrasion observed on the specimens. Every one of the Ambrona specimens and all but three of the Torralba specimens showed microscopic evidence of abrasion...This finding is remarkable considering that replicated specimens were intentionally selected to exclude those showing ...visible signs of sedimentary abrasion. (Shipman and Rose 1983:468)

Additional work confirming the conclusions drawn from examining the photographs of alleged worked bone from Torralba has been recently produced by Klein and Allwarden (in press). These researchers found that 36-37 percent of the bones from Torralba exhibit gross sedimentary abrasion and rounding (many of the specimens have been characterized as "bone pebbles"). Thus, I think it can be confidently asserted that it is no longer feasible to defend either the earlier claims by the excavators (Freeman and Butzer 1966; Freeman 1975a, 1978) or even the slightly more tempered claims (Freeman and Howell 1981, 1982) that Torralba was a high-integrity site of "living floors" on which the bones and tools were found roughly where they were dropped by ancient big-game hunters.

An additional claim illustrates the willingness of researchers to accept everything found at Torralba as resulting from hominid agencies. I am referring to the "worked" elephant tusk "artifacts" recovered at Torralba; I had the opportunity to examine these specimens in Berkeley in the spring of 1983. It is quite true that there is regularity in the pattern of breakage. In addition, the tusk tips frequently have scratch marks and other indications of abrasion. It must be recalled, however, that elephant tusks are in fact used as "tools" by elephants in removing bark from trees and in many other tasks, so that modified elephant tusks are not uncommon in nature. Although the breakage pattern at Torralba is quite distinctive, I have frequently observed the identical pattern on the ends of tusks from paleontologically excavated complete elephant carcasses. (A good example of this kind of breakage is found on the mammoth skeleton on display in the Natural History Museum at the University of Alaska, Fairbanks.)

Even more remarkable is the tusk tip shown in Figure 3, where the "distinctive" breakage pattern noted on the Torralba specimens is duplicated, but in addition an even more spectacular set of modifications characterize the specimen. A deep groove has been worn into the tip of the tusk, which when seen in plan view looks very much like an abstract human figure. Figure 3 shows a side view of this specimen; the remains of an earlier tip break are still visible behind the "head" of the rounded area at the right. This specimen was recovered by Mike English, a game warden in Kruger National Park, South Africa. He found this tusk near a waterhole; Figure 4 illustrates typical behavior of the young male elephants he was studying.

As with many other claims for distinctive breakage, it is quite likely that this modification to elephant tusks occurs when the tips are knocked off while

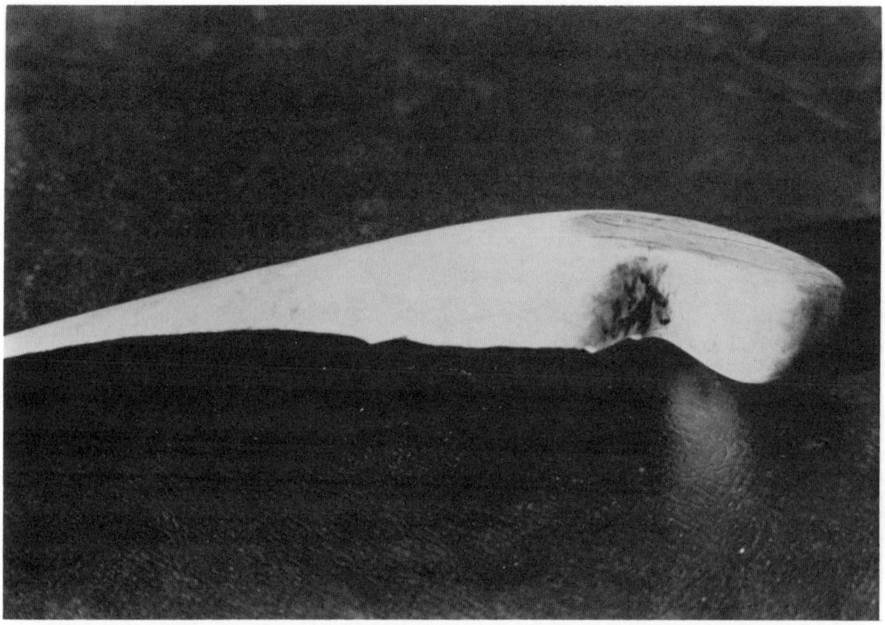

Figure 3. Worn and "worked" tip of a young elephant's tusk. (Photo by Binford)

the animal is alive. I have been told that tusk breakage is common among young males during the time when they are beginning to compete among themselves and before they leave their mothers. And most of the alleged tusk "artifacts" at Torralba are from young elephants! I strongly doubt that hominids produced these artifacts.

The skepticism expressed above regarding the integrity of the Torralba deposits is totally consistent with the results of the analysis thus far presented, specifically that the "worked" bone is consistently associated either with unidentified bone fragments indicative of natural breakage or with faunal groups that do not covary with hominid-produced tools. Further, the analysis clearly suggests that the depositional events at Torralba are more complex than the single episodes of hominid hunting and butchering that are postulated

Figure 4. Young male elephants "sparring" with one another at a waterhole. (Photo by Binford)

by the excavators. The evidence for this conclusion is as follows:

(1) There is strong and nearly ubiquitous evidence of sedimentary modification of the bones (Shipman and Rose 1983; Klein and Allwarden in press). The findings of these researchers are totally consistent with my own observations.

(2) The different species present at Torralba segregate among themselves and independently associate with stone tools. This pattern is not consistent with the claims that different species were killed in single episodes and butchered in mass processing activities organized by multiple labor groups who performed different aspects of the butchering process in different places (Freeman and Butzer 1966; Freeman 1975a, 1978).

(3) The pattern revealed in this analysis shows that the stone tools only rarely covary with elephant remains, as opposed to their relatively frequent associations with bovid, equid, and cervid remains. This pattern contradicts the imaginatively presented picture of game drives at Torralba with elephants as a major food target.

Analysis Step Two: Looking at Stone Tools

Freeman has argued that the important unit for observation relevant to inferences about tool use is not the entity, the tool or item recovered, but the multiple utilized or modified edges on each item. He has therefore reported the frequencies not of tools but of edges. Thus, if a single item exhibits simple utilization on one edge and modification indicative of a scraper on another, this single item would be tabulated under two separate classes of variables -- one tabulation for utilized pieces and another independent tabulation for scrapers. For instance, from the entire Torralba collection only 55 tools were classified as sidescrapers (extrapolated from Freeman 1975a:668; Howell 1966:135). Nevertheless, in the published "counts" (Freeman 1978:72, Table 2), 73 sidescraper edges are tabulated among the levels reported. Clearly, if we are interested in the depositional context of the artifacts, which are, of course, the units lost or discarded by ancient hominids, Freeman's data do not report any units of relevance. His choice of variables is only useful if one is interested in tool histories -- that is, when the artifacts as unit entities were components of a technologically aided set of activities. Put another way, it is artifacts that can be used and modified in different ways when they were part of a cultural system but it is entities that are discarded or lost.

Freeman's data-reporting techniques are not appropriate for investigations of the context of loss, abandonment, or natural reorganization of these items after they left the technological system. Properties of edges are relevant only in the dynamic context, not in the depositional context. Their study only makes sense if one assumes from the beginning that the depositional and subsequent archaeological contexts are one and the same and that both contexts are directly referable to the dynamic tool-use context of the past. Since we have good reason to doubt the validity of these assumptions, the study of edges makes little sense as a basis for "seeing" activity areas. Nevertheless, I have conducted an analysis on these data in search of patterning that might be relevant to the problems being considered here.

As in the previous analysis, I have used a chi-square normalization of the raw count data presented by Freeman (1978). These normalized data were then subjected to a singular value decomposition resulting in a principal component multivariate view of the interaction among variables in the matrix. This solution accounts for 60 percent of the total variance, with Dimension 1 representing 30 percent, Dimension 2 representing 18 percent, and Dimension 3 representing 12 percent. The remaining 40 percent was distributed among many dimensions, all of which appeared to represent noise. This fact alone suggests that both the method of tabulation and the postdepositional history of the site have perhaps contributed to a rather unstructured set of relationships.

The results shown in Table 4 summarize some components of patterns that are well documented in the Paleolithic archaeological record and other characteristics that are difficult to understand. Perhaps the most noticeable pattern is the fact that lithic waste (positive Dimension 1) varies inversely with a very recognizable group of associated tools. The pattern of independent distribution of lithic waste is well known in the Paleolithic from the Oldowan through some Mousterian facies. Thus, this situation is familiar in one regard, but the particular group of tools with which waste is inversely correlated is atypical. Most students of the Paleolithic would recognize the group of tools mutually associated negatively in Dimension 1 as a variant of what is known from the Mousterian as a "denticulate" (Bordes 1963) assemblage, which when accompanied by bifaces and some "Upper Paleolithic" tools may be known as "Mousterian of Acheulian tradition."

In my experience there is considerable lithic waste in denticulate-dominated assemblages, indicating that the tools generally seem to have been produced and discarded at the same place. This also characterizes the "Mousterian of Acheulian tradition," but in these latter assemblages, denticulates may be associated with considerable quantities of nonlocal raw

Table 4. Analysis Step Two: three-component solution, lithic items only.

Dimension 1		Dimension 2		Dimension 3	
Lithic waste	7.48	Sidescrapers	3.38	Abrupt retouch	2.02
Utilized flakes	2.69	Endscrapers	2.81	Denticulates	1.27
Abrupt retouch	2.37	Cores	1.42	Burins	1.14
Choppers	0.29	Choppers	1.19	Cores	0.92
		Abrupt retouch	0.82	Utilized flakes	0.84
		Bifaces	0.46	Sidescrapers	0.80
				Retouched edges	0.53
- - - - - - - - - - - - - -		- - - - - - - - - - - - -		- - - - - - - - - - - - -	
Burins	-0.24	Retouched edges	-0.06	Perforators	-0.23
Retouched edges	-0.57	Lithic waste	-0.10	Choppers	-0.40
Endscrapers	-1.03	Notches	-0.45	Endscrapers	-0.45
Cores	-1.04	Burins	-0.45	Lithic waste	-1.42
Perforators	-2.13	Utilized flakes	-1.44	Bifaces	-1.69
Bifaces	-2.58	Denticulates	-2.52	Notches	-2.42
Sidescrapers	-2.75	Perforators	-2.64		
Notches	-2.99				
Denticulates	-3.99				

material, frequently in the form of hand-axes accompanied by flakes derived from retouching. Another earlier denticulate-dominated assemblage is from the site of Isernia, which dates from approximately 700,000 years ago. There, however, the frequencies of lithic waste are not reported, and the site is probably fluvially structured. Other "pre-Mousterian" assemblages dominated by denticulates and notches are reported. For instance, assemblages from the Evenosian "culture" (de Lumley 1975:790, 798), which was recognized by Bordes (1968:92) long ago as perhaps being ancestral to the "Denticulate Mousterian," are said to be characterized by much lithic waste. The assemblage from the roughly contemporaneous site of Cantolouette (Guichard 1965) is also dominated by denticulates and notches and has abundant lithic waste.

It is of course possible that at Torralba we are seeing some new "variant" of a denticulate-dominated assemblage (see Freeman's [1975a:670] summary of the Torralba assemblage) in which lithic debris is inversely related to denticulates and notches, especially since the excavators report that the majority of the raw materials used in tool production at Torralba were introduced from lithic sources located some distance from the site (Howell 1965:91). On the other hand, a significant proportion of the hand-axes were made of locally available limestone (Freeman 1975a:668), while the majority of the shaped tools, including denticulates, are said to be manufactured of flint and quartzite (materials not found at the site). This pattern is exactly opposite to that known from most Acheulian sites, where the large tools commonly are not associated with the debris of their manufacture and are made of materials exotic to the site, and where small tools are frequently associated with the debris of primary reduction. It is possible that the large numbers of composite tools, those noted by Freeman as having multiple types of edges, indicate that the archaeological remains at Torralba were actually used as a lithic source by groups of hominids who occasionally visited the site. The subsequent hominid modification of many of the tools would obscure their original form, and certainly cores would be greatly reduced by such reuse. This type of hominid behavior is suspected as early as the Oldowan, and if it can be substantiated for the Torralba materials, it would challenge the excavators' views that the deposits reflect the remains of well-planned, cooperative game drives. The presence of hand-axes of locally available limestone undermines the conclusion that at Torralba the hominids planned ahead and "geared up" for the hunt. During the Acheulian, bifaces appear to be one of the more regularly transported items, but this does not appear to be true at Torralba.

Still another possibility is that since the reported tool frequencies are not counts of items, but rather counts of edges on items, the first dimension in its negative aspect is primarily reporting the types of edges that in fact tend to occur on single items most frequently. That we are seeing some by-product of the classificatory procedure is a distinct possibility, since such covariation among edges on a single piece would always be associated with provenience and since the "counts" represent independently tabulated properties of single entities. It is the entities that have provenience. This possibility must not be dismissed since waste, utilized flakes, abruptly retouched pieces, and choppers are all classes for which it is hard to imagine multiple edge properties.

Turning now to Dimension 2, we note that sidescrapers, endscrapers, cores, choppers, abruptly retouched edges, and bifaces are grouped in opposition to perforators, denticulates, utilized flakes, burins, notches, and lithic waste. We see here the commonly occurring opposition of scrapers on the one hand and denticulates/notches on the other, which occurs most frequently in Mousterian and "pre-Mousterian" assemblages. This familiar pattern is demonstrated in these data both by the inversely varying pattern on Dimension 2 and by the independent distribution of scrapers and denticulate/notches (i.e., they appear on different dimensions). The association of scrapers with choppers (also known as core scrapers or core axes) and cores is a fairly well known pattern. In fact, this is one of the few patterns of tool-type association that correlates with a rather unambiguous faunal context. At the well excavated and documented Middle Stone Age elephant butchery site of Mwanganda's village in Malawi (Clark and Haynes 1970), stone tools were unambiguously associated with and clustered around three major segments of a single elephant carcass. One focus of artifact clustering was an elephant femur, and another was an upper thorax section represented by ribs, vertebrae, and the skull and mandible. A short distance away were the humerus and a distal radius shaft section. The assemblage was mainly composed of scrapers followed by roughly equal numbers of small core scrapers and notch/bec forms. Chopper/core scrapers were the most common large tools. By far the most numerous items were unmodified flakes (lithic waste), which amounted to 77 percent of the assemblage. It would be interesting to see the suite of edges isolated in positive Dimension 2 as being similar to this well-documented context of tool use in elephant butchering. As in the case of the denticulate association, however, lithic waste is inversely correlated.

The functional association of shaped tools remains provocative. As has been noted by the excavators, most of the lithic materials appear to have been introduced, and only some large tools appear to be made of locally available materials. It must be emphasized that this pattern is the reverse of that known from most Acheulian sites, where the large tools are more commonly found at some distance from the sources of materials of which they were made. If subsequent work continues to sustain this emerging pattern, then it would appear that the tool users at Torralba were ill prepared to deal with the conditions they encountered on the site. They had to make large tools on the spot (these are the tools that the excavators assume were related to hominid hunting activities), and according to the reports of many composite tools, it may have been necessary to exploit archeological specimens for recycling and further reduction and modification. Such a scenario does not fit well with the idea of planned game drives!

In its negative aspect Dimension 2 is quite provocative. If we view the "perforators" as reused or specialized denticulates, we thus see the negative aspect as a "denticulate" manifestation without bifaces and scrapers but with utilized flakes, notches, and lithic waste. The distribution of this dimension throughout the levels may well indicate the location of archaeological remains that have some integrity.

Dimension 3 groups some edge forms that are not easily interpreted. Perhaps the most important feature is that notches dominate the negative aspect while denticulates appear on the positive aspect. Occurring with notches are the bifaces, lithic waste, endscrapers, choppers, and perforators, which represent a suite of artifacts that probably do not normally exhibit multiple edge forms. Varying with the denticulates and inversely with the above group are abruptly retouched edges, burins, cores, utilized flakes, sidescrapers, and retouched edges. These types of edges are more easily imagined as perhaps occurring on a single piece with some other edge form. The question of whether the method of classification is responsible for this dimension is not easily solved since the combinations of edges and their frequencies on single pieces are not reported. Before venturing a "functional" interpretation of this dimension I feel we must await a more detailed report of the site data.

Despite the analytical limitations imposed by the use of the published data categories, we can make the following observations. (1) Notches and denticulates, followed by the majority of the other shaped pieces, tend to vary independently of lithic waste and utilized flakes. (2) Sidescrapers tend to vary inversely with denticulates, notches, and perforators when lithic waste is associated with the latter artifacts. (3) Sidescrapers and cores always occur in association with each other when lithic waste is absent. We have noted that the results of this analysis reflect well-known patterns from other sites, especially with regard to the generally inverse relationships between scrapers on one hand and denticulates/notches on the other. Torralba is one of the few sites where faunal material is apparently associated with these opposing patterns. The faunal data may provide some context for these patterns and thereby offer clues to their meaning.

Analysis Step Three: Faunal Remains

In addressing this material I have once again used the data presented by Freeman (1978). In his original tabulation he independently listed parts considered "worked" and parts considered "unworked." Because the patterned relationships among animal parts may give some indication of site integrity, as well as of the possible differential distribution of animal parts caused by hominid movement and/or introduction of anatomical segments from elsewhere,

I have summed the frequencies of worked and unworked classes for the same anatomical units. The modified data derived from Freeman's (1978:72-73) tabulations and used in this analysis are summarized in Table 5.

As in previous cases I have normalized the data by calculating chi-square values for the cells in the matrix. These were then used as input for a singular value decomposition producing a principal component multivariate solution on completely normalized data. A four-component solution that accounts for 63 percent of the variance was accepted. The remaining 37 percent is distributed among a series of low-valued dimensions that appear to represent noise. Dimension 1 accounts for 21 percent of the total variance, Dimension 2 for 17 percent, Dimension 3 for 13 percent, and Dimension 4 for 12 percent. The results of these operations are summarized in Table 6.

The most striking feature of the above patterns of relationships is that Dimension 1, which accounts for 21 percent of the variance, tells us very clearly that elephant remains vary inversely with the remains of the other large animals (bovids, cervids, and equids). If we look primarily at the variables heavily loaded on Dimension 2, the positive aspect tells us that teeth and skulls of bovids, equids, and elephants are associated and are accompanied by cervid feet, elephant feet, and elephant limbs. These parts behave as a suite and vary inversely with the feet and limb bones of bovids and equids as well as cervid antlers and equid pelves. Keeping in mind that most of the tabulated items are fragmentary rather than complete anatomical units, we must consider the possibility that Dimension 2 represents the contrast between a lag deposit (positive aspect) and a transported bone assemblage (negative aspect). The former has large or dense bones, such as teeth, while the latter has the remains of small species and fragments of generally smaller bones. This conclusion is just a suggestion at this point; however, the associations of covariantly related parts make little sense as segregations related to butchering or other organized meat processing in light of recent research at elephant butchery locations (Frison and Todd 1986; Scott 1980).

Dimension 3 has some of the same properties as Dimension 2, in that its positive aspect has bovid feet associated with elephant skulls, ribs, and feet, followed by equid skull parts. Since we already know that most of the

Table 5. Faunal Remains by Occupation.

	Occupation									
Variables	1	2	3	4	5	6	7	8	9	10
Elephant teeth	16	0	5	8	4	2	0	1	10	5
Elephant limbs	42	2	9	24	4	0	4	0	23	3
Elephant ribs	29	1	12	17	5	4	3	7	27	4
Elephant skulls	4	0	2	10	4	2	1	2	12	0
Elephant tusks	30	0	6	14	4	2	5	2	18	6
Elephant feet	11	0	4	11	3	1	0	1	18	0
Elephant scapulae	8	0	1	2	0	0	0	0	5	0
Elephant pelves	9	0	1	1	1	1	1	0	5	2
Elephant vertebrae	33	1	2	11	6	1	3	0	22	6
Equid teeth	32	6	9	21	19	3	2	2	23	22
Equid skulls	5	0	3	2	3	1	1	1	4	2
Equid limbs	14	1	6	13	4	1	7	3	7	18
Equid scapulae	7	0	2	4	2	1	0	0	2	3
Equid feet	7	0	4	6	2	0	1	2	6	4
Equid vertebrae	11	0	2	3	1	0	0	0	2	4
Equid pelves	4	0	6	5	2	2	1	1	0	4
Bovid teeth	0	4	1	8	6	2	1	0	1	0
Bovid skulls	5	1	0	9	2	0	2	0	1	0
Bovid limbs	9	0	0	5	0	3	2	2	3	6
Bovid feet	1	0	1	2	1	5	5	2	1	3
Cervid antlers	15	0	5	3	3	0	5	1	5	8
Cervid skulls	2	0	2	2	1	0	1	0	1	2
Cervid limbs	2	1	0	7	1	2	0	0	1	4
Cervid scapulae	4	0	1	2	1	1	1	0	1	1
Cervid feet	2	1	3	4	10	2	1	0	0	2
Cervid vertebrae	3	0	2	1	2	0	0	1	0	0
Unidentified bones	149	14	47	126	36	34	21	10	56	66

tabulated items are fragmentary, the absence of teeth could be due to the fact that small fragments are easily transported by natural agencies. The association varies inversely with equid vertebrae, equid teeth, unidentified bone, cervid antlers, equid scapulae, and elephant limbs. The negative aspect is dominantly an equid association of parts that might be expected to remain at the site of a carcass, with cervid antler (which I have been told is mostly shed red deer antler at Torralba; Howell, personal communication 1984) and fragments of elephant limbs in a palimpsest association.

Dimension 4 groups cervid feet and vertebrae with equid teeth, skulls, pelves, and feet, again joined by cervid antlers. These are small and light cervid bones and could easily represent materials naturally deposited into areas

Table 6. Analysis Step Three: four-component solution, faunal remains only.

Dimension 1		Dimension 2		Dimension 3		Dimension 4	
Bovid teeth	4.79	Bovid teeth	5.02	Bovid feet	3.98	Cervid feet	3.19
Cervid feet	4.35	Cervid feet	3.15	Elephant skulls	3.51	Cervid vertebrae	2.25
Bovid feet	3.59	Equid teeth	2.26	Elephant ribs	3.29	Equid teeth	2.05
Unidentifiable bones	3.14	Elephant feet	2.11	Elephant feet	2.14	Equid skulls	1.77
Cervid limbs	2.41	Elephant skulls	1.71	Equid skulls	1.00	Cervid antlers	1.77
Equid pelves	2.05	Bovid skulls	1.54	Bovid teeth	0.75	Equid pelves	1.34
Equid limbs	2.10	Elephant limbs	1.38	Cervid feet	0.47	Equid feet	1.19
Equid teeth	1.39	Elephant vertebrae	0.92	Bovid limbs	0.41	Cervid skulls	0.95
Bovid skulls	1.04	Cervid vertebrae	0.42	Equid pelves	0.29	Equid limbs	0.89
Bovid limbs	0.63	Elephant scapulae	0.41			Elephant teeth	0.86
Cervid skulls	0.61	Cervid limbs	0.38			Elephant ribs	0.77
Cervid vertebrae	0.35	Elephant teeth	0.25			Equid vertebrae	0.54
Cervid scapulae	0.34					Equid scapulae	0.41
Equid scapulae	0.09					Elephant pelves	0.18
- - - - - -		- - - - - -		- - - - - -		- - - - - -	
Equid skulls	-0.01	Equid scapulae	-0.03	Cervid scapulae	-0.07	Elephant tusks	-0.13
Cervid antlers	-0.13	Elephant ribs	-0.09	Elephant teeth	-0.14	Elephant vertebrae	-0.14
Equid feet	-0.72	Equid skulls	-0.11	Elephant tusks	-0.29	Elephant feet	-0.16
Equid vertebrae	-1.05	Cervid scapulae	-0.54	Bovid skulls	-0.31	Cervid scapulae	-0.17
Elephant skulls	-1.19	Elephant pelves	-0.61	Elephant scapulae	-0.39	Elephant skulls	-0.20
Elephant teeth	-1.43	Equid feet	-0.62	Elephant pelves	-0.42	Elephant scapulae	-0.36
Elephant pelves	-1.73	Cervid skulls	-0.68	Cervid skulls	-0.50	Bovid feet	-1.00
Elephant tusks	-2.64	Unidentifiable bones	-0.71	Cervid vertebrae	-0.63	Bovid limbs	-1.16
Elephant ribs	-2.99	Elephant tusks	-0.78	Cervid limbs	-0.68	Bovid teeth	-1.29
Elephant scapulae	-3.17	Equid vertebrae	-0.82	Equid feet	-0.69	Elephant limbs	-1.31
Elephant feet	-3.57	Equid pelves	-1.38	Elephant vertebrae	-0.91	Cervid limbs	-2.08
Elephant vertebrae	-3.61	Cervid antlers	-2.63	Equid limbs	-0.93	Bovid skulls	-2.19
Elephant limbs	-3.81	Bovid limbs	-2.89	Elephant limbs	-1.07	Unidentifiable bones	-2.29
		Equid limbs	-3.31	Equid scapulae	-1.10		
		Bovid feet	-4.08	Cervid antlers	-1.17		
				Unidentifiable bones	-1.26		
				Equid teeth	-1.59		
				Equid vertebrae	-2.49		

where perhaps equid carcasses once existed, since both equid teeth and axial skeletal parts are represented. In its negative aspect this dimension isolates unidentified bone and bovid skulls, limbs, teeth, and feet, along with elephant limbs and cervid limbs. This group has some properties of a bovid carcass associated with fragments of cervid and elephant limbs varying in a coordinated manner.

While this is a very "noisy" solution, several facts are clear. (a) Elephant remains vary inversely with bovids, equids, and cervids. (b) In several combinations what might be seen as bovid, equid, or cervid carcass associations occur independently, with carcass remnants of only equids and cervids occurring together. (c) In all dimensions except the first, some elements are always inversely related to other elements from the same type of animal. Clearly, this implies movement of parts of individual animals, if we imagine that complete carcasses on the site served as the sources of the bones. Although these implications are certainly not conclusive, at least some of this movement is probably referable to modification of land surfaces by water, solifluction, and other natural processes. The inversely related patterns suggest that many bones could actually have been accumulated by hominids and introduced to the site as anatomical segments or bone sets. This possibility cannot be ruled out with positive evidence, and the noisy patterns seen here are consistent with such a possibility. The argument that all the bones derived from animals killed at the site is not demonstrable. This becomes a crucial point when it is recalled that stone tools predominantly associate with cervids, equids, and bovids, and not with elephants, which behave in this analysis more like carcass remnants.

This analysis was very frustrating. We know something about the transport probabilities of bones, and we have several pilot studies of elephants and other faunas to serve as guides for the recognition of fluvial transport and sorting under a variety of conditions (see Frison and Todd 1986). Unfortunately, we cannot use any of this knowledge because of the way in which the Torralba faunas are reported. We know that many bones are abraded and rolled, yet these important facts are not reported. We know that most bones are broken and fragmentary, yet the bones are summarized in terms of sets of anatomical parts as known in living animals without consideration of the problem of formation processes. The units in which bones occur in these deposits are not anatomical units per se but instead bone fragments, which must be studied in terms of properties that could have affected their sorted and restructured presence in these deposits. It is these data, summarized in classes that are not relevant for such analyses, that have served as the basis for all the published interpretations as well as the arguments forcefully put forth regarding big-game hunting at Torralba. This can only be described as unfortunate.

Analysis Step Four: Combined Tool and Faunal Variables

My original idea in doing a sequence of analyses was simply to enable recognition of patterning by looking at stone tools and bones independently that might not be as obvious when both types of remains were viewed together. My final step in pattern recognition was to study the total matrix of both bones and tools as originally presented and studied by Freeman (1978:72-73).

In his original tabulations, Freeman independently listed faunal parts considered "worked" and parts considered "unworked." Because I am interested in the patterned covariation among the parts of animals as some indication of site integrity as well as the possible differential distribution of animal parts caused by hominid activity, I have retained Freeman's distinctions and have studied, using my own techniques, the identical matrix of data studied by Freeman in his analysis. As in previous cases I normalized the data by calculating chi-square values for the cells in the matrix. These were then used as input for a singular value decomposition producing a principal component multivariate solution. The six-component solution summarized in Table 7 accounts for 79 percent of the total variance, distributed as follows: Dimension 1, 19 percent; Dimension 2, 15 percent; Dimension 3, 13 percent; Dimension 4, 12 percent; Dimension 5, 10 percent; and Dimension 6, 10 percent.

In order to discuss the patterning revealed by the independent dimensions in the principal component solution I have prepared a different form of table, one that displays the patterning more graphically. Table 8 summarizes the Dimension 1 patterning in this manner.

Here we again see the pattern recognized when we studied both categories of remains (Table 3), namely that the elephant bone varies inversely with that from all other species. Those parts of other species loading along with elephant are small or light parts that might be expected to be readily transported by natural agencies. Covarying with the non-elephant species are a suite of tool classes that have consistently high loadings (see Table 7, Dimension 1). As we shall see, this is consistent with the earlier results, where tools were predominantly shown to covary with the non-elephant species (Table 3). Turning to the negative aspect of Dimension 1, we see a grouping of all the elephant bone and a much more weakly associated suite of tool classes (see Table 7, Dimension 1) headed by sidescrapers, bifaces, and endscrapers. Clearly we are being told two important facts. Elephant remains vary inversely with the other species, and sidescrapers and bifaces rather unambiguously go with elephants and are inversely related to the species/tool relationships indicated by the positive component. It cannot be overemphasized that both bifaces and sidescrapers are infrequent tool forms at Torralba relative to denticulates, notches, and retouched and utilized forms (see Freeman 1975a:670). If this were all we knew about the Torralba site, we would be forced to recognize two separate activities involving different species and different tool assemblages. One is the set of activities in which cervids, equids, and bovids were involved and the hominids were using a tool kit similar to that described earlier as being associated with an unambiguous butchering site at which the butchers seem mainly to have produced the tools at the carcasses. On the other hand, the association of the elephant remains with scraper edges (which are suspected as commonly being compound-edge tools) and bifaces (which are said to be made mainly of local materials) together with endscrapers, notches, denticulates, perforators, and cores -- but in the absence of waste -- is a strange association. If these elephant hunts were well planned, one wonders why the butchers were not equipped for the task ahead. They had to make their hand-axes on the spot of poor material,

Table 7. Analysis Step Four: master loadings

Case	Dimension 1 Variables		Dimension 2 Variables		Dimension 3 Variables	
1	-3.74	elephant feet	-4.98	bovid teeth	-3.64	worked unidentified bone
2	-3.69	unworked elephant limb	-4.94	denticulate	-3.43	worked elephant rib
3	-3.67	worked elephant limb	-3.08	cervid feet	-2.57	unworked unidentified bone
4	-3.42	elephant vertebra	-3.08	notch	-2.40	worked elephant tusk
5	-3.31	worked elephant rib	-2.10	bovid skull	-2.32	equid vertebra
6	-2.86	elephant scapula	-2.06	perforator	-1.96	worked equid teeth
7	-2.44	sidescraper	-1.89	equid pelvis	-1.88	equid pelvis
8	-2.40	elephant skull	-1.86	worked unidentified bone	-1.60	equid scapula
9	-2.35	unworked elephant rib	-1.56	cervid vertebra	-0.89	cervid scapula
10	-2.31	worked elephant tusk	-1.11	unworked unidentified bone	-0.84	endscraper
11	-2.30	unworked elephant tusk	-0.82	biface	-0.79	cervid vertebra
12	-1.47	elephant teeth	-0.74	cervid limb	-0.69	abrupt retouch
13	-1.34	bovid skull	-0.70	retouched edge	-0.65	elephant scapula
14	-1.29	biface	-0.67	worked equid limb	-0.61	cervid limb
15	-1.24	elephant pelvis	-0.60	core	-0.60	cervid antler
16	-1.22	worked unidentified bone	-0.47	cervid skull	-0.56	bovid skull
17	-0.98	endscraper	-0.36	worked equid teeth	-0.54	cervid skull
18	-0.80	notch	-0.30	elephant skull	-0.35	elephant pelvis
19	-0.59	equid feet	-0.25	equid feet	-0.32	unworked bovid limb
20	-0.40	equid vertebra	-0.24	worked elephant rib	-0.30	bovid feet
21	-0.32	cervid vertebra	-0.23	equid skull	-0.25	unworked equid limb
22	-0.12	cervid scapula	-0.11	worked elephant limb	-0.16	sidescraper
23	-0.11	denticulate	-0.05	burin	-0.10	elephant teeth
24	-0.11	perforator	0.05	cervid scapula	-0.06.	core
25	-0.09	core	0.07	chopper	-0.05	chopper
26	-0.09	equid skull	0.09	utilized flake	0.06	worked bovid feet
27	0.12	equid scapula	0.22	unworked equid teeth	0.13	cervid feet
28	0.22	worked bovid limb	0.23	abrupt retouch	0.13	elephant vertebra
29	0.26	burin	0.28	bovid feet	0.28	equid feet
30	0.41	cervid antler	0.38	equid scapula	0.34	equid skull
31	0.65	worked equid teeth	0.59	elephant feet	0.60	worked equid limb
32	0.67	cervid skull	0.79	worked bovid limb	0.91	waste
33	0.69	chopper	0.90	endscraper	0.93	biface
34	0.76	unworked bovid limb	0.90	worked elephant tusk	0.99	burin
35	0.87	retouched edge	1.06	sidescraper	1.22	bovid teeth
36	0.95	unworked equid teeth	1.07	cervid antler	1.36	notch
37	1.08	worked equid limb	1.14	unworked bovid limb	1.43	elephant feet
38	1.21	bovid teeth	1.33	unworked equid limb	1.45	unworked elephant rib
39	1.44	unworked unidentified bone	1.49	equid vertebra	1.60	retouched edge
40	1.54	bovid feet	1.64	elephant scapula	1.60	denticulate
41	1.59	cervid limb	2.00	elephant pelvis	1.96	unworked elephant tusk
42	1.72	equid pelvis	2.01	elephant teeth	2.07	utilized flake
43	2.46	cervid feet	2.31	unworked elephant tusk	2.08	worked elephant rib
44	2.60	unworked equid limb	2.36	unworked elephant rib	2.24	unworked equid teeth
45	2.99	abrupt retouch	2.89	elephant vertebra	2.87	elephant skull
46	5.36	utilized flake	2.90	unworked elephant limb	2.94	unworked elephant limb
47	10.18	waste	5.17	waste	3.60	perforator

Table 7. Analysis Step Four: master loadings. (continued).

	Dimension 4		Dimension 5		Dimension 6	
Case	Variables		Variables		Variables	
1	-4.42	cervid feet	-3.84	worked elephant rib	-2.62	cervid limb
2	-3.40	unworked equid teeth	-2.72	worked bovid limb	-2.09	biface
3	-3.31	bovid teeth	-2.14	unworked equid limb	-1.81	bovid teeth
4	-1.71	unworked elephant rib	-2.00	sidescraper	-1.73	waste
5	-1.57	worked elephant tusk	-1.21	elephant teeth	-1.53	bovid skull
6	-1.47	abrupt retouch	-1.08	elephant skull	-1.41	worked elephant limb
7	-1.32	elephant vertebra	-1.04	unworked unidentified bone	-1.34	notch
8	-1.28	worked unidentified bone	-1.01	perforator	-1.04	core
9	-1.27	elephant teeth	-0.91	cervid vertebra	-0.94	retouched edge
10	-0.95	elephant skull	-0.89	cervid limb	-0.89	worked equid teeth
11	-0.77	chopper	-0.86	utilized flake	-0.85	unworked elephant rib
12	-0.69	retouched edge	-0.82	worked elephant limb	-0.70	worked unidentified bone
13	-0.66	core	-0.71	worked elephant tusk	-0.61	elephant skull
14	-0.61	worked elephant limb	-0.46	equid feet	-0.57	unworked bovid limb
15	-0.59	burin	-0.35	abrupt retouch	-0.45	sidescraper
16	-0.58	sidescraper	-0.28	notch	-0.37	endscraper
17	-0.55	equid scapula	-0.24	waste	-0.36	denticulate
18	-0.46	elephant pelvis	-0.16	unworked equid teeth	-0.29	chopper
19	-0.44	elephant feet	-0.11	worked equid teeth	-0.16	elephant teeth
20	-0.31	cervid limb	-0.11	equid scapula	-0.12	unworked elephant limb
21	-0.31	equid skull	-0.11	equid vertebra	-0.10	unworked elephant tusk
22	-0.29	cervid scapula	-0.04	denticulate	-0.05	worked bovid feet
23	-0.26	cervid vertebra	-0.03	endscraper	-0.03	worked unidentified bone
24	-0.22	bovid skull	-0.02	elephant feet	0.06	worked elephant tusk
25	-0.10	elephant scapula	-0.02	chopper	0.02	elephant feet
26	0.10	unworked elephant limb	0.05	bovid feet	0.15	abrupt retouch
27	0.17	utilized flake	0.16	elephant scapula	0.18	bovid feet
28	0.19	endscraper	0.25	equid skull	0.24	elephant scapula
29	0.23	equid vertebra	0.26	bovid teeth	0.26	elephant vertebra
30	0.33	bovid feet	0.37	equid pelvis	0.35	utilized flake
31	0.68	waste	0.43	core	0.37	worked equid limb
32	0.76	cervid skull	0.63	unworked elephant rib	0.38	cervid scapula
33	0.92	unworked bovid limb	0.67	elephant pelvis	0.49	equid scapula
34	0.94	worked equid teeth	0.69	unworked elephant tusk	0.61	elephant pelvis
35	0.98	unworked elephant tusk	0.75	cervid feet	0.84	equid vertebra
36	0.98	worked bovid limb	0.78	cervid scapula	0.89	perforator
37	1.20	bovid teeth	0.79	bovid skull	0.99	burin
38	1.25	biface	0.87	elephant vertebra	1.03	cervid skull
39	1.43	perforator	1.20	biface	1.24	unworked equid limb
40	1.44	unworked equid limb	1.23	unworked bovid limb	1.34	worked elephant rib
41	1.45	equid feet	1.36	cervid skull	1.59	equid feet
42	1.53	cervid antler	1.49	worked unidentified bone	1.81	unworked equid teeth
43	1.88	notch	1.73	retouched edge	1.94	equid pelvis
44	1.99	worked elephant rib	1.99	burin	2.09	equid skull
45	2.03	worked unidentified bone	2.11	cervid antler	2.49	cervid feet
46	2.39	denticulate	2.42	unworked elephant limb	2.53	cervid antler
47	2.51	worked equid limb	2.74	worked equid limb	3.01	cervid vertebra

Cervid	Equid	Bovid	Elephant	Tools
(Positive)				
feet limbs skulls antlers	unworked limbs pelves worked limbs unworked teeth worked teeth scapulae	feet teeth unworked limbs worked limbs		waste utilized flakes abrupt retouch retouched edges choppers burins
- - - - -	- - - - - - -	- - - - - - -	- - - - - - - - -	- - - - - - - - -
(Negative)				
vertebrae scapulae	feet vertebrae skulls	skulls	feet unworked limbs worked limbs vertebrae worked ribs scapulae skulls unworked ribs worked tusks unworked tusks teeth pelves	sidescrapers bifaces endscrapers notches denticulates perforators cores

and they quite possibly made use of a variety of tools picked up on the site. (This suggestion will be clarified when the other dimensions are considered.) The normal Acheulian pattern of frequently transported hand-axes does not appear to apply at Torralba. Similarly, the common pattern of scrapers inversely related to notches and denticulates does not seem to apply when elephant remains are associated.

Dimension 2 reports an independent pattern present at Torralba (Table 9). There are several striking features to this component. Perhaps the most important is that there is generally an inverse relationship between the alleged worked bones and their unworked counterparts for both species and anatomical parts. In addition, all of the unidentified bones (both worked and unworked) go with the "worked" bones on the negative aspect of Dimension 2. The exceptions to this generalization are "worked" elephant tusks, which I have already suggested are the naturally broken tips of young elephant tusks, known

Table 9. Analysis Step Four: Dimension 2, combined tool and faunal analysis.

Cervid	Equid	Bovid	Elephant	Tools
(Positive)				
antlers scapulae	vertebrae unworked limbs scapulae unworked teeth	unworked limbs worked limbs feet	unworked limbs vertebrae unworked ribs unworked tusks teeth pelves scapulae worked tusks feet	waste sidescrapers endscrapers abrupt retouch utilized flakes choppers
- - - - -	- - - - - - - -	- - - - - - - -	- - - - - - - - -	- - - - - - - - -
(Negative)				
feet vertebrae limbs skulls	pelves worked limbs worked teeth feet skulls	teeth skulls	skulls worked ribs worked limbs	denticulates notches perforators bifaces retouched edges cores burins

to occur while the animals are still alive. The other exception is worked bovid limbs, which appear with the "unworked" bones. One way to view this pattern is that Dimension 2 separates the largely untrampled or geologically unmodified levels from the heavily trampled and/or geologically modified deposits. For those who wish to see the variability at Torralba as referable exclusively to hominid activity, a case could be made that the negative aspect of this component, upon which the majority of the tools with significant loadings are grouped, represents major hominid activities during which many bones were broken. This suggestion is found to be plausible when it is realized that it is skull parts and limb bones that are predominant for all species, including elephant. Clearly some bias in anatomical parts seems to be represented, and it is head parts that in other sites are frequently abundant when denticulates and notches dominate the tool assemblage. The activity-related interpretation is also consistent with patterns recognized earlier in which lithic waste tends to vary independently of most of the tools.

The previously observed connection between sidescrapers, endscrapers, and elephants appears here again, with what clearly seem to be elephant carcass suites or death-site remains. The presence of waste and utilized flakes makes it somewhat easier to see aspects of the positively associated variables as being related to carcass processing. No matter how we interpret the patterning, the case for "activity areas" is very hard to sustain. The elephant carcass material is inversely related to remains of other species, making it difficult to argue that the differences in tools represent tools appropriate to sequential processing steps in the butchering of a single animal. In addition, it must be kept in mind that we are studying different stratigraphic levels at Torralba. If these two sets of interrelated tools and body parts represented phases in a processing sequence, we would not expect them to be inversely related to one another among the various levels at Torralba.

Dimension 3 (Table 10) offers further insight into the suggestions made regarding Dimension 2. Here we see some very interesting associations. First, the positive aspect groups head and leg parts of both equids and bovids and a suite of elephant bones that have some of the properties of a skull and a leg association joined by ribs. The tools that are linked to this association are perforators and utilized flakes, along with retouched flakes and a heavy tool -- in this case, bifaces -- plus waste. These artifact forms are all "at home" with the notches and denticulates with which they are also associated. Perhaps a similar tool kit is associated with a similar processing task, regardless of the species being processed. This could then be seen as a similarly organized suite of tools, as was seen in negative Dimension 2. In the previous case, however, the main difference was that equids and cervids were the associated species. When we consider the negative aspect of Dimension 3, the occurrence of tool suites appears to be even more plausible. We see a clear carcass grouping of body parts from cervids and equids associated with a skull and a leg set from bovids and another carcass set -- ribs, scapulae, pelves, and teeth -- of elephants. As was the case in negative Dimension 1 and positive Dimension 2, sidescrapers and endscrapers seem to go with carcass indicators, also accompanied here by abruptly retouched edges, cores, and choppers. Choppers appear with the scraper forms in all previous suites in

Table 10. Analysis Step Four: Dimension 3, combined tool and faunal analysis.

Cervid	Equid	Bovid	Elephant	Tools
(Positive)				
feet	unworked teeth	teeth	unworked limbs	perforators
	worked limbs	worked teeth	skulls	utilized flakes
	skulls		worked ribs	denticulates
	feet		unworked tusks	retouched edges
			unworked ribs	notches
			feet	burins
			vertebrae	bifaces
				waste
- - - - -	- - - - - - -	- - - - - - -	- - - - - - - -	- - - - - - - -
(Negative)				
scapulae	vertebrae	skulls	worked ribs	endscrapers
vertebrae	worked teeth	unworked limbs	worked tusks	abrupt retouch
limbs	pelves	feet	scapulae	sidescrapers
antlers	scapulae		pelves	cores
skulls	unworked limbs		teeth	choppers

apparent association with carcass indicators, while cores and abruptly retouched edges have appeared in association in two of the three examples seen thus far. As in the case of Dimension 2, it must be kept in mind that these inversely related suites of tools and faunal associations occur independently across the different levels at the Torralba site.

We have begun to recognize a pattern, and the characteristics of Dimension 4 (Table 11) generally support what we have seen thus far -- notches and denticulates varying inversely with choppers, cores, and sidescrapers. Similarly, we see that with the denticulate tool suite (positive aspect), head parts of cervids, teeth and limbs of equids, and teeth and limbs of bovids go together and only associate with worked elephant ribs, unworked tusks, and unworked elephant limbs. In addition, there is the provocative loading of "worked equid teeth" with denticulates. I have previously noted that in some Mousterian assemblages equid upper teeth may be "worked;" their crowns have been crushed and "flakes" have even been removed from their lateral surfaces. In both Mousterian assemblages (Binford 1984:159-165) and the recently studied fauna from Zhoukoudian (Binford and Stone 1986) upper

Table 11. Analysis Step Four: Dimension 4, combined tool and faunal analysis.

Cervid	Equid	Bovid	Elephant	Tools
(Positive)				
antlers skulls	worked limbs feet unworked limbs worked teeth vertebrae pelves	worked limbs unworked limbs feet	worked ribs unworked tusks unworked limbs	denticulates notches perforators bifaces waste utilized flakes endscrapers
- - - - -	- - - - - - -	- - - - - - -	- - - - - - - -	- - - - - - - -
(Negative)				
feet limbs scapulae vertebrae	unworked teeth scapulae skulls	teeth skulls	unworked ribs worked tusks vertebrae teeth skulls worked limbs pelves feet scapulae	abrupt retouch choppers retouched edges cores burins sidescrapers

teeth of equids are sometimes burned, particularly the fourth premolars and first molars. I have suggested that this results from the roasting and subsequent impact breakage of skulls while they are resting on their teeth on a hard surface. The associations indicated by positive Dimension 4 have strong analogies to other known hominid-processed faunas in which heads and lower legs were being accumulated and processed for "marginal food." In short, the denticulate/notch "tool kits" predominantly associate with non-elephant parts and, significantly, with parts of heads and limbs/feet (see Table 11). Various elephant parts may infrequently appear, but their presence varies inversely across independent levels with elephant carcass indicators.

Turning to the negative aspect of Dimension 4, elephant now dominates the associated species. This repeats the now established pattern in which elephant remains vary independently both of the other species and of the frequently occurring tool "edge" classes. Thus, elephant tends to vary inversely with tool frequencies in general (Tables 3 and 11). As in previous cases, when an elephant carcass assemblage of bones is indicated, cores, choppers, and sidescrapers, sometimes accompanied by abruptly retouched pieces, are related. We appear to have the repetition of a very robust and internally consistent pattern. Elephant remains vary independently of the other fauna as a group. Various combinations of the other species occur in different stratigraphic levels, and elephants may sometimes coincide with varying combinations of the other species. Elephant may occur with equids, but only rarely is there any interaction except independence noted between bovids and elephants. The vast majority of the animal/tool interactions involve the non-elephant species. Among these positive animal/tool associations, when denticulates and notches (sometimes accompanied by bifaces and other small tools) dominate the tool suite, the related anatomical parts are heads and limbs, with axial parts rare. On the other hand, when sidescrapers, endscrapers, cores, choppers, and abruptly retouched edges dominate the tool associations, the dominant anatomical part frequencies, regardless of species, are suggestive of the presence of carcasses. Since elephants generally vary independently of the other species, the latter tool group strongly associates with bone suites that appear to represent elephant carcass remains.

Dimensions 5 and 6 continue as variations on this theme. Summaries of these components are presented in Tables 12 and 13. Given the rather clear pattern demonstrated by the initial dimensions to be extracted (which account for most of the variance among the tool-yielding levels at Torralba), it is not surprising that these dimensions document situations in which these robust patterns are somewhat more confused. The last two meaningful dimensions can most likely be seen as some measure of the frequency with which independent events were compounded on single occupation surfaces at Torralba.

Positive Dimension 5 shows the rare situation in which all species are represented. When this situation occurs, the only tools related are low-count forms (burins) and large tools (bifaces and cores). One suspects that this pattern is perhaps related to postdepositional disturbance. The negative aspect of this dimension, however, exhibits a return to the robust pattern previously outlined. Carcasses seem indicated for the equids, and these relate to a suite of elephant parts that could represent naturally sorted carcass

Table 12. Analysis Step Four: Dimension 5, combined tool and faunal analysis.

Cervid	Equid	Bovid	Elephant	Tools
(Positive)				
antlers skulls scapulae feet	worked limbs pelves skulls	unworked limbs skulls teeth feet	unworked limbs vertebrae unworked tusks pelves unworked ribs scapulae	burins retouched edges bifaces cores
(Negative)				
vertebrae limbs	unworked limbs feet unworked teeth worked teeth scapulae vertebrae	worked limbs	worked ribs teeth skulls worked limbs worked tusks feet	sidescrapers perforators utilized flakes abrupt retouch notches waste choppers endscrapers denticulates

remnants, with the lighter parts having been removed by natural agents. With what is most likely naturally sorted carcass remains of equids and elephants we see the familiar group of tools -- sidescrapers, perforators, utilized flakes, abruptly retouched edges, notches, and waste. Viewing this dimension as the above-described pattern of carcass-related anatomical parts and tools, with the added dimension of some natural sorting, makes a great deal of sense.

Dimension 6 seems to represent the other half of the pattern seen in negative Dimension 5, namely processing locations for skulls and limbs that have also been modified by natural sorting agents (Table 13). The positive dimension now loads primarily small and low-count tools and what appear to be carcass remains from cervids, equids, and elephants. This rare interaction of carcass indicators for multiple species must be viewed as carcasses with which hominids were rarely if ever involved since only small tools are related, and these have low loading values. They may represent tools easily transported by natural agents. In any event, the only pattern that could possibly be seen as a "mass kill" occurs in positive Dimension 6. Ironically, tools are only

Table 13. Analysis Step Four: Dimension 6, combined tool and faunal analysis.

Cervid	Equid	Bovid	Elephant	Tools
(Positive)				
vertebrae	skulls	feet	worked ribs	burins
antlers	pelves		pelves	perforators
feet	unworked teeth		vertebrae	utilized flakes
skulls	feet		scapulae	abrupt retouch
scapulae	unworked limbs		feet	
	vertebrae		worked tusks	
	scapulae			
	worked limbs			
- - - - -	- - - - - - -	- - - - - - -	- - - - - - - -	- - - - - - - -
(Negative)				
limbs	worked teeth	teeth	worked limbs	bifaces
		skulls	unworked ribs	waste
		unworked limbs	skulls	notches
		worked feet	teeth	cores
			unworked limbs	retouched edges
			unworked tusks	sidescrapers
				endscrapers
				denticulates
				choppers

marginally involved, and these are small and hence could be naturally sorted forms.

The negative dimension associates elephants and bovids in a rare pattern of interaction in which both appear to represent head and limb associations. The elephant suite exhibits a clear bias for heavy and "lag" deposit remains, however, such as tusks and teeth. These could represent highly modified carcass remains (lag deposits) forming a background for bovid processing events carried out by hominids. This suggestion would render more understandable the rare association of both notches and denticulates with sidescrapers linked to cores, waste, and retouched flakes.

I consider the results of this pattern recognition study to be exciting and provocative. We have isolated some robust patterns in the Torralba data. The interaction between hominids and faunal remains seems clear. In fact, the results are not in conflict with the results that Freeman obtained (see Table 1); however, because of his procedures he could not see the detail observable here. His first three factors all linked skull and limb parts of a variable suite of largely non-elephant species to notches, denticulates, utilized flakes, lithic waste, perforators, and becs. These three factors accounted for the majority of the variance in his data. Factor IV then loaded what is clearly the elephant carcass suite of bones, and as we found, endscrapers and sidescrapers were strongly linked to the carcass indicators. Freeman had before him patterns demonstrating that one suite of the most abundant tools went with the non-elephant species, which were primarily represented by head and leg parts. Independent of the above relationships, elephant carcasses varied with scrapers. It is very hard to see how killing an elephant and initially processing it could result in a secondary processing area dominated by cervid, equids, and bovids. Freeman simply believed the earlier intuitive interpretations (Freeman and Butzer 1966) and accommodated his pattern-recognition results to these beliefs.

CONCLUSIONS

Much of the bone from Torralba has been heavily modified by natural agents; we strongly suspect that water and solifluction and other, biologically produced modifications have occurred, such as trampling, particularly by elephants. This means that the excavated surfaces, while probably variable in integrity (Binford 1981:19), cannot rationally be assumed to represent high-integrity "living floor" surfaces where the total content can be directly referable to hominid actions. Instead, careful taphonomic studies and informed considerations of formation processes are required even for remains and tools unambiguously produced by hominids. Thus far such studies have not been conducted. In fact, the popularized interpretations of the Torralba data have essentially pretended that these problems were nonexistent. This situation immediately calls into question the interpretations made under bad assumptions.

Claims for high frequencies of cut marks (Howell et al. 1963; Howell and Freeman 1982), patterned breakage of bone by hominids for purposes of bone tool production (Biberson and Aguirre 1965; Howell 1965:84; Freeman 1975a:675), and the citation of special anatomical parts, such as the alleged tools manufactured from the tips of elephant tusks and thought to have been hafted (Freeman 1975a:675; Howell and Freeman 1982:12), are common. Recent middle-range research has cast grave doubt on the accuracy of such interpretations. As pointed out here, investigations by Shipman and Rose (1983) show that very few of the specimens selected specifically by the investigators as "good" examples of bones exhibiting cut marks could be accepted as such when modern identification techniques were used. During the course of these investigations much evidence indicative of abrasion and other natural modifications was noted on the bones. These findings strongly support Klein and Allwarden's (in press) work, which reports a high order of modification by natural agents and the observation that many bones are clearly rolled or modified by solifluction. My own taphonomic research supports these conclusions, particularly emphasizing that the "distinctive breakage" patterns are frequently observable on bones broken by natural agents or are a consequence of marrow bone breakage by hominids where no tool manufacture was intended or carried out. Finally, the "hafted elephant tusk" spear points have been noted to simply result from natural tip breakage among living elephants. Thus, most of the early and spectacular claims for unique and provocative actions by hominids using bone tools turn out to be enthusiastic arguments made with little knowledge of patterns in nature. In short, middle-range studies did not anchor such interpretations; instead, "what else could it be" arguments warranted romantic views of the Torralba hominids.

Other reconstructions of ancient life have also been based on Torralba data. For instance, manuports (natural but allegedly hominid-introduced stones) have been "identified" and interpreted at Torralba.

> Some manuport accumulations could be deliberate stockpiles of ammunition made for future eventualities; a social group sophisticated enough to conduct a drive like those suggested would probably have been capable of moving the prey quite precisely in a given direction, taking advantage of the terrain. Stockpiles of raw material for tool manufacture as well as ammunition caches might have been made at several suitable "natural

abbatoirs" along the valley. (Freeman 1975a:680)

I had intended to add manuports to the list of "tools" presented by Freeman for study in this analysis. In fact, scattered in the literature are counts of manuports for all the levels used in this analysis. Nevertheless, still other reports give conflicting counts, so I concluded that I could not be sure that the reported counts that seemed to be consistent did in fact derive from the same years of excavation as those represented by Freeman's (1978) tables. During the course of this study, however, I did analyze a matrix of tools and fauna with counts of manuports from similar proveniences. The results were striking. Manuports varied with no other variables and behaved inversely with both tools and fauna on the first component. This situation was mainly caused by the report of 1700+ manuports from Level I. Level I at Torralba is the only stratum described as being heavily disturbed, and it even yields "patterned ground" phenomena known to result from freeze/thaw conditions operating on deposits (Butzer 1965:1719). One wonders how one could identify manuports in what is described as a "coarse, subangular to subrounded gravel" deposit (Butzer 1965:1719). I think it is clear that the investigators have been overzealous in their identification of manuports. Secondary and low variance components did group manuports with tools of the denticulate/notch suite discussed in this report; however, much of the variance in the occurrence of manuports remained unaccounted for. It seems quite likely to me that manuports do exist at Torralba, but their indiscriminate identification has largely obscured this important fact. I think the identification of ammunition piles is unwarranted since the only levels where manuports appeared to vary with the tools were levels yielding less than fifty such items, and most contained between one and ten.

Another area of unwarranted interpretation concerns the evidence for fire at Torralba. Early in the history of the site's excavation the recognition of diffuse charcoal in the levels at Torralba was taken as evidence for purposeful fires used by the hominids "to drive unwieldy elephants into the mud" (Howell 1965:84, 94-95). There seems to be no realization here that fire is a natural and constant part of ecosystems. The presence of dispersed charcoal in a geological deposit does not demand "heroic" interpretation. This is particularly true in sediments collected in lakes and bogs, where floating carbon is concentrated and then deposited as the water level retreats.

The age structure of the elephants at Torralba was initially said to be biased in favor of young elephants, which were thought to be "most vulnerable to the hunt" (Biberson 1964:216). This view has more recently been revised and the claim has been made that "adult specimens make up 70 percent of recognizable individual elephants" (Freeman 1975a:677). If this is true, it would be hard to argue for a "catastrophic" death situation for the elephants involved. Nevertheless, the citation of age profiles of whatever accuracy as evidence for hunting is strange since I know of no data available at that time that described what a natural elephant death population might look like. Now such data is accumulating (Soffer 1985), and I have been informed by Gary Haynes (personal communication, 1986) that the Torralba elephants are similar to known natural elephant death populations. Arguments favoring hunting interpretations based on age profiles, particularly "catastrophic" game driving techniques, seem to be totally unwarranted for Torralba.

Freeman has advanced the interpretation of the Torralba data beyond the specific data-centered approach cited above. He has entered into the domain of pattern recognition studies and as a consequence has presented a much more elaborate picture of past life at Torralba than that presented in the early scenario of big-game hunters using game drives. His analytical goal was to warrant his claim that the vast quantities of meat obtained in the game drives were sequentially processed, leaving in the archaeological record structured and redundant patterns referable to "activity areas."

This report has been a restudy of the same data used by Freeman. While better data would certainly be desirable, it is these data which have been used in constructing a picture of past life at Torralba. I have been concerned with the methodological issue regarding the degree to which earlier interpretations are warranted.

Results of this Study

This study was conducted using analytical strategies to ensure that all the assumptions of the pattern recognition techniques used were in fact met. It was conducted in a step-wise fashion.

Preliminary analysis was conducted on a matrix of summed frequencies for the separate species represented at Torralba as well as the sum of tool counts, each tabulated by "occupation" level. This gross overview approach established certain patterns that were later found to permeate the results of all subsequent analyses. Most striking perhaps was the recognition that artifacts dominantly covary with the non-elephant species -- cervids, equids, and bovids. This suite varies *inversely* with elephant remains. Independent of this dominant pattern was a second pattern in which elephant bones varied with equid remains and both covaried with tools. This suite was inversely related to bovid and cervid remains. In addition, bovids alone were linked to tools and bovids and tools together varied inversely with equids and cervids.

These patterns immediately alert us to several important implications. First, the differential distributions and associations of tools at Torralba are species-related. If Freeman's arguments regarding "activity areas" were correct, we would expect the tools to be partitioned differentially within species groups. This is definitely not the case. Second, the predominant interactions between tools and species are with the cervid, equid, and bovid group and only secondarily with elephant remains. Third, equid remains are the only ones that are associated with tools twice in the solution, and cervid remains are never independently associated with tools. This immediately suggests not only that the dominant hominid involvement was with the cervid, equid, and bovid groups but that, among these, equid remains were the most common companion of hominid presence at the site.

The two following analyses, which considered tools and fauna independently, illustrate important patterns within each class. Both, however, share some important properties. Perhaps the most noticeable feature was that both were rather "noisy" solutions. It should be recalled that in the faunal study the distinctions between "worked and unworked" bones were ignored and the classes were collapsed. In my experience, when one obtains "noisy" solutions the important variables underlying the structure in the materials being explored are not well represented among the variables chosen for study.

In the tool study we learned that lithic waste varies inversely with most of the tools. The next most important independent pattern is that sidescrapers, endscrapers, cores, choppers, abruptly retouched edges, and bifaces vary inversely with lithic waste, notches, burins, utilized flakes, denticulates, and perforators. This provocative pattern is well known in Europe; denticulates and notches tend to vary inversely with scraper forms in Mousterian assemblages. This pattern is inherent in the Torralba data, and viewing Torralba as a whole, it is the denticulate/notch suite of tools that dominates the numerical frequencies (see Freeman 1975a:670). We note a third independent pattern, one in which abruptly retouched pieces, denticulates, and burins (as dominant elements) vary inversely with notches, bifaces, and lithic waste. This is clearly a breakdown of the "denticulate/notch" pattern into a provocative subpattern. This result was encouraging in that a familiar pattern emerges, suggesting that some structural facts regarding the past organization of hominid technologies remain at Torralba.

In the faunal study the pattern seen in the gross study of fauna and tools was reinforced and elaborated. As in the early study, elephant remains varied *inversely* with the remains of cervids, equids, and bovids. Independent of this overriding pattern was an association of heads of bovids and elephants accompanied by elephant limb bones. This suite varied inversely with feet and limbs of bovids and equids and with some cervid parts. This relationship tells us that, while the inverse relationships between elephants and other species is continuing, there is some interaction between elephant skull parts and bovid skull parts. Although additional independent patterns continue to document the overall independence between elephants and other species, interspecies relationships are suggested. The nature of these interactions is unclear, however, since taphonomically ambiguous clues indicate that the interactions could be caused by either natural or hominid agents.

In the final analysis, an integrated matrix of both tool-edge frequencies and species categories broken down by anatomical units was searched for patterning. In this study the distinctions made by the excavators between "worked" and "unworked" bone was maintained. This study yielded a more robust solution than those previously obtained, suggesting immediately that there was significant structured interaction between fauna and tools and also between "worked" and "unworked" bones. The patterns we had seen in earlier studies were clarified, and some really exciting results were obtained. There was repeated confirmation of the previously recognized pattern in which denticulates and notches dominate a suite of tools that varies quite independently of tool sets dominated by side- and endscrapers. We also noted that in a comparative environment of tools the faunal remains break out into a dichotomy of what appear to be carcass remnants versus aggregated head and limb parts from different species. *The scraper-dominated tool suite is regularly associated with the carcass indicators, regardless of species, while the denticulate/notch-dominated tool suite is consistently related with the head and limb pattern, regardless of species* (except elephant). These are truly exciting results, since never before have we seen the *context* of the pattern of independent variation between denticulated and notched tools versus the scraper class of tools so familiar to most European archaeologists.

Even more important for the interpretation of the Torralba site is the fact that carcass indicators vary independently among species, and inversely within species, with the head and limb associations. This has immediate implications for the validity of Freeman's argument that activity areas are the organizational framework standing behind the patterns noted at Torralba. Simply stated, no activity area differentiation is indicated. *The differentiations are independent of one another and commonly are inversely related within a species class, and as such they could only arise as independent episodes of hominid activity at the site.* It is also quite clear that no mass coincidences of carcasses ever occurred at the Torralba location. They vary independently of one another among the species and inversely among the differentiated activities indicated. *Game drives could not account for the accumulations at Torralba.*

Finally, *the dominant tool suite at Torralba, the denticulate/notch suite, is interactively related to non-carcass indicators.* The vast majority of episodes of hominid use at Torralba refer to the accumulation of marginal animal parts from carcasses, most of which were located elsewhere. Head and limb parts appear to have been aggregated and processed at Torralba. In fact, no anatomical parts indicative of bovid carcasses are reported from Torralba. The exploitation of carcasses using a scraper-dominated tool kit is rare, but when it occurs, primarily equids, fewer cervids, and minimal numbers of elephants appear to be involved. These tools are frequently "compound" tools made from the same suite of raw materials as the denticulate/notch tool suite. One can speculate that the hominids were opportunistically "surprised" by the presence of exploitable carcasses and reworked their tool kits into appropriate forms for butchering or, alternatively, scavenged tools from already present archaeological assemblages, occasionally manufacturing larger tools from poor but immediately available materials. *It cannot be overstressed that the involvement by hominids with carcass remnants at the site of Torralba is a very minor aspect of the overall behaviorally relevant materials found there.* In turn, the total artifact inventory from all ten occupations at Torralba amounts to only 785 pieces (Freeman 1975a); as I have suggested previously (Binford 1981:15-18), hominids were not the major agents responsible for the presence of faunal remains at Torralba.

The exciting patterned interaction demonstrated here between tools and fauna is totally inconsistent with the heroic interpretations that have been popularized. Instead, the patterns are consistent with (a) minimal and episodically sporadic hominid presence; (b) accumulation of marginal parts of animal carcasses for processing, thus far repeatedly seen in the Lower Paleolithic and strongly suggestive of scavenging; and (c) occasional exploitation of carcasses of a variety of species independently distributed in the Torralba deposits. This ad hoc exploitation was performed by poorly equipped hominid groups. Judging from the extremely low frequencies of carcass-related tools, we cannot view this behavior as anything but expedient exploitation of resources that "present themselves," and it carries no positive evidence that is in anyway suggestive of planned hunting.

REFERENCES

Biberson, P. 1964. Torralba et Ambrona. Notes sur deux stations Acheuléenes de chasseurs d'elephants de la Vielle Castille. *Diputaciòn Provincial de Barcelona, Instituto de Prehistoria y Arqueologia, Monografias* 6:201-248.

Biberson, P., and E. Aguirre. 1965. Expériences de taille d'outils préhistoriques dan des os d'eléphant. *Quaternaria* 7:165-183.

Binford, L. R. 1981. *Bones: Ancient Men and Modern Myths*. New York: Academic Press.

Binford, L. R. 1983. *Working at Archaeology*. New York: Academic Press.

Binford, L. R. 1984. *Faunal Remains from Klasies River Mouth*. New York: Academic Press.

Binford, L. R. 1985. Human ancestors: Changing views of their behavior. *Journal of Anthropological Archaeology* 4:292-327.

Binford, L. R., and S. R. Binford. 1966. A preliminary analysis of functional variability in the Mousterian of Levallois facies. *American Anthropologist* 68(2):238-295.

Binford, L. R., and N. M. Stone. 1986. Zhoukoudian: A closer look. *Current Anthropology* 27(5):453-475.

Bordes, F. 1963. Le Mousterien a denticules. *Arheoloski Vestnik*. Acta Archaeologica XIII-XIV:43-49. Ljubljana: Slavik Academy of Arts and Sciences.

Bordes, F. 1968. *The Old Stone Age*. World University Library, New York: McGraw-Hill.

Butzer, K. W. 1965. Acheulian occupation sites at Torralba and Ambrona, Spain: Their geology. *Science* 150:1718-1722.

Clark, J. D., and C. V. Haynes. 1970. An elephant butchery site at Mwanganda's Village, Karonga, Malawi, and its relevance for paleolithic archaeology. *World Archaeology* 1:390-411.

de Lumley, H. 1975. Cultural evolution in France in its paleoecological setting during the Middle Pleistocene. In: *After the Australopithecines*, ed. K. W. Butzer and G. L. Isaac, 745-808. The Hague: Mouton.

Freeman, L. G., Jr. 1975a. Acheulian sites and stratigraphy in Iberia and the Mahgreb. In: *After the Australopithecines*, ed. K. W. Butzer and G. L. Isaac, 661-743. The Hague: Mouton.

Freeman, L. G., Jr. 1975b. By their works you shall know them: Cultural developments in the Paleolithic. In: *Hominisation und Verhalten*, ed. Kurth/Eibl-Eibesfeld, 234-251. Stuttgart: Gustav Fischer.

Freeman, L. G., Jr. 1978. The analysis of some occupation floor distributions from Earlier and Middle Paleolithic sites in Spain. In: *Views of the Past: Essays in Old World Prehistory and Paleoanthropology*, ed. L. G. Freeman, Jr., 57-115. The Hague: Mouton.

Freeman, L. G., Jr. 1981. The fat of the land: Notes on Paleolithic diet in Iberia. In: *Omnivorous Primates: Gathering and Hunting in Human Evolution*, ed. R. S. O. Harding and G. Telecki, 104-165. New York: Columbia University Press.

Freeman, L. G., Jr., and K. W. Butzer. 1966. The Acheulean Station of Torralba (Spain): A Progress Report. *Quaternaria* 8:9-21.

Freeman, L. G., Jr., and F. C. Howell. 1981. Acheulian occupation at Ambrona (Spain). Paper presented at the forty-sixth annual meeting of the Society for American Archaeology, San Diego.

Freeman, L. G., Jr., and F. C. Howell. 1982. Acheulian hunters on the Spanish Meseta: Torralba and Ambrona reconsidered. Paper presented at the annual meeting of the American Anthropological Association, Washington, D. C.

Frison, G. C., and L. C. Todd. 1986. *The Colby Mammoth Site: Taphonomy and Archaeology of a Clovis Kill in Northern Wyoming*. Albuquerque: University of New Mexico Press.

Gowlett, J. 1984. *Ascent to Civilization: The Archaeology of Early Man*. New York: Knopf.

Guichard, J. 1965. Un facies original de L'Acheuleen: Cantalouette (commune de Creysse). *L'Anthropologie* 69:413-464.

Harpending, H., and A. Rogers. 1985. Antana: A Package for Multivariate Data Analysis. Ms. available from senior author, Department of Anthropology, Pennsylvania State University, State College, Pennsylvania.

Howell, F. C. 1965. *Early Man*. New York: Time-Life Books.

Howell, F. C. 1966. Observations on the Earlier Phases of the European Lower Paleolithic. *American Anthropologist* 68(2, Part 2):88-201.

Howell, F. C., and L. G. Freeman, Jr. 1982. Ambrona: An early Stone Age site on Spanish Meseta. *The L. S. B. Leakey Foundation News* 22:11-13.

Howell, F. C., K. W. Butzer, and E. Aguirre. 1963. Noticia Preliminar Sobre el Emplazamiento Acheulense de Torralba (Soria). *Excavaciones Arqueologicas en Espana* 10:1-38.

Jolly, C. J., and F. Plog. 1976. *Physical Anthropology and Archaeology*. New York: Knopf.

Klein, R. G., and K. Allwarden. In press. An overview of the Torralba fauna. In: *Torralba: An Acheulian Butchering Site on the Spanish Meseta*, ed. L. G. Freeman, Jr., and F. C. Howell. New York: Academic Press.

Pfeiffer, J. E. 1978. *The Emergence of Man* (third edition). New York: Harper & Row.

Schiffer, M. B. 1976. *Behavioral Archaeology*. New York: Academic Press.

Scott, K. 1980. Two Hunting Episodes of Middle Paleolithic Age at La Cotte de Saint-Brelade, Jersey (Channel Islands). *World Archaeology* 12(2):137-152.

Shipman, P. 1984. Ancestors: Scavenger Hunt. *Natural History* 93(4):20-27.

Shipman, P., and J. Rose. 1983. Evidence of butchery and hominid activities at Torralba and Ambrona; An evaluation using microscopic techniques. *Journal of Archaeological Science* 10:465-474.

Soffer, O. 1985. *The Upper Paleolithic of the Central Russian Plain*. New York: Academic Press.

Bodies, Brawn, Brains and Noses: Human Ancestors and Human Predation

Erik Trinkaus

Department of Anthropology
University of New Mexico, Albuquerque, New Mexico

> In some scrawny troop of beleaguered not-yet-men on some scrawny, forgotten plain a radian particle from an unknown source fractured a never-to-be-forgotten gene, and a primate carnivore was born. For better or for worse, for tragedy or for triumph, for ultimate glory or ultimate damnation, intelligence made alliance with the way of the killer, and Cain with his sticks and his stones and his quickly running feet emerged on the high savannah.
>
> Abel stayed behind in the bush. (Ardrey 1961:269)

It has long been assumed that many of the unique biological and behavioral characteristics of modern humans evolved in response to the selective pressures of a predominantly predatory existence. In this scenario, for which Ardrey's prose provides but one example, the habitual pursuit, capture and dispatch of large vertebrate prey emerged early in hominid evolution. It appeared, perhaps at the time of the evolutionary split of the hominids from the lineage leading to the extant African apes, perhaps at the time of the emergence of the genus *Homo*. Regardless of the exact time of its emergence, it is frequently assumed or inferred from prehistoric data that relatively early in the existence of the genus *Homo*, and certainly by the Middle Pleistocene, humans were acting out versions of "Rambo meets the megafauna". If such a scenario is an accurate reflection of human adaptive evolution, it may be appropriate to view systematic predation an as important factor in the evolution of many of the biological features that distinguish us from our primate relatives, and perhaps as having a role in extant human behavior. If not, it may be more appropriate to see predation, to the extent that it occurs among living peoples, as but one aspect of our eclectic means of exploiting the environment.

At issue here is the extent to which Pliocene and Pleistocene hominid anatomy was "adapted" for a predatory existence. This issue has two aspects. First, to what extent could these prehistoric hominids have engaged in successful predation, given the behavioral limitations of their anatomies? Second, to what extent can past alterations in human anatomy be seen as reflections of increases in the incidence of predatory behavior? The second of these questions is potentially more answerable than the first, but both of them require a knowledge of what aspects of human anatomy (and physiology) are relevant to a predatory existence. In the past, complex cognitive abilities (presumably to allow successful, group-oriented pursuit of prey), precise manipulative abilities (presumably to permit the fabrication and wielding of weapons given our lack of biological [e.g., teeth and claws] weapons) and bipedalism (presumably to free the hands to wield those weapons) have been invoked as indications of past human predation. Although likely to have been important in the emergence of human predatory behavior, these biobehavioral complexes are neither necessary nor sufficient indicators of such behavior. Yet, by taking them along with other indicators of changing human biology and behavior during the Pliocene and Pleistocene, it *may* be possible to provide answers to the above questions.

Even though the emphasis here is on prehistoric human biology, it *must* be kept in mind that the ultimate resolution of the nature of the evolution of human predatory behavior depends upon the full integration of interpretations from human paleontology and from paleolithic archeology. Given the complex biocultural nature of the evolution of human subsistence during the Pleistocene and part of the Pliocene, neither academic sphere will be able to provide a resolution to these issues without consideration of data and interpretations from the other.

To assess the role of hunting in hominid evolution, I will review the paleoanthropological evidence for human behavioral patterns during the Pliocene and Pleistocene, emphasizing those which may have been related to the pursuit, capture and consumption of medium to large vertebrate prey. The groups to be discussed will be: the Australopithecines (including *A. afarensis, A. africanus, A. robustus* and *A. boisei*), *Homo habilis, H. erectus*, "archaic *H. sapiens*" (especially the Neandertals of the northwestern Old World), and anatomically modern *H. sapiens* (Figure 1). The last four biological groups are here assumed to have been responsible for, respectively, the Oldowan, the Acheulian, the Middle Paleolithic and the Upper Paleolithic, even though strict concordance between these biological and cultural typological units, especially around times of transition, did not always exist.

THE ANCESTRAL BACKGROUND

Throughout the following discussion it should be kept in mind, even though it should no longer need to be emphasized, that nonhuman primates occasionally engage in predatory behavior. This has been documented primarily for baboons (*Papio anubis* [Harding 1973; Strum 1975]) and common chimpanzees (*Pan troglodytes* [Hasegawa et al. 1983; Nishida et al. 1979; Telecki 1975, 1981]), but most primates have been known to prey on small vertebrates (Harding 1981; Telecki 1975). In addition, among chimpanzees there is evidence of rudimentary planning and social cooperation associated with predation as well as social consumption of the resultant carcass (Telecki 1975) (whether that social interaction is interpreted as sharing or tolerated scrounging [Isaac 1978]). This is combined with the evidence among chimpanzees for the selection, modification and transportation of materials for use as tools to facilitate food extraction (Beck 1980; Boesch and Boesch 1982; Goodall 1964; Nishida and Uehara 1982; Sugiyama and Koman 1979; Telecki 1974; White-sides 1985). It is thus evident that some level of predation as well as occasional tool use is not unique to the Hominidae. Much, if not all, of the evolution of these activities in the hominid lineage was therefore

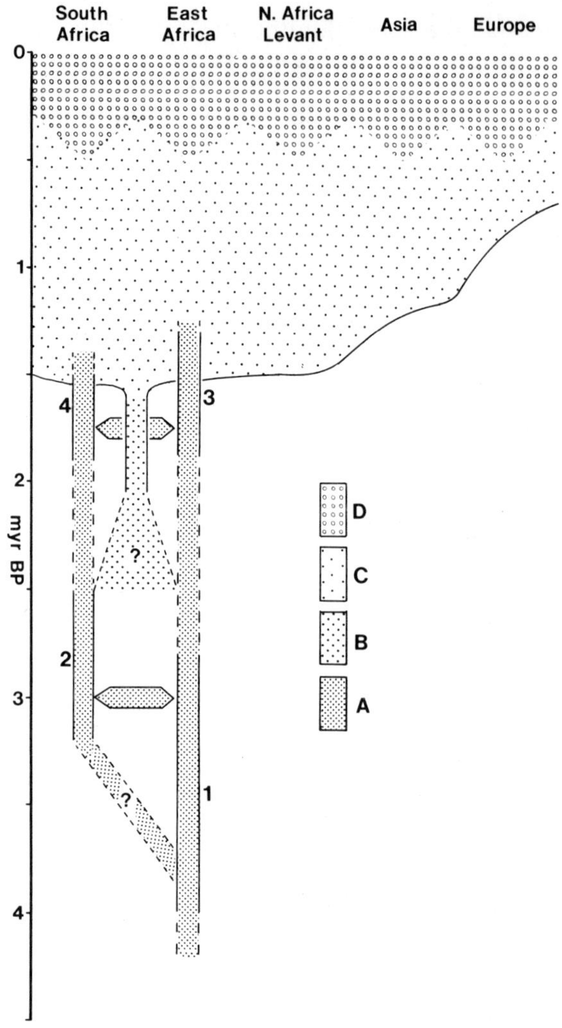

Figure 1. Phylogeny of the Hominidae. A: *Australopithecus*; B: *Homo habilis*; C: *Homo erectus*; D: *Homo sapiens*. The species of *Australopithecus* are: 1: *A. afarensis*; 2: *A. africanus*; 3: *A. boisei*; 4: *A. robustus*. *Homo sapiens* includes both "archaic *H. sapiens*" and anatomically modern humans; the boundary between them is between 50 and 30 kyr BP depending upon the geographic region. The Neandertals of central and western Asia and Europe immediately precede early anatomically modern humans in those regions and emerged gradually from more

primarily a shift in emphasis rather than the introduction of completely novel behavioral patterns.

THE AUSTRALOPITHECINES

Recent discoveries and especially analyses of fossil remains of australopithecines, as well as of their taphonomic contexts, have shown them to have been, in a behavioral sense, much closer to living apes and probably Miocene ancestral Hominoidea than to later Pleistocene and extant humans. Although clearly members of the Hominidae, they present (from our modern perspective) a mosaic of functional anatomical features.

Most characteristics of their postcranial skeleton experienced stasis throughout the australopithecine lineage; only a slight increase in body size (McHenry 1982) appears to separate the postcranial remains of the Pliocene

---

Figure 1 (*continued*) archaic members of "archaic *H. sapiens*" between ca. 150 and 75 kyr BP. The species within the genus *Homo* are considered lineal chronospecies, even though some geographical groups may have died out without issue. In the genus *Australopithecus*, the east African *A. afarensis* to *A. boisei* and the south African *A. africanus* to *A. robustus* "lineages" should be seen as regional representatives of a general *Australopithecus* lineage during the Pliocene and Lower Pleistocene. At any one point in time during that period, there was undoubtedly an irregular continuum of morphology across eastern and southern Africa (indicated by the horizontal arrows between Pliocene and Lower Pleistocene species of *Australopithecus*), with far steeper geographical clines than are known for the genus *Homo*; the regional differences in morphology were apparently sufficient to warrant the identification of contemporaneous geographical species of *Australopithecus*, but the morphological discontinuities were far less than the current incomplete fossil record suggests. *A. afarensis*, as the oldest known (perhaps going back as far as 5.5 myr BP [Kramer 1986]) and most pleisomorphous hominid species, is considered as the ancestral species of *Australopithecus*; the time of its spread into southern Africa (represented by the appearance of *A. africanus*) is not known. The genus *Homo*, represented first by *H. habilis*, is best seen as emerging from the east and south African continuum of *Australopithecus* populations between 3.0 and 2.0 myr BP, probably in the vicinity of 2.5 myr BP.

(*A. afarensis* and *A. africanus*) species from the early Pleistocene (*A. robustus* and *A. boisei*) ones. It is possible that there was a modest encephalization between the earlier and later species (Falk 1985a; McHenry 1982), and there were significant changes in facial morphology and the development of cranial masticatory muscle attachment areas, probably due to the changing morphology and orientation of the anterior dentition between *A. afarensis* and the other species of *Australopithecus* and to the increasing molarization of the premolars and molars and perhaps patterns of use of the anterior dentition in *A. africanus, A. robustus* and especially *A. boisei* (Kimbel et al. 1984; Rak 1983).

The postcranial morphology of these early hominids indicates a locomotor repertoire that included bipedalism combined with arboreal climbing and perhaps suspensory locomotion. Their inferiorly directed foramina magna, lumbar lordosis, short and broad ilia, relative femoral head size, large bicondylar angles, ovoid femoral condyles, talar facets for the plantar calcaneonavicular ligament, large calcaneal tuberosities, fully adducted and robust halluces, and dorsal extension of metatarsal head articulations (Johanson et al. 1982; Kimbel et al. 1984; Lamy 1986; Latimer 1983; Latimer et al. 1982; Lovejoy 1978; Lovejoy et al. 1973; Robinson 1972; Susman et al. 1984) indicate that they were fully orthograde bipedal primates. This conclusion is supported by the pattern of weight distribution through the foot indicated by the footprints from Laetoli (especially the G trails) (Day and Wickens 1980; M. D. Leakey and Hay 1979; Tuttle 1985). However, the minimal development of the posterior sacral and iliac tuberosities, the predominantly posterior orientations of the ilia, the minimal development of a pubic extension of the acetabular subchondral bone, their long femoral necks, the absence of lateral femoral condylar dominance, the single medial ligamentous attachment for the lateral meniscus, the pronounced breadth of the lateral malleolus and associated peroneal musculature hypertrophy, the ovoid and rotated calcaneal tuberosities, the large dimensions of the naviculocuboid facets, the predominantly proximal (rather than proximodorsal) orientations of the proximal pedal phalangeal metatarsal articulations, the elongation and diaphyseal curvature of the pedal phalanges, and the probably short hind limbs relative to trunk size (among other features) (Deloison 1985; Gomberg and Latimer 1984; Jungers 1982;

Latimer et al. 1982; Senut and Tardieu 1985; Stern and Susman 1983; Susman et al. 1984; Tardieu 1983; Trinkaus pers. observ.; Tuttle 1981) indicate that they practiced a form of bipedalism significantly different from that of extant humans. Their bipedalism undoubtedly combined greater habitual hip flexion, knee rotation, midtarsal rotation and toe grasping than is usual or possible for extant humans. All of these features would be useful to a primate engaged in frequent movement along vertical or inclined arboreal substrates but not to one committed to terrestrial bipedalism.

The morphology of their upper limb remains imply significant use of them in locomotion, probably in climbing and, to some degree, suspension. The diagnostic features include more cranially oriented scapular glenoid fossae, humeral lateral epicondyle hypertrophy, ulnar elongation and diaphyseal curvature, an enlarged and elongate pisiform bone, curved manual phalanges, third carpometacarpal articulations designed to resist predominantly axial joint reaction forces, and first and fifth carpometacarpal joints that favored stability at the expense of mobility (Bush 1980; Bush et al. 1982; Day and Scheuer 1973; Howell and Wood 1974; Marzke 1983; McHenry 1983; Senut 1981; Senut and Tardieu 1985; Stern and Susman 1983; Susman et al. 1984; Trinkaus pers. observ.; Tuttle 1981).

When combined with their relatively small body size and likely vulnerability to predation on the ground, these anatomical traits suggest a locomotor regime that included terrestrial bipedalism, climbing, some suspensory locomotion, and arboreal bipedalism. It is very possible, in fact, that this mosaic locomotor repertoire was merely the result of a small-bodied orthograde hominoid adding bipedality to its habitual patterns of movement so as to be able to move efficiently along large diameter arboreal substrates and on the ground. It may have been little more than the response of a small-bodied ape to pressures similar to those that promoted knuckle-walking in the large-bodied African Hominoidea.

The locomotor anatomy of the australopithecines thus suggests an adaptation in which both terrestrial and arboreal resources were important. Certainly, the pronounced postcanine megadontia, mandibular robusticity and masticatory musculature hypertrophy of all of the *Australopithecus* species

(Kimbel et al. 1984; McHenry 1984; Rak 1983), and especially of the Lower Pleistocene ones, implies habitual and prolonged mastication of tough, fibrous and relatively low quality foods, such as would be found most frequently in more open country environments.

Although australopithecines probably engaged in at least as much tool use as extant apes, it is uncertain whether habitual tool use was part of their behavioral repertoire. There are no unequivocal cases of tools associated with members of the genus *Australopithecus* in which there were no contemporaneous and regionally sympatric members of the genus *Homo* who could have been responsible for the implements. More importantly, the hand remains of *Australopithecus* (including members of at least *A. afarensis, A. africanus* and *A. robustus*) (Bush 1980; Marzke 1983; Marzke and Shackley 1986; McHenry 1983; Susman et al. 1984) indicate few if any modifications for tool manipulation. Their first and fifth carpometacarpal joints retain palmar interlocking elements that would have restricted axial rotation and hence limited opposition, their third metacarpals lack styloid processes and hence the ability to resist proximoulnarly direct joint reaction forces such as are generated in many human grip positions, and their distal phalanges lack any enlargement of the apical tuft for gripping with the finger tip. They did have thumb to ulnar digit length proportions similar to those of recent humans, and their second-to-third carpometacarpal articulations permitted more axial rotation of the second digit than is usual among nonhuman hominoids. It is unknown whether the former is an ancestral or derived condition, but it and the latter would have permitted slightly more grip positions, especially for activities such as throwing, than is possible for extant apes (Marzke and Shackley 1986).

It appears that at least the later australopithecines were slightly more encephalized than extant apes (Falk 1985a; McHenry 1976, 1982), but it is uncertain whether their brains had undergone significant reorganization in the direction of recent humans (Falk 1980, 1983, 1985a, b; Holloway 1978, 1981, 1983). Interestingly, recent data (Bromage and Dean 1985; B. H. Smith 1986) indicate that the australopithecines (and possibly also *H. habilis*) had growth rates similar to those of extant apes and considerably faster than those of modern humans. They thus appear to have lacked the energetically, demographically and socially expensive prolonged developmental periods necessary in the context of information-based recent human cultural systems; the implication is that the level of information processing among them was significantly less than that of recent humans.

These anatomical considerations of the australopithecines fit well with recent taphonomic analyses (e.g., Behrensmeyer 1978; Brain 1981) of the contexts of australopithecine remains, which indicate that those early hominids had little if anything to do with the accumulation of the faunal remains with which they were associated (contra Dart 1957) and were more likely the prey than the predators.

THE EMERGENCE OF THE GENUS HOMO

In the Lower Pleistocene, between about 2.0 and 1.6 myr BP, there is evidence for a hominid that was significantly different from and contemporaneous with members of the genus *Australopithecus*. Even though it is not currently possible to determine the morphological limits of this group (and hence which specimens should be included within it), there is little doubt as to its existence.[1] Since it includes the holotype of the species *H. habilis*, OH 7 (L. S. B. Leakey et al. 1964), it is best to use that species designation for this group.[2]

H. habilis contrasts with contemporaneous and preceding members of the genus *Australopithecus* primarily with respect to the lower limbs, the hands and the brain. There were undoubtedly additional contrasts, especially in the dentition and aspects of cranial base morphology (Tobias 1985; Wood 1985), but uncertainties in assigning specimens to this taxon and the absence of functional interpretations of those possible differences preclude making statements about them at this time.

The lower limbs of *H. habilis* indicate the first full commitment to terrestrial bipedalism in hominid evolution (Day et al. 1975; Day and Napier 1966; Rose 1984; Senut and Tardieu 1985; Tardieu 1983). The ilium shows the same lateral orientation of its anterior portion as those of more recent humans, a definite posterior projection of the posterior superior ilium and a clear iliac pillar. The ischium is robust, and the acetabulum is large relative to overall pelvic dimensions. The femoral head is large and positioned on a short femoral neck, and the distal femur shows lateral condylar dominance. The proximal tibia shows dual attachments for the medial ends of the lateral meniscus, indicating limited axial rotation at the knee. And the distal hallucial phalanx is large and robust, indicating full ground reaction force through the hallux during toe-off. Even though some arboreal activity was undoubtedly part of the locomotor repertoire of members of *H. habilis* for predator avoidance, sleeping and foraging, there is no evidence in their lower limbs for a continued commitment to arborealism, such as is seen in *Australopithecus*.

Just as importantly, the lower limb remains of *H. habilis* are exceptionally robust compared to those of modern humans. This is seen in the thick and dense iliac pillar (Rose 1984) and in femoral diaphyseal morphology, which includes a rounded nonpilastric diaphysis with a proximomedial shaft swelling and marked cortical thickening (Kennedy 1983; Trinkaus 1984a). The pronounced cortical thicknesses of the iliac pillar and femoral diaphysis in particular indicate resistance to both high peak levels of force (from muscular contraction and body momentum) and fatigue. Given the tendency of bone to respond through remodelling primarily to habitual patterns of stress,[3] it is most likely that the pronounced cortical hypertrophy seen in *H. habilis* lower limbs was a response to habitually prolonged periods of bipedal locomotion, hence an indication of elevated endurance.

The lower limb morphology of *H. habilis* thus suggests that they were spending a considerable portion of their activity periods moving bipedally across the landscape, presumably during the food quest. This need not imply, however, that they were carrying food to a central location or implements to food sources any more than chimpanzees do. Alterations in energy consumption patterns during locomotion and the development of a more efficient cooling of the body through respiration, which are associated with bipedalism (Carrier 1984), may well have provided sufficient selective pressure, especially in an equatorial region, to explain the shift to full terrestrial bipedalism once hominids were exploiting widely dispersed terrestrial and arboreal resources.

The hand remains of *H. habilis* consist essentially of the partial hand skeleton of OH 7; they exhibit two important features related to manipulation (Susman and Creel 1979). The first carpometacarpal joint lost the palmar metacarpal projection seen in the same articulation of *Australopithecus*, indicating greater mobility, and especially axial rotation as used in pollical opposition, at a level similar to that of modern humans. The distal pollical phalanx is broad and has a robust apical tuft, generally comparable to those of more recent humans. These alterations indicate increased dexterity and use of the thumb in powerful gripping. A powerful grip is also indicated by pronounced tendinous insertions of the flexor digitorum superficialis muscle on the middle hand phalanges of the OH 7 hand (Susman and Creel 1979).

These manual changes were associated with the first appearance of lithic implements, the Oldowan industry (M. D. Leakey 1971). These early stone tools reflect merely the discovery of a single principle relative to the use of stone hammers by chimpanzees, namely that one can obtain a sharp edge (on either the core or flake) that enables the exploitation of additional resources (Roche 1980; Toth 1985). Yet that innovation alone would have allowed an expansion in the use of resources in the landscape. In fact, it is at this time that we find the first evidence that hominids were processing portions of large vertebrate carcasses using lithic implements (Bunn pers. comm.; Potts and Shipman 1981), presumably for dietary purposes, and it is likely that those simple tools, or similar ones in wood or bone (Brain 1982), were used to extract previously inaccessible plant foods.

These contrasts with the australopithecines were associated with the first major increase in encephalization during hominid evolution; the actual amount of encephalization that occurred is uncertain, given disagreements as to which specimens should be included in *H. habilis*, problems associated with estimating the endocranial capacities of several specimens (including OH 7) and the absence of appropriate body size indicators for all of the specimens providing reliable endocranial capacity estimates. Yet, body weight estimates (e.g., McHenry 1984) suggest little if any increase in body size, especially when

H. habilis is compared to *A. robustus* and *A. boisei* remains. This is associated with an endocranial capacity increase of between 25% and 50% relative to the same australopithecines (between 40% and 70% when compared to the slightly smaller *A. africanus* remains) (Blumenberg 1985), depending upon the composition of one's *H. habilis* cranial sample. In addition, the endocranial morphology of KNM-ER 1470 (Falk 1983) suggests the first frontal lobe reorganization in the direction of recent humans.

Despite these clear alterations in locomotor and manipulative anatomy, encephalization and technology, the simplest explanation of these changes is that members of *H. habilis* were using their new abilities merely to extend the range of food resources available to them. The available paleontological evidence suggests only that the expansion involved increased manual dexterity, greater ability to perceive and recall how to obtain access to those resources, and considerable but probably poorly organized (i.e., opportunistic) movement across the landscape during the food quest. The contemporaneous archeological record indicates only that they were making tools and engaged in some processing of carcasses, probably in a variety of the available ecozones (Binford 1981, 1985; Potts 1982, 1984; Toth 1985). The associated faunal remains do not permit determination of whether the processed carcasses were hunted or scavenged, or even what percentage of them were processed by hominids; the distribution of skeletal elements among them (Potts 1982), however, strongly implies that they were scavenged, probably relatively late in the sequential exploitation of the original carcasses.

THE EXPANDING HORIZONS OF HOMO ERECTUS

The period around 1.6 myr BP saw the appearance of a new hominid species, *H. erectus*, and a new technological complex, the Acheulian (Brown et al. 1985; Clark and Kurashina 1979; Isaac and Curtis 1974; M. D. Leakey 1971; R. E. F. Leakey and Walker 1985a). Aspects of both of these taxonomic units underwent significant evolution during the 1.1 to 1.3 myr of their existences, but their general patterns appear to have remained relatively constant for this considerable period of time.

The earliest *H. erectus* are distinguishable from the preceding *H. habilis*, given current fossil evidence, primarily with respect to craniofacial morphology.[4] The endocranial capacities of the earliest sufficiently complete specimens, KNM-ER 3733 and 3883 and WT 15000 (R. E. F. Leakey and Walker 1985a; Walker and Leakey 1986), are only slightly greater than those of the largest *H. habilis* specimens, and subsequently there was only a (admittedly hard to document [Rightmire 1981]) gradual increase in encephalization during *H. erectus* existence (Holloway 1978; Weidenreich 1943; Wolpoff 1984).

The comparable postcranial elements of early *H. erectus* (primarily the innominate bone and femur) cannot be distinguished from those of *H. habilis* (Brown et al. 1985; Kennedy 1983; R. E. Leakey et al. 1978; R. E. F. Leakey and Walker 1985a; Rose 1984; Trinkaus 1984a), implying the same habitually prolonged activity patterns inferred for the earlier species. There was in fact stasis in *H. erectus* lower limb morphology, and hence in locomotor patterns, throughout the existence of the species (Day 1971, 1982; R. E. Leakey et al. 1978; Rose 1984; Trinkaus 1984a; Weidenreich 1941). The few upper limb elements preserved for *H. erectus* (Brown et al. 1985; R. E. Leakey et al. 1978; R. E. F. Leakey and Walker 1985a, b; Napier 1959; Weidenreich 1941), like those of *H. habilis*, indicate mobility patterns similar to those of recent humans combined with pronounced muscular hypertrophy.

The craniofacial alterations included reductions in the postcanine dentition and associated maxilla and mandible, a vertical shortening of the face, formation of a full supraorbital torus with a supratoral sulcus, the emergence of an occipital torus and associated structures, and the anterior projection of the skeletal nasal aperture relative to the neighboring maxillary and zygomatic regions (Brown et al. 1985; R. E. Leakey et al. 1978; R. E. F. Leakey and Walker 1985a; Rightmire 1979; Weidenreich 1937, 1943; Wolpoff 1971, 1985). The first four changes have been attributed to increasing paramasticatory use of the anterior dentition associated with reduced emphasis on the postcanine teeth (Wolpoff 1980, 1985). The last is best seen as part of an increasing emphasis on thermal efficiency during prolonged and elevated activity levels.

The anterior projection of the skeletal nasal aperture in *H. erectus* consisted of the sharp angulation between the largely vertical nasoalveolar clivus and the internal nasal floor and development of a prominent anterior nasal spine combined with a definite anterior projection of the lateral margins of the nasal aperture and of the nasal bones (Clarke et al. 1970; R. E. F. Leakey and Walker 1985a, b; Sartono 1971; Weidenreich 1943, 1945; Franciscus and Trinkaus in prep.). This provides, for the first time in human evolution, an indication of an anteriorly positioned external nose with inferiorly directed nares. Such a structure would have enabled members of *H. erectus* to retrieve moisture from exhaled air, by creating turbulence within a structure whose temperature was normally below that of the body core and hence condensing out moisture from that exhaled air. The resultant moisture could then be used for humidifying the next breath without using additional body moisture from the nasopharangeal membranes (Carey and Steegman 1981; Franciscus and Trinkaus in prep.). Such a system would have been more efficient at conserving body fluids than the flatter, more ape-like, noses of the australopithecines and probably of *H. habilis*.[5] This would have been important for a diurnal primate exploiting resources in open country, especially in relatively arid regions. Furthermore, it is at this time that significant basicranial flexion first appears in hominid evolution (e.g., KNM-ER 3733 [Laitman 1985]), implying the emergence of oral tidal respiration and the associated increased potential for voluntary pulmonary ventilation during sustained elevated activity. These shifts in nasal and basicranial morphology, in conjunction with the stasis in locomotor anatomy across the *H. habilis* to *H. erectus* transition, imply that *H. erectus* maintained the extensive movement across the landscape associated with *H. habilis* and combined it with increased activity levels in more arid country and/or during more thermally stressful periods of the day.

The contemporaneous archeological record, that subsumed by the Acheulian technological complex, indicates a continuation of the pattern seen in the Oldowan with gradual changes throughout its 1.0+ myr duration (Bordes 1984a, b; Butzer and Isaac 1975; Jelinek 1977; M. D. Leakey 1971; Villa 1983). Initially it differed from the Oldowan primarily in the introduction of tools made of flakes, hence resultant implements that were conceptually one step removed from the original raw materials; these include principally the hand axe (or biface) and the related cleaver. Through time, there was an increasing differentiation between small and large tools, frequently in terms of raw material used, and various technological advances, such as soft-hammer retouch of working edges and prepared-core techniques of knapping, appeared and became more prevalent. There is no evidence that any of these lithic implements were hafted, even though towards the end of this time period, well within the Middle Pleistocene, there is evidence of the use of wooden (probably thrusting) spears (Oakley et al. 1977). The source of the associated faunal remains is the subject of an ongoing debate (e.g., Freeman 1975 versus Klein and Allwarden n. d.; Shipman and Rose 1983; Klein this volume and Binford this volume; Shipman et al. 1981 versus Binford and Todd 1982); what is clear is that *H. erectus* was only one of the agents that collected and processed portions of the carcasses associated with tools and/or human remains during this time period. It is only at the end of, or slightly more recent than, this time period that there is evidence of humans being the sole processing agents of whole carcasses, and those carcasses are frequently the remains of relatively small mammals (Binford 1985; de Lumley 1969).

One adaptively significant change occurred during this time period; hominids spread, for the first time, out of their eastern and southern African homeland. They appear around 1.5 myr BP in the highlands of eastern Africa (Clark and Kurashina 1979), and *H. erectus* had reached Indonesia by around 1.2 myr BP (Leinders et al. 1985). Therefore, assuming that hominids reached southeast Asia through gradual extensions of ranges and group fissioning, *H. erectus* had spread across most of the tropical and subtropical Old World before 1.0 myr BP. Subsequently, by the initial Middle Pleistocene humans had spread into the temperate zones of Eurasia (Cremaschi et al. 1983; Villa 1983; Yi and Clark 1983), although there is no conclusive evidence for them in those more northern regions during glacial stadials until the end of the Middle Pleistocene, after the period of *H. erectus*. It is most likely that they moved into more northern regions along with many other mammalian species as the climate ameliorated during interglacial/interstadial periods of the Middle

Pleistocene, and either retreated southward or died out locally with the onset of subsequent stadial conditions. Evidence of fire, although frequently equivocal during this time period, does not appear until well into the Middle Pleistocene (Binford and Ho 1985; de Lumley 1966; Perlès 1977). This geographical spread of *H. erectus* indicates increasing abilities to deal with more varied environments and especially with greater seasonality of plant resource availability.

The gradual changes in human technology and increases in encephalization during the time span of *H. erectus* imply significant changes in their abilities to exploit varied resources, abilities that, although hard to define, undoubtedly were responsible for their geographical expansions between 1.6 and 1.2 myr BP and subsequent to 700 kyr BP. Yet, the stasis in their locomotor and manipulative anatomies indicate that the high levels of opportunistic terrestrial (and arboreal) foraging characteristic of *H. habilis* continued throughout this time period.

THE NEANDERTALS AND OTHER "ARCHAIC HOMO SAPIENS"

The human populations we identify as *H. erectus* evolved gradually into forms of late archaic humans referred to as "archaic *H. sapiens*" [6] (Trinkaus 1982a, 1986). Ongoing arguments as to the placement of various Middle Pleistocene specimens into one or the other of these taxonomic groups (e.g., Stringer 1984) merely emphasize the gradual nature of this transition and the arbitrariness of any firm boundary between the groups. Yet, by the end of the Middle Pleistocene, during oxygen isotope stages 6 and 7 (Berggren et al. 1980), sufficient changes had taken place to warrant the definition of a separate hominid group.

This group extended from somewhere between 300 and 500 kyr BP (depending upon where one puts the arbitrary boundary with late *H. erectus*) to the emergence of early modern humans between 30 and 50 kyr BP. It includes the Neandertals of Europe and western Asia plus their immediate predecessors in those regions and similar grade humans from Africa and eastern Asia. Most of these hominids were making lithic implements that have been, or could be, referred to as Middle Paleolithic (*sensu lato*) (Bordes 1984a, b; Trinkaus 1986).

The anatomical pattern of these hominids includes a mixture of features that are usually associated with both modern humans and earlier members of the genus *Homo* (Smith and Spencer 1984; Trinkaus 1983a, 1986). Their endocranial capacities were well within modern human ranges of variation, but their continued platycephaly indicates that on the average they had cerebral growth which was slow relative to skeletal growth (compared to modern humans) (Trinkaus and LeMay 1982). The maintenance of prognathism and large anterior teeth, in the Neandertals but not in their African and east Asian colleagues, documents continued use of the anterior dentition for paramasticatory purposes; this interpretation is supported by analyses of Neandertal facial biomechanics (Rak 1986; F. H. Smith 1983) and anterior dental attrition (Heim 1976; Ryan 1980; Trinkaus 1983a). In addition, a series of morphological features of the masto-occipital region, which distinguish the Neandertals and their northwestern Old World predecessors (Hublin 1978), as well as their hypertrophied cervical vertebral spinous processes (Heim 1976; Trinkaus 1983a), indicate enhanced stabilization of the mandible and of the whole head for such nondietary use of the anterior dentition (Trinkaus 1986).

Their lower limb remains maintained the pattern of robusticity seen in their predecessors, especially the pronounced cortical thickness of the femoral and tibial diaphyses (Endo and Kimura 1970; Heim 1982a; Lovejoy 1982; Lovejoy and Trinkaus 1980; Trinkaus 1976, 1983a, b, 1984a, b,; Trinkaus and Ruff unpub. data), although the iliac pillar hypertrophy seen in *H. habilis* (Rose 1984) and *H. erectus* (Day 1971) gradually reduced during the Middle Pleistocene (Day 1982; Stringer 1986) to reach athletic modern human levels by the time of the Neandertals in the early Upper Pleistocene (Endo and Kimura 1970; Heim 1982a; Schaefer 1957; Trinkaus 1983a). In addition, the proximal pedal phalanges of the Neandertals exhibit a mediolateral diaphyseal enlargement (Trinkaus 1983a), a response to irregular locomotion over the landscape. Humans thus continued to be engaged in habitually high levels of prolonged locomotor activity, locomotion that involved endurance, strength and a high frequency of irregular movement. They therefore appear to have spent a considerable portion of their activity periods involved in relatively unplanned, opportunistic foraging.

In addition, the faces of the Neandertals and other late archaic humans exhibit the prominent nasal apertures that emerged in *H. erectus*. Some of them, particularly among the European and Near Eastern Neandertals (Boule 1911-13; Heim 1976; F. H. Smith 1976; Trinkaus 1983a), possessed among the largest hominid nasal apertures known. It is possible that the configurations of their nasal regions were influenced extensively by the spatial and biomechanical demands of their large, anteriorly placed and heavily used anterior teeth (Rak 1986; Trinkaus 1982b; Wolpoff 1980). It is more likely, however, that their large nasal apertures, voluminous nasal cavities and projecting external noses were in large part reflections, not of cold adaptation as has been suggested (e.g., Coon 1962; Wolpoff 1980), but of a need to get rid of excess body heat during bouts of elevated activity.

The strength apparent in their lower limbs was present in their upper limbs as well (Trinkaus 1983a, b, 1986), but there it was principally muscular, not diaphyseal, and hence oriented toward peak strength rather than endurance. This muscular hypertrophy is evident particularly in the attachments for their rotator cuff muscles on the scapulae, the pectoralis major and latissimus dorsi muscles on the humeri, the pronator muscles on the ulna, and the carpal flexor, digital flexor and opponens muscles on the carpals, metacarpals and distal phalanges. In addition, their laterally bowed radii increased the moment arms of the pronator muscles and more effectively resisted muscular tension on the interosseus membrane, their relatively short proximal pollical phalanges increased thenar muscle effectiveness during the power grip, and the pronounced expansion of the distal phalangeal apical tufts indicate that they were powerfully gripping objects across their finger tips. In combination with this muscular strength, there is evidence, in the form of proximal ulnar articular orientations, to indicate that, on the average, they loaded their arms in more flexed positions than do more recent humans.

This upper limb robusticity implies that they were habitually using considerably more biological strength than do modern humans to perform manipulative tasks, and their facial and dental size and morphology and dental attrition patterns second that interpretation. When viewed from this perspective, the Middle Paleolithic archeological record can only support that view. Lithic implements, although now predominantly scrapers of considerable variety (Bordes 1984a, b), appear to have been hand-held and generally multipurpose tools; evidence for hafting is rare and equivocal indicating that most leverage was humanly generated. Bone tools were crude and expedient, modified little if at all prior to use (e.g., Gábori-Csánk 1968; Martin 1907-10; Rigaud 1982; Tode 1953). Effective projectiles have not been identified, and only wooden thrusting spears appear to have been in use as weapons (e.g., Jacob-Friesen 1956). These late archaic humans would therefore have had to exert considerably greater force to accomplish tasks comparable to those facing more recent humans, and their fossil remains indicate that they did so.

Despite their apparent continuation of an endurance- and strength-oriented and largely opportunistic foraging pattern, these late archaic humans were able to survive the rigors of the last and penultimate glacial periods (oxygen isotope periods 4 and 6), even though they were restricted to low and mid latitudes across Eurasia. Their frequent use of fire, despite limited pyrotechnology (Perlès 1976), and of natural shelters was undoubtedly partly responsible for their ability to do so. However, this also meant that some of these archaic humans experienced increasing seasonality of resources and especially greater periods without available plant foods. It should therefore not be surprising that their lithic debris is usually mixed with abundant faunal remains. Yet, as with their predecessors, it is frequently difficult to determine whether they had first and exclusive access to the carcasses, although they appear to have been able to get primary access to small- to medium-sized prey (Binford 1982b, 1984, 1985; Klein 1976, 1978; Straus 1982, 1983). Humans certainly were not the only occupants of the shelters and caves from which most of their archeological and paleontological remains derive.

There is increasingly more data which indicate that aspects of the life cycle were, on the average, different among the Neandertals (and their predecessors) than among modern humans. Studies of Neandertal ontogeny (e.g., Heim 1982b; Legoux 1970; Wolpoff 1979; Dean et al. 1986) imply that Neandertal growth rates were accelerated relative to, although overlapping with, those of modern humans. And estimates of the age at death of older Neandertals (Heim 1976; Thompson and Trinkaus 1981; Trinkaus 1983a; Trinkaus and Thompson 1987) suggest that significant postreproductive survival was extremely rare or absent among them. These interpretations imply that information was still not as important a part of the cultural adaptation as among modern humans, given the lesser amount of time available to absorb information during development and the rarity of grandparenting for trans-multiple-generational communication. The absence of clear notational or mnemonic devices among Middle Paleolithic "art" objects (e.g., Bordes 1969; Davis 1974; Marshack 1982) only serves to confirm their less developed use of information about the environment during subsistence and other activities. This, of course, accords well with the evidence for opportunistic foraging present in their locomotor anatomies.

THE EMERGENCE AND EVOLUTION OF ANATOMICALLY MODERN HUMANS

Between 50 and 30 kyr BP across the Old World, these late archaic humans were replaced, through complex and geographically variable phylogenetic processes (Smith and Spencer 1984; Trinkaus 1986), by robust versions of extant humanity. These early modern humans exhibit total morphological patterns within the ranges of variation of recent (postglacial) humans, differing on the average from later humans in having faces and teeth at the large ends of recent human ranges of variation and postcrania that fall among the more robust of recent humans (Ferembach 1976; Frayer 1978; Matiegka 1934, 1938; McCown and Keith 1939; Rightmire 1984; Singer and Wymer 1982; F. H. Smith 1984; Vallois and Billy 1965; Vandermeersch 1981; Wolpoff et al. 1984). They clearly lacked the muscularity of the upper limbs, the diaphyseal hypertrophy of the lower limbs, and the robusticity and prognathism of the facial skeleton characteristic of archaic members of the genus *Homo*.

The shift in facial morphology is probably a reflection of a marked reduction in paramasticatory use of the dentition, a decrease which is evident as well in reductions in anterior dental attrition despite smaller incisors and canines among these early modern humans. The craniofacial change thus appears to have been the product, at least in part, of technological improvements that eliminated the need for frequent use of the teeth as a vise. Those technological advances show up early in the Upper Paleolithic, in the form of lithic assemblages with frequent emphasis on blades, a polished bone technology, and evidence of hafting (Bordes 1984a, b; Clark 1983; Delporte 1984; Klein 1977; Leroy-Prost 1974; Otte 1979; Singer and Wymer 1982). The first two provided longer and more task specific tools, which in themselves provided increased leverage, and together they permitted the development of composite tools with replaceable cutting edges and greater nonbiological leverage than was available to previous, archaic humans. The paleontologically evident reductions in upper limb muscularity and muscle moment arms, as well as the shift to more habitually extended positions, follow naturally from these technological changes.

The loss at this time of the locomotor hypertrophy that characterized all archaic members of the genus *Homo* implies the emergence of planned, less opportunistic foraging patterns, ones that must have been based in extensive knowledge of the environment as are those of recent human hunter-gatherers (Binford 1982a; Gould 1980). Other indications of the emergence of a fully information-based subsistence system around this time are the presence of modern human growth and development patterns, the postreproductive survival of individuals (Skinner 1981), the loss of the platycephaly of archaic *Homo* (Stringer 1978) indicating accelerated postnatal cerebral growth (Trinkaus and LeMay 1982), and the appearance of information-coding systems in the archeological record in the form of representational art and repetitive marks on bones (Conkey 1983; Pfeiffer 1982). It was also at this time that humans first inhabited high-latitude regions and hence were able to cope effectively with the stresses of an arctic existence.

After the appearance of early modern humans in the middle of the last glacial, they continued to evolve within regions of the Old World. Primarily, they gradually lost the moderate cranial and postcranial robusticity of early modern humans, their dentitions became progressively smaller, and stature decreased (Billy 1972; Frayer 1978, 1984; Wolpoff 1980). These changes reflect gradual reductions in the physical demands made upon human anatomy, reductions that can only be viewed as the consequences of an accelerating rate of increase in technological sophistication, cultural complexity and communication within human groups during the Upper Paleolithic.

IMPLICATIONS FOR THE EVOLUTION OF HUMAN PREDATORY BEHAVIOR

What does all of this tell us about the extent to which Pliocene and Pleistocene hominid anatomy was "adapted" for a predatory existence? Does it allow us to answer the two questions posed above, namely: to what extent could they have engaged in successful predation, and to what extent were paleontologically documented anatomical alterations reflections of increases in the incidence of predatory behavior?

The answer to the first question must be that at least all of the members of the genus *Homo*, from a strictly anatomical perspective, could have been successful predators, but it is unlikely that the australopithecines engaged in more than occasional predation on small mammals. This follows primarily from the evidence for considerable commitment to arboreality among the australopithecines, especially in their upper limb anatomy. Although they could have manipulated tools effectively with their hands, the minimal modifications in their carpometacarpal regions for gripping postures and clear adaptations for use of the hands for arboreal locomotion suggests that the use of predatory implements (a necessity for so small-bodied a primate to capture more than the occasional small vertebrate) was not habitual. The existence of hands apparently with the full range of grips present in modern humans among all members of the genus *Homo*, combined with their commitments to terrestrial bipedality, implies that they had the capabilities to follow and capture mammals of at least medium size. Furthermore, given the energetic and thermoregulatory advantages of bipedalism (Carrier 1984), it is possible that they were able to run prey to exhaustion and needed few implements to dispatch them. These considerations, however, do not mean that the observed anatomical changes associated with the origin and evolution of the genus *Homo* were necessarily related to or associated with increases in human predatory behavior.

All that the prehistoric record associated with *H. habilis* indicates is that those hominids were terrestrial, had a rudimentary lithic technology, spent considerable time opportunistically foraging, and engaged in some processing of carcasses. It is, in fact, most likely that these early members of the genus *Homo* merely integrated consumption of animal products into a technologically aided, opportunistic foraging system, in much the same way that they undoubtedly expanded the range of available plant foods through the use of tools. There is no necessary reason to infer from the data that they had any priority to the animal carcasses that they processed, and it appears likely that most of the animal remains were scavenged (Binford 1981; Potts and Shipman 1981; Potts 1982). Nor is it necessary to infer that they posed any serious competition to contemporaneous carnivores/scavengers of the order Carnivora (contra Potts [1984] and Walker [1984]), even though such competition (in a general ecological sense) is possible given the presence of both cutmarks and carnivore toothmarks on the same faunal specimens (Potts 1982). We simply do not know whether these hominids utilized carcasses during seasons of prey abundance or scarcity. They may well have been using their apparently increased cognitive abilities to fit themselves effectively around more specialized consumers, without posing direct competition to any of the existing guilds.

The anatomical changes associated with the origin of *H. erectus* suggest that they may have incorporated more animal products into their diet than was common for their predecessors. Principally, the emergence of upper respiratory regions adapted for conserving body moisture and enhancing pulmonary air flow, in the context of a locomotor system adapted for endurance, implies that they were frequently engaged in high, as well as prolonged, activity levels. Such movement is best seen as associated with the pursuit of mobile prey, such as mammals.[7] Could the running of prey to exhaustion, such as was proposed by Carrier (1984), have become part of the hominid subsistence repertoire around the time of *H. erectus* origins?

Unfortunately, taphonomic analyses of faunal remains that were processed by early *H. erectus* are insufficient to indicate whether those hominids had priority to, hence were responsible for, the carcasses in question. However, the geographical spread of *H. erectus* into increasingly seasonal environments with more and more variance in plant food availability, suggests that they did indeed incorporate increasing amounts of animal products into their diet subsequent to their origins around 1.6 myr BP. Furthermore, the significant encephalization during the 1.0+ myr of *H. erectus* existence plus the increasing, though modest, conceptual complexity of their technologies imply that they were using improving cognitive and technological capabilities to exploit more varied environments and resources. Yet, it needs to be stressed that their locomotor anatomies indicate still highly opportunistic foraging patterns and that the size of the prey that they might have been hunting need not have been very large.

The data for "archaic *H. sapiens*", although more complete and more temporally compressed than that for earlier species of *Homo*, indicate primarily a continuation of the same pattern. The modest reductions in postcranial robusticity, the continued encephalization, the technological shifts associated with the emergence of the Middle Paleolithic, and the occupation of mid-latitude periglacial environments all imply greater abilities to deal effectively with seasonal variability in resource availability. Yet, the continued postcranial hypertrophy, evidence for paramasticatory use of the dentition, apparently shorter life cycle periods than those of modern humans, rarity of composite tools, and absence of evidence for information processing all demonstrate that their foraging remained opportunistic, poorly organized, energetically inefficient, and dependent upon biological compensations for cultural limitations. The available faunal data support this interpretation; they indicate that these hominids were preying primarily upon small- to medium-sized and/or weak individuals, avoiding large and dangerous prey and scavenging considerable proportions of the animal products in their diets (Binford 1982b, 1984; Klein 1976, 1978; Straus 1982).

It is only with the emergence of anatomically modern humans that we find evidence, both paleontological and archeological, for effective, systematic and organized predation as a significant part of hominid subsistence. The relevant shifts are primarily those with which more archaic humans have been contrasted, namely various indications of reductions in postcranial strength and endurance, decreases in the use of the anatomy to compensate for technological limitations and expansions of the transmission of information (presumably about the environment and its resources). Yet, it remains uncertain, given the current state of faunal analyses, to what extent the earliest modern humans, those that followed immediately after the Neandertals and other late archaic humans, were highly successful predators (Spiess 1979; Straus 1983; Klein 1977); it is primarily for later Upper Paleolithic populations that we have unquestionable archeological evidence of systematic, planned exploitation of larger mammals on a regular basis (Audouze et al. 1981; Frison this volume; Fisher this volume; Klein 1977, 1984, this volume; Leroi-Gourhan and Brézillon 1972; Soffer 1985; Straus 1983, 1985; Todd this volume).

CONCLUSION

This consideration of the Pliocene and Pleistocene paleoanthropological record and its implications for the evolution of human predation reveals a complex process, one in which predation played an increasingly important role during the course of human evolution. Yet, it appears that human predation as known ethnographically and ethnohistorically is a recent phenomenon in human evolution, associated with anatomically modern humans.

It also appears that relatively few of the anatomical characteristics of recent humans can be seen solely as the results of selective pressures generated by a predatory existence. Although the introduction of animal products into the human diet permitted expansion into increasingly seasonal environments, most of the distinguishing human characteristics, such as bipedality, manual dexterity and elaborate technology, and marked encephalization can be viewed as having been promoted by the demands of an opportunistic foraging system in which hominids fit themselves around more specialized mammalian groups. Predation, especially in equatorial regions, need only be seen as one minor facet of that foraging system. It is only the projecting noses of members of *Homo erectus* and *Homo sapiens* that appear to have evolved in direct response to the pressures of a predatory existence.

ACKNOWLEDGEMENTS

The interpretations presented here have resulted from innumerable discussions with paleoanthropological colleagues and personal examinations of the hominid fossil record. I would like to express my gratitude to the many colleagues whose ideas I have absorbed and subsequently modified and to the many curators who have let me examine original fossils in their care. Portions of this work, primarily those concerned with the Upper Pleistocene, have been supported by grants from the Wenner-Gren Foundation, the National Science Foundation, the L. S. B. Leakey Foundation, Harvard University, the University of New Mexico and the Centre National de la Recherche Scientifique. I would also like to than M. Houghteling for suggesting Rambo imagery, R. Franciscus for many profitable discussions about noses, and R. G. Klein, M. H. Nitecki, D. Pilbeam, R. B. Potts and R. H. Tuttle for helpful comments on the manuscript.

END NOTES

[1] The discussion of *H. habilis* is here based primarily on the OH 7, 10, 16 and 24 and KNM-ER 1470, 1481, 1590, 1802, 3228 and 3732 specimens. Other Lower Pleistocene remains have been included within *H. habilis* (e.g., OH 8 and 13, KNM-ER 1805 and 1813, and STw 53); some are probably attributable to *Australopithecus* and others to an additional hominid species that was present in eastern and southern Africa at the same time as *H. habilis* and early *H. erectus*. However, it is extremely difficult to achieve a consensus as to which specimens might belong in such an additional species and even

whether it existed (see R. E. Leakey et al. 1978; Walker 1984; Wood 1985). Even if it did exist, it is unlikely to have exhibited predatory behaviors outside of the range indicated by the undoubted remains of *Australopithecus* and early members of the genus *Homo*.

[2] Specimen designations for Pliocene and early Pleistocene hominids use standard site abbreviations. Those referred to here include: OH (Olduvai Hominid; from Olduvai Gorge, Tanzania), KNM-ER (Kenya National Museums - East Rudolf; from Koobi Fora, Kenya), WT (West Turkana, Kenya), STw (Sterkfontein, South Africa), and SK (Swartkrans, South Africa).

[3] Cortical bone hypertrophy, such as is produced by remodeling that involves endosteal and subperiosteal bone apposition and increased intracortical density, appears to be due primarily to normal fracture repair processes acting upon microfractures within cortical bone (Chamay and Tschantz 1972; Goodship et al. 1979; Lanyon 1982; Lanyon et al. 1982; Liskova and Hert 1971; Tschantz and Rustihauser 1967). Such a process of remodeling leading to cortical hypertrophy can be initiated by any stress upon a bone that leads to strains which exceed the elastic limit of the bone and hence produce plastic deformation within portions of the diaphyseal cross-section, leading to microfractures (Currey 1970). Short term, peak loads can be responsible for remodeling-promoting deformations, but repetitive intermittent mechanical loading, especially given the low resistance of bone to fatigue (Carter et al. 1981; Carter and Spengler 1982), is much more likely to produce significant hypertrophy of diaphyseal cortical bone (Chamay 1970; Liskova and Hert 1971; Lanyon 1982). For these reasons, the diaphyseal cortical hypertrophy seen in *H. habilis* and more recent archaic members of the genus *Homo* is best interpreted as a response (whether predominantly genetically based or developmentally induced, given the known plasticity of mammalian bone) to high levels of repetitive, moderate level, mechanical loading of the locomotor skeleton. Among terrestrial bipedal hominids, such loading of the lower limb is most likely to be induced by relatively continuous movement across the landscape, such as would be associated with opportunistic foraging for widely

dispersed and relatively unpredictable (at least to the hominids involved) resources.

4 The sample of initial members of *H. erectus*, dated to approximately 1.5 myr BP (±0.1 myr) is here taken to include (at least) KNM-ER 730, 736, 737, 992, 1808, 3733, 3883, SK 84 and 847, and WT 15000 (Clarke et al. 1970; R. E. Leakey et al. 1978; R. E. F. Leakey and Walker 1985a, b; Napier 1959; Walker and Leakey 1986). More recent specimens include those from African and Asian sites such as Lantien, Ngandong, Olduvai Upper Bed II to Bed IV, Salé, Sangiran, Ternifine, Thomas, Trinil and Zhoukoudian, and may include additional specimens from European Middle Pleistocene sites (Stringer 1984; Wolpoff 1984).

5 Only three specimens of *H. habilis* known in the fossil record preserve portions of the nasal region, OH 24, KNM-ER 1470 and KNM-ER 3732 (M. D. Leakey et al. 1971; Day et al. 1975; R. E. Leakey et al. 1978; R. E. F. Leakey and Walker 1985a). KNM-ER 1470 possesses minimally projecting nasal bones, and the last two, but not OH 24, have slightly everted lateral nasal margins. The inferior nasal margin of KNM-ER 1470 is too damaged to provide a clear indication of its morphology, although its largely flat or even slightly concave and vertically set nasoalveolar clivus suggests that the angle between it and the internal nasal floor may have been relatively acute. The inferior nasal margin of OH 24 appears to have been rounded, similar to those of australopithecines and nonhuman hominoids, and it possesses a prominent crest of bone that may be identifiable as a hominine anterior nasal spine but in fact is nonprojecting and hence more ape-like (M. D. Leakey et al. 1971). It is unclear, however, whether this nasal aperture morphology indicates a projecting external nose or merely a slight shift in nasal configuration relative to that present in members of the genus *Australopithecus*.

Furthermore, one of the specimens discussed in note 1 as representative of a possible third species of hominid in the Lower Pleistocene of east Africa (KNM-ER 1813) exhibits lateral nasal aperture eversion and an angulated inferior nasal margin (R. E. Leakey et al. 1978). Two other crania mentioned,

KNM-ER 1805 and STw 53, have nonprojecting nasal bones and appear to lack both prominent anterior nasal spines and angulated inferior nasal aperture margins. Interestingly, KNM-ER 1813 (and the similar OH 13) overlapped the time range of early *H. erectus* (M. D. Leakey 1971; R. E. Leakey et al. 1978). This might imply that KNM-ER 1813 and OH 13 belong in *H. erectus* (which would indicate a major extension of the known range of endocranial volumes for that species to include specimens with endocranial capacities less than 550 cc). Alternatively, it might suggest that a projecting external nose evolved in parallel in two Lower Pleistocene hominid lineages, and therefore, that two hominid species were developing elevated activity levels during this time period. This indicates that conventional scenarios of Lower Pleistocene hominid evolution, including a few aspects of the one present in this review, underestimate the complexity of the situation, either by assuming too few synchronic and roughly sympatric hominid species or by limiting the possible ranges of variation with the recognized species.

[6] The term "archaic *Homo sapiens*" is used here for convenience and following current usage. It should be recognized, however, that the distinctions between the total morphological patterns of the specimens included in "archaic *Homo sapiens*" and those in "anatomically modern" *H. sapiens* are at least as great as those which are habitually used by human paleontologists to define distinct species of Plio-Pleistocene hominids. Yet, the complex nature of regional geographic hominid evolution during the late Middle and early Upper Pleistocene (Smith and Spencer 1984; Trinkaus and Smith 1985; Trinkaus 1986) makes the definition of a distinct species for these hominids (which would have to be referred to as "*H. neanderthalensis*") difficult and unjustifiable at this time. Furthermore, such species distinctions inevitably become arbitrary, given the difficulties inherent in applying the nonevolutionary Linnean taxonomic systems to a dynamic evolutionary sequence.

[7] Walker has suggested that the emergence of *H. erectus* was associated with a marked increase in the consumption of animal products by hominids. His arguments have been based on dental occlusal microwear (1981), a possible

case of hypervitaminosis A (KNM-ER 1808) (Walker et al. 1982), and considerations of possible competition with carnivore guilds and the temporal association of the emergence of *H. erectus* with changes in that guild (1984). The first needs further substantiation, the second may be an isolated case, and the last does not necessarily follow from the probable ways in which early hominids may have obtained access to carcasses (see text).

REFERENCES CITED

Ardrey, R. 1961. *African Genesis*. New York: Dell Pub.

Audouze, F., D. Cahen, L.H. Keeley, and B. Schmider. 1981. Le site magdalenien du Buisson Campin a Verberie (Oise). *Gallia Préh*. 24:99-143.

Beck, B.B. 1980. *Animal Tool Behavior*. New York: Garland Press.

Behrensmeyer, A.K. 1978. The habitat of Plio-Pleistocene hominids in East Africa: Taphonomic and micro-stratigraphic evidence; pp. 165-189. In: *Early Hominids of Africa*, ed. C. Jolly. New York: St. Martin's.

Berggren, W.A., L.H. Burckle, M.B. Cita, H.B.S. Cooke, B.M. Funnell, S. Gartner, J.D. Hays, J.P. Kennett, N.D. Opdyke, L. Pastouret, N.J. Shackleton, and Y. Takayanagi. 1980. Towards a Quaternary time scale. *Quart. Res*. 13:277-302.

Billy, G. 1972. L'évolution humaine au paléolithique supérieur. *Homo* 23:2-12.

Binford, L.R. 1981. *Bones*. New York: Academic Press.

Binford, L.R. 1982a. The archaeology of place. *J. Anthrop. Archaeol*. 1:5-31.

Binford, L.R. 1982b. Comment on R. White: Rethinking the Middle/Upper Paleolithic Transition. *Curr. Anthrop*. 23:177-181.

Binford, L.R. 1984. *Faunal Remains from Klasies River Mouth*. New York: Academic Press.

Binford, L.R. 1985. Human ancestors: Changing views of their behavior. *J. Anthrop. Archaeol*. 4:292-327.

Binford, L.R., and C.K. Ho. 1985. Taphonomy at a distance: Zhoukoudian, "The Cave Home of Beijing Man"? *Curr. Anthrop*. 26:413-442.

Binford, L.R., and L.C. Todd. 1982. On arguments for the "butchering" of giant geladas. *Curr. Anthrop*. 23:108-110.

Blumenberg, B. 1985. Population characteristics of extinct hominid endocranial volume. *Am. J. Phys. Anthrop*. 68:263-279.

Boesch, C., and H. Boesch. 1982. Optimization of nut cracking with hammers by wild chimpanzees. *Behavior* 83:265-286.

Bordes, F. 1969. Os percé moustérien et os gravé acheuléen du Pech de l'Azé. *Quaternaria* 11:1-6.

Bordes, F. 1984a. *Leçons sur le Paléolithique II*. Paris: C.N.R.S.

Bordes, F. 1984b. *Leçons sur le Paléolithique III*. Paris: C.N.R.S.

Boule, M. 1911-13. L'homme fossile de La Chapelle-aux-Saints. *Ann. Paléont.* 6:111-172; 7:21-56, 85-192; 8:1-70.

Brain, C.K. 1981. *The Hunters or the Hunted?* Chicago: Univ. Chicago Press.

Brain, C.K. 1982. The Swartkrans site: Stratigraphy of the fossil hominids and a reconstruction of the environment of early Homo; pp. 676-706. In: *L'Homo erectus et la Place de l'Homme de Tautavel Parmi les Hominidés Fossiles*, ed. H. de Lumley. Paris: C.N.R.S.

Bromage, T.G., and M.C. Dean. 1985. Re-evaluation of the age at death of immature fossil hominids. *Nature* 317:525-527.

Brown, F., J. Harris, R. Leakey, and A. Walker. 1985. Early *Homo erectus* skeleton from west Lake Turkana, Kenya. *Nature* 316:788-792.

Bush, M.E. 1980. The thumb of *Australopithecus afarensis* (abstr.). *Am. J. Phys. Anthrop.* 52:210.

Bush, M.E., C.O. Lovejoy, D.C. Johanson, and Y. Coppens. 1982. Hominid carpal, metacarpal and phalangeal bones recovered from the Hadar formation: 1974-1977 collections. *Am. J. Phys. Anthrop.* 57:651-667.

Butzer, K.W., and G.L. Isaac, eds. 1975. *After the Australopithecines*. The Hague: Mouton.

Carey, J.W., and A.T. Steegman, Jr. 1981. Human nasal protrusion, latitude, and climate. *Am. J. Phys. Anthrop.* 56:313-319.

Carrier, D.R. 1984. The energetic paradox of human running and hominid evolution. *Curr. Anthrop.* 25:483-495.

Carter, D.R., W.E. Caler, D.M. Spengler, and V.H. Frankel. 1981. Fatigue behavior of adult cortical bone: The influence of mean strain and strain range. *Acta orthop. scand.* 52:481-490.

Carter, D.R., and D.M. Spengler. 1982. Biomechanics of fracture; pp. 305-334 In: *Bone in Clinical Orthopaedics*, ed. G. Sumner-Smith. Philadelphia: Saunders.

Chamay, A. 1970. Mechanical and morphological aspects of experimental overload and fatigue in bone. *J. Biomech.* 3:263-270.

Chamay, A., and P. Tschantz. 1972. Mechanical influences in bone remodeling. Experimental research on Wolff's Law. *J. Biomech.* 5:173-180.

Clark, J.D. 1983. The significance of culture change in the early later Pleistocene in northern and southern Africa. In: *The Mousterian Legacy*, ed. E. Trinkaus, *Brit. Arachaeol. Rep.* S164:1-12.

Clark, J.D., and H. Kurashina. 1979. Hominid occupation of the east-central highlands of Ethiopia in the Plio-Pleistocene. *Nature* 282:33-39.

Clarke, R.J., F.C. Howell, and C.K. Brain. 1970. More evidence for an advanced hominid at Swartkrans. *Nature* 225:1219-1222.

Conkey, M.W. 1983. On the origins of Paleolithic art: A review and some critical thoughts. In: *The Mousterian Legacy*, ed. E.Trinkaus, *Brit. Archaeol. Rep.* S164:201-227.

Coon, C.S. 1962. *The Origin of Races*. New York: Knopf.

Cremaschi, M., C. Peretto, and B. Sala, eds. 1983. *Isernia La Pineta*. Bologna: Calderini Editore.

Currey, J.D. 1970. The mechanical properties of bone. *Clin. Orthop.* 73:210-231.

Dart, R.A. 1957. The osteodontokeratic culture of *Australopithecus prometheus. Transv. Mus. Mem.* 10.

Davis, S. 1974. Incised bones from the Mousterian of Kebara Cave (Mount Carmel) and the Aurignacian of Ha-yonim Cave (Western Galilee), Israel. *Paléorient* 2:181-182.

Day, M.H. 1971. Postcranial remains of *Homo erectus* from Bed IV, Olduvai Gorge, Tanzania. *Nature* 232:383-387.

Day, M.H. 1982. The *Homo erectus* pelvis: Punctuation or gradualism? pp. 411-421. In: *L'Homo erectus et la Place de l'Homme de Tautavel parmi les Hominidés Fossiles*, ed. H. de Lumley. Paris: C.N.R.S.

Day, M.H., R.E.F. Leakey, A.C. Walker, and B.A. Wood. 1975. New hominids from East Rudolf, Kenya, I. *Am. J. Phys. Anthrop.* 42:461-476.

Day, M.H., and J.R. Napier. 1966. A hominid toe bone from Bed I, Olduvai Gorge, Tanzania. *Nature* 211:929-930.

Day, M.H., and J.L. Scheuer. 1973. SKW 14147: A new hominid metacarpal from Swartkrans. *J. Hum. Evol.* 2:429-438.

Day, M.H., and E.H. Wickens. 1980. Laetoli Pliocene hominid footprints and bipedalism. *Nature* 286:385-387.

Dean, M.C., C.B. Stringer, and T.G. Bromage. 1986. Age at death of the Neanderthal child from Devil's Tower, Gibraltar and the implications for studies of general growth and development in Neanderthals. *Am. J. Phys. Anthrop.* 70:301-309.

Deloison, Y. 1985. Comparative study of calcanei of primates and *Pan-Australopithecus-Homo* relationship; pp. 143-147. In: *Hominid Evolution: Past, Present and Future*, ed. P.V. Tobias. New York: Liss.

Delporte, H. 1984. L'Aurignacien de La Ferrassie. In: *Le Grande Abri de La Ferrassie*, ed. H. Delporte. *Études Quart.* 7:145-234.

de Lumley, H. 1966. Les fouilles de Terra Amata à Nice. *Bull. Mus. Anthrop. Préh. Monaco* 13:29-51.

de Lumley, H., ed. 1969. Une Cabane Acheuléenne dans la Grotte du Lazaret (Nice). *Mém. Soc. Préh. Franç.* 7.

Endo, B., and T. Kimura. 1970. Postcranial skeleton of the Amud man; pp. 231-406. In: *The Amud Man and his Cave Site*, ed. H. Suzuki and F. Takai. Tokyo: Academic Press of Japan.

Falk, D. 1980. A reanalysis of the South African Australopithecine natural endocasts. *Am. J. Phys. Anthrop.* 53:525-539.

Falk, D. 1983. Cerebral cortices of East African early hominids. *Science* 221:1072-1074.

Falk, D. 1985a. Hadar AL 162-28 endocast as evidence that brain enlargement preceded cortical reorganization in hominid evolution. *Nature* 313:45-47.

Falk, D. 1985b. Apples, oranges, and the lunate sulcus. *Am. J. Phys. Anthrop.* 67:313-315.

Ferembach, D. 1976. Les restes humains de la Grotte de Dar-es-Soltane 2 (Maroc) Campagne 1975. *Bull. Mém. Soc. Anthrop. Paris* Série XIII, 3:183-193.

Franciscus, R.G., and E. Trinkaus. In prep. Nasal morphology and the emergence of *Homo erectus*. *Am. J. Phys. Anthrop.*

Frayer, D.W. 1978. Evolution of the dentition in Upper Paleolithic and Mesolithic Europe. *Univ. Kansas Pub. Anthrop.* 10.

Frayer, D.W. 1984. Biological and cultural change in the European late Pleistocene and early Holocene; pp. 211-250. In: *The Origins of Modern Humans*, ed. F.H. Smith and F. Spencer. New York:Liss.

Freeman, L.G. 1975. Acheulian Sites and stratigraphy in Iberia and the Maghreb; pp. 661-744. In: *After the Australopithecines*, ed. K.W. Butzer and G.L. Isaac. The Hague: Mouton.

Gábori-Csánk, V. 1968. *La Station du Paléolithique Moyen d'Érd, Hongrie*. Budapest: Akademiai Kiado.

Gomberg, D.N., and B.M. Latimer. 1984. Observations on the transverse tarsal joint of *A. afarensis*, and some comments on the interpretation of behavior from morphology (abstr.). *Am. J. Phys. Anthrop*. 63:164.

Goodall, J. 1964. Tool using and aimed throwing in a community of free living chimpanzees. *Nature* 201:1264-1266.

Goodship, A.E., L.E. Lanyon, and H. McFie. 1979. Functional adaptation of bone to increased stress. *J. Bone Joint. Surg.* 61A:539-546.

Gould, R.A. 1980. *Living Archaeology*. Cambridge: Cambridge Univ. Press.

Harding, R.S.O. 1973. Predation by a troop of olive baboons (*Papio anubis*). *Am. J. Phys. Anthrop*. 38:587-592.

Harding, R.S.O. 1981. An order of omnivores: Nonhuman primate diet in the wild; pp. 191-214. In: *Omnivorous Primates*, ed. R.S.O. Harding and G. Telecki. New York: Columbia Univ. Press.

Hasegawa, T., M. Hiraiwa, T. Nishida, and H. Takasaki. 1983. New evidence on scavenging behavior in wild chimpanzees. *Curr. Anthrop*. 24:231-232.

Heim, J.L. 1976. Les Hommes Fossiles de La Ferrassie I. *Arch. Inst. Paléont. Hum*. 35.

Heim, J.L. 1982a. Les Hommes Fossiles de La Ferrassie II. *Arch. Inst. Paléont. Hum*. 38.

Heim, J.L. 1982b. *Les Enfants Néandertaliens de La Ferrassie*. Paris: Masson.

Holloway, R.L. 1978. Problems of brain endocast interpretation and African hominid evolution; pp. 379-401. In: *Early Hominids of Africa*, ed. C.J. Jolly. New York: St. Martin's.

Holloway, R.L. 1982. Revisiting the South African Taung Australopithecine endocast: The position of the lunate sulcus as determined by the stereoplotting technique. *Am. J. Phys. Anthrop*. 56:43-58.

Holloway, R.L. 1983. Cerebral brain endocast pattern of *Australopithecus afarensis* hominid. *Nature* 303:420-422.

Howell, F.C., and B.A. Wood. 1974. Early hominid ulna from the Omo basin, Ethiopia. *Nature* 249:174-176.

Hublin, J.J. 1978. *Le Torus Occipital transverse et les Structures Associées: évolution dans le Genre Homo*. Thèse de Troisième Cycle, Univ. Paris VI.

Isaac, G.L. 1978. The food-sharing behavior of protohuman hominids. *Sci. Am*. 238(4):90-108.

Isaac, G.L., and G.H. Curtis. 1974. Age of early Acheulian industries from the Peninj group. *Nature* 249:624-627.

Jacob-Friesen, K.H. 1956. Eiszeitliche Elefantenjäger in der Luneburger Heide. *Jarhb. Rom.-Germ. Zentralmus*. Mainz 3:1-22.

Jelinek, A.J. 1977. The Lower Paleolithic: Current evidence and interpretations. *Ann. Rev. Anthrop*. 6:11-32.

Johanson, D.C., C.O. Lovejoy, W.H. Kimbel, T.D. White, S.C. Ward, M.E. Bush, B.M. Latimer, and Y. Coppens. 1982. Morphology of the Pliocene partial hominid skeleton (AL288-1) from the Hadar formation, Ethiopia. *Am. J. Phys. Anthrop.* 57:403-452.

Jungers, W.L. 1982. Lucy's limbs: skeletal allometry and locomotion in *Australopithecus afarensis*. *Nature* 297:676-678.

Kennedy, G.E. 1983. A morphometric and taxonomic assessment of a hominine femur from the Lower Member, Koobi Fora, Lake Turkana. *Am. J. Phys. Anthrop*. 61:429-436.

Kimbel, W.H., T.D. White, and D.C. Johanson. 1984. Cranial morphology of *Australopithecus afarensis*: A comparative study based on a composite reconstruction of the adult skull. *Am. J. Phys. Anthrop*. 64:337-388.

Klein, R.G. 1976. The mammalian fauna of the Klasies River Mouth sites, Southern Cape Province. *S. Afr. Archaeol. Bull*. 31:75-98.

Klein, R.G. 1977. The ecology of early man in southern Africa. *Science* 197:115-126.

Klein, R.G. 1978. Stone age predation on large African bovids. *J. Archaeol. Sci.* 5:195-217.

Klein, R.G. 1984. Mammalian extinctions and stone age people in Africa; pp. 553-573. In: *Quaternary Extinctions: A Prehistoric Revolution*, ed. P.S. Martin and R.G. Klein. Tucson: Univ. Arizona Press.

Klein, R.G., and K. Allwarden. n.d. An overview of the Torralba fauna. Unpub. manuscript, Univ. Chicago.

Kramer, A. 1986. Hominid-pongid distinctiveness in the Miocene-Pliocene fossil record: The Lothagam mandible. *Am. J. Phys. Anthropol*. 70:457-473.

Laitman, J.T. 1985. Evolution of the hominid upper respiratory tract: The fossil evidence; pp. 281-286. In: *Hominid Evolution: Past, Present and Future*, ed. P.V. Tobias. New York: Liss.

Lamy, P. 1986. The settlement of the longitudinal plantar arch of some african Plio-Pleistocene hominids: A morphological study. *J. Hum. Evol*. 15:31-46.

Lanyon, L.E. 1982. Mechanical function and bone remodeling; pp. 273-304. In: *Bone in Clinical Orthopaedics*, ed. G. Sumner-Smith. Philadelphia: Saunders.

Lanyon, L.E., A.E. Goodship, C.J. Pye, and J.H. MacFie. 1982. Mechanically adaptive bone remodelling. *J. Biomech*. 15:141-154.

Latimer, B.M. 1983. The anterior foot skeleton of *Australopithecus afarensis* (abstr.). *Am. J. Phys. Anthrop*. 60:217.

Latimer, B.M., C.O. Lovejoy, D.C. Johanson, and Y. Coppens. 1982. Hominid tarsal, metatarsal and phalangeal bones recovered from the Hadar formation: 1974-1977 collections. *Am. J. Phys. Anthrop*. 53:701-719.

Leakey, L.S.B., P.V. Tobias, and J.R. Napier. 1964. A new species of the genus *Homo* from Olduvai Gorge. *Nature* 202:7-9.

Leakey, M.D. 1971. *Olduvai Gorge 3*. Cambridge: Cambridge Univ. Press.

Leakey, M.D., R.J. Clarke, and L.S.B. Leakey. 1971. New hominid skull from Bed I, Olduvai Gorge, Tanzania. *Nature* 232:308-312.

Leakey, M.D., and R.L. Hay. 1979. Pliocene footprints in the Laetoli Beds at Laetoli, northern Tanzania. *Nature* 278:317-323.

Leakey, R.E., M.G. Leakey, and A.K. Behrensmeyer. 1978. The hominid catalogue; pp. 86-187. In: *Koobi Fora Research Project 1*., ed. M.G. Leakey and R.E. Leakey. Oxford: Clarendon Press.

Leakey, R.E.F., and A.C. Walker. 1985a. Further hominids from the Plio-Pleistocene of Koobi Fora, Kenya. *Am. J. Phys. Anthrop*. 67:135-163.

Leakey, R.E.F., and A.C. Walker. 1985b. *Homo erectus* unearthed. *Nat. Geog*. 168(5):624-629.

Leinders, J.J.M., F. Aziz, P.Y. Sondaar, and J. De Vos. 1985. The age of the hominid-bearing deposits of Java: State of the art. *Geol. Mijnbouw* 64:167-173.

Legoux, P. 1970. Étude odontologique de l'enfant néandertalien du Pech de l'Azé. In: *L'Enfant du Pech de l'Azé, Arch. Inst. Paléonto. Hum.* 33:53-87.

Leroi-Gourhan, A., and M. Brézillon. 1972. Fouilles de Pincevent. *Gallia Préh. Suppl.* 7.

Leroy-Prost, C. 1974. Les pointes en matière osseuse de l'Aurignacien. *Bull. Soc. Préh. Franc.* 71:449-458.

Liskova, M., and J. Hert. 1971. Reaction of bone to mechanical stimuli part 2. Periosteal and endosteal reaction of tibial diaphysis in rabbit to intermittent loading. *Folia Morph. (Prague)* 19:301-317.

Lovejoy, C.O. 1978. A biomechanical review of the locomotor diversity of early hominids; pp. 403-429. In: *Early Hominids of Africa*, ed. C.J. Jolly. New York: St. Martin.

Lovejoy, C.O. 1982. Diaphyseal biomechanics of the locomotor skeleton of Tautavel man with comments on the evolution of skeletal changes in Late Pleistocene man; pp. 447-470. In: *L'Homo erectus et la Place de l'Homme de Tautavel parmi les Hominidés Fossiles*, ed. H. de Lumley. Paris: C.N.R.S.

Lovejoy, C.O., K.G. Heiple, and A.H. Burnstein. 1973. The gait of *Australopithecus*. *Am. J. Phys. Anthrop.* 38:757-780.

Lovejoy, C.O., and E. Trinkaus. 1980. Strength and robusticity of the Neandertal tibia. *Am. J. Phys. Anthrop.* 53:456-470.

Marshack, A. 1982. Non-utilitarian fragment of bone from the Middle Paleolithic layer; p. 117. In: *Excavation in the Bacho Kiro Cave (Bulgaria)*, ed. J.K. Kozlowski. Warsaw: Panstwowe Wydawnictwo Naukowe.

Martin, H. 1907-10. *Recherches sur l'Évolution du Moustérien dans le Gisement de La Quina I.* Paris: Librairie C. Reinwald.

Marzke, M.W. 1983. Joint function and grips of the *Australopithecus afarensis* hand, with special reference to the region of the capitate. *J. Hum. Evol.* 12:197-211.

Marzke, M.W., and M.S. Shackley. 1986. Tools and the morphology of hominid hands (abstract). *Am. J. Phys. Anthrop.* 69:237.

Matiegka, J. 1934. *Homo predmostensis, Fosilní Clovek z Predmosti na Morave I.* Prague: Nákladem Ceské Akademie Ved a Umení.

Matiegka, J. 1938. *Homo predmostensis, Fosilní Clovek z Predmosti na Morave II.* Prague: Nákladem Ceské Akademie Ved a Umení.

McCown, T.D., and A. Keith. 1939. *The Stone Age of Mount Carmel II.* Oxford: Clarendon Press.

McHenry, H.M. 1976. Early hominid body weight and encephalization. *Am. J. Phys. Anthrop.* 45:77-83.

McHenry, H.M. 1982. The pattern of human evolution: Studies on bipedalism, mastication, and encephalization. *Ann. Rev. Anthrop.* 11:151-173.

McHenry, H.M. 1984. The capitate of *Australopithecus afarensis* and *A. africanus*. *Am. J. Phys. Anthrop.* 62:187-198.

McHenry, H.M. 1984. Relative cheek-tooth size in *Australopithecus*. *Am. J. Phys. Anthrop.* 64:297-306.

Napier, J.R. 1959. Fossil metacarpals from Swartkrans. *Fossil Mamm. Afr.* 17:1-18.

Nishida, T., and S. Uehara. 1982. Natural history of a tool using behavior by wild chimpanzees in feeding on wood boring ants. *J. Hum. Evol.* 11:73-99.

Nishida, T., S. Uehara, and R. Nyundo. 1979. Predatory behavior among wild chimpanzees of the Mahale Mountains. *Primates* 20:1-20.

Oakley, K.P., P. Andrews, L.H. Keeley, and J.D. Clark. 1977. A reappraisal of the Clacton spearpoint. *Proc. Preh. Soc.* 43:13-30.

Otte, M. 1979. Le paléolithique supérieur ancien en Belgique. *Mongr. Archéol. Nat.* 5.

Perlès, C. 1976. Le feu; pp. 728-733. In: *Préhistoire Française*, ed. H. de Lumley. Paris: C.N.R.S.

Perlès, C. 1977. *Préhistoire du Feu*. Paris: Masson.

Pfeiffer, J.E. 1982. *The Creative Explosion*. New York: Harper and Row.

Potts, R.B. 1982. *Lower Pleistocene Site Formation and Hominid Activities at Olduvai Gorge, Tanzania*. Ph.D. Thesis, Harvard Univ.

Potts, R.B. 1984. Home bases and early hominids. *Am. Sci.* 72:338-347.

Potts, R.B., and P. Shipman. 1981. Cutmarks made by stone tools on bones from Olduvai Gorge, Tanzania. *Nature* 291:577-580.

Rak, Y. 1983. *The Australopithecine Face*. New York: Academic Press.

Rak, Y. 1986. The Neanderthal: A new look at an old face. *J. Hum. Evol.* 15:151-164.

Rigaud, J.P. 1982. *Le Paléolithique en Périgord: Les Données du Sud-Ouest Sarladais et Leurs Implications*. Thèse de Doctorat d'État, Univ. Bordeaux I.

Rightmire, G.P. 1979. Cranial remains of *Homo erectus* from Beds II and IV, Olduvai Gorge, Tanzania. *Am. J. Phys. Anthrop.* 51:99-116.

Rightmire, G.P. 1981. Patterns in the evolution of *Homo erectus. Paleobiol.* 7:241-246.

Rightmire, G.P. 1984. *Homo sapiens* in sub-Saharan Africa; pp. 295-325. In: *The Origins of Modern Humans*, eds. F.H. Smith and F. Spencer. New York: Liss.

Robinson, J.T. 1972. *Early Hominid Posture and Locomotion*. Chicago: Univ. Chicago Press.

Roche, H. 1980. *Premiers Outils Taillés d'Afrique*. Paris: Société d'Ethnographie.

Rose, M.D. 1984. A hominine hip bone, KNM-ER 3228, from east Lake Turkana, Kenya. *Am. J. Phys. Anthrop.* 63:371-378.

Ryan, A.S. 1980. *Anterior Dental Microwear in Hominid Evolution*: Comparisons with Humans and Nonhuman Primates. Ph.D. Thesis, Univ. Michigan.

Sartono, S. 1971. Observations on a new skull of *Pithecanthropus erectus (Pithecanthropus VIII)* from Sangiran, central Java. *Koninkl. Ned. Akad. Wed.* 74:185-194.

Schaefer, U. 1957. Homo neanderthalensis King. 1. Das Skelet aus dem Neandertal. *Z. Morph. Anthrop.* 48:268-297.

Senut, B. 1981. *L'Humérus et ses Articulations chez les Hominidés Plio-Pleistocènes*. Paris: C.N.R.S.

Senut, B., and C. Tardieu. 1985. Functional aspects of Plio-Pleistocene hominid limb bones: Implications for taxonomy and phylogeny; pp. 193-201. In: *Ancestors: The Hard Evidence*, ed. E. Delson. New York: Liss.

Shipman, P., W. Bosler, and K.L. Davis. 1981. Butchering of giant geladas at an Acheulian site. *Curr. Anthrop.* 22:257-268.

Shipman, P., and J. Rose. 1983. Evidence of butchery and hominid activities at Torralba and Ambrona: An evaluation using microscopic techniques. *J. Archaeol. Sci.* 10:465-474.

Singer, R., and J. Wymer. 1982. *The Middle Stone Age at Klasies River Mouth in South Africa*. Chicago: Univ. Chicago Press.

Skinner, M. 1981. Dental attrition in immature hominids of the Late Pleistocene: Implications for adult longevity (abstr.). *Am. J. Phys. Anthrop.* 54:278-279.

Smith, B.H. 1986. Dental development in *Australopithecus* and early *Homo*. *Nature* 323:327-330.

Smith, F.H. 1976. The Neandertal remains from Krapina. *Dept. Anthrop. Univ. Tenn. Rep. Invest.* 15.

Smith, F.H. 1983. A behavioral interpretation of changes in craniofacial morphology across the archaic/modern *Homo sapiens* transition. In: *The Mousterian Legacy*, ed. E. Trinkaus. *Brit. Archaeol. Rep.* S164:141-163.

Smith, F.H. 1984. Fossil hominids from the Upper Pleistocene of central Europe and the origin of modern Europeans; pp. 137-209. In: *The Origins of Modern Humans*, ed. F.H. Smith and F. Spencer. New York: Liss.

Smith, F.H., and F. Spencer, eds. 1984. *The Origins of Modern Humans*. New York: Liss.

Soffer, O. 1985. *The Upper Paleolithic of the Central Russian Plain*. Orlando: Academic Press.

Spiess, A.E. 1979. *Reindeer and Caribou Hunters*. New York: Academic Press.

Stern, J.T., Jr., and R.L. Susman. 1983. The locomotor anatomy of *Australopithecus afarensis*. *Am. J. Phys. Anthrop.* 60:279-317.

Straus, L.G. 1982. Carnivores and cave sites in Cantabrian Spain. *J. Anthrop. Res.* 38:75-96.

Straus, L.G. 1983. From Mousterian to Magdalenian: Cultural evolution viewed from Vasco-Cantabrian Spain and Pyrenean France. In: *The Mousterian Legacy*, ed. E. Trinkaus. *Brit. Archaeol. Rep.* S164:73-112.

Straus, L.G. 1985. Stone age prehistory of northern Spain. *Science* 230:501-507.

Stringer, C.B. 1978. Some problems in Middle and Upper Pleistocene hominid relationships. In: *Recent Advances in Primatology*, ed. D.J. Chivers and K.A. Joysey. London: Academic Press. 3:395-418.

Stringer, C.B. 1984. The definition of *Homo erectus* and the existence of the species in Africa and Europe. *Cour. Forsch.-Inst. Senckenberg.* 69:131-143.

Stringer, C.B. 1986. An archaic character in the Broken Hill innominate E.179. *Am. J. Phys. Anthrop.* 71:115-120.

Strum, S.C. 1975. Primate predation: An interim report on the development of a tradition in a troop of olive baboons. *Science* 187:755-757.

Sugiyama, J., and J. Koman. 1979. Tool using and making behavior in wild chimpanzees at Bossou, Guinea. *Primates* 20:513-524.

Susman, R.L., and N. Creel. 1979. Functional and morphological affinities of the subadult hand (O.H. 7) from Olduvai Gorge. *Am. J. Phys. Anthrop.* 51:311-332.

Susman, R.L., J.T. Stern Jr., and W.L. Jungers. 1984. Arboreality and bipedality in the Hadar Hominids. *Folia Primat.* 43:113-156.

Tardieu, C. 1983. *L'Articulation du Genou*. Paris: C.N.R.S.

Teleki, G. 1974. Chimpanzee subsistence technology: Material and skills. *J. Hum. Evol.* 3:575-594.

Telecki, G. 1975. Primate subsistence patterns: Collector-predators and gatherer-hunters. *J. Hum. Evol.* 4:125-184.

Telecki, G. 1981. The omnivorous diet and eclectic feeding habits of chimpanzees in Gombe National Park, Tanzania; pp. 303-343. In: *Omnivorous Primates*, ed. R.S.O. Harding and G. Telecki. New York: Columbia Univ. Press.

Thompson, D.D., and E. Trinkaus. 1981. Age determination for the Shanidar 3 Neanderthal. *Science* 212:575-577.

Tobias, P.V. 1985. Single characters and the total morphological pattern redefined: The sorting effected by a selection of morphological features of the early hominids; pp. 94-101. In: *Ancestors: The Hard Evidence*, ed. E. Delson. New York: Liss.

Tode, A. 1953. Die Untersuchung der palaolithischen Frielandstation von Salzgitter-Lebenstedt 8: Archaologische Erkenntnisse. *Eiszeitalt. Gegenw.* 3:192-215.

Toth, N. 1985. The Oldowan reassessed: A close look at early stone artifacts. *J. Archaeol. Sci.* 12:101-120.

Trinkaus, E. 1976. The evolution of the hominid femoral diaphysis during the Upper Pleistocene in Europe and the Near East. *Z. Morph. Anthrop.* 67:291-319.

Trinkaus, E. 1982a. Evolutionary continuity among archaic *Homo sapiens*. In: *The Transition from Lower to Middle Palaeolithic and the Origin of Modern Man*, ed. A. Ronen. *Brit. Archaeol. Rep.* S151:301-314.

Trinkaus, E. 1982b. The evolution of the Neandertal face. Paper presented at 81st Mtng. Am. Anthrop. Assoc.

Trinkaus, E. 1983a. *The Shanidar Neandertals*. New York: Academic Press.

Trinkaus, E. 1983b. Neandertal postcrania and the adaptive shift to modern humans. In: *The Mousterian Legacy*, ed. E. Trinkaus. *Brit. Archaeol. Rep.* S164:165-200.

Trinkaus, E. 1984a. Does KNM-ER 1481A establish *Homo erectus* at 2.0 myr BP? *Am. J. Phys. Anthrop.* 64:137-139.

Trinkaus, E. 1984b. Western Asia; pp. 251-293. In: *The Origins of Modern Humans*, ed. F.H. Smith and F. Spencer. New York: Liss.

Trinkaus, E. 1986. The Neandertals and the Origins of Modern Humans. *Ann. Rev. Anthrop.* 15:193-218.

Trinkaus, E., and M. LeMay. 1982. Occipital bunning among later Pleistocene hominids. *Am. J. Phys. Anthrop.* 57:27-35.

Trinkaus, E., and F.H. Smith. 1985. The fate of the Neandertals; pp. 325-333. In: *Ancestors: The Hard Evidence*, ed. E. Delson. New York: Liss.

Trinkaus, E., and D.D. Thompson. 1987. Femoral diaphyseal histomorphometric age determinations for the Shanidar 3, 4, 5 and 6 Neandertals and Neandertal longevity. *Am. J. Phys. Anthrop.* 72:123-129.

Tschantz, P., and E. Rutishauser. 1967. La surcharge mécanique de l'os vivant. Les déformations plastiques initiales et l'hypertrophie d'adaptation. *Ann. Anat. path.* 12:223-248.

Tuttle, R.H. 1981. Evolution of hominid bipedalism and prehensile capacities. *Phil. Trans. Roy. Soc. Lond.* B292:89-94.

Tuttle, R.H. 1985. Ape footprints and Laetoli impressions: A response to the SUNY claims; pp. 129-133. In: *Hominid Evolution: Past, Present and Future*, ed. P.V. Tobias. New York: Liss.

Vallois, H.V., and G. Billy. 1965. Nouvelles recherches sur les hommes fossiles de l'Abri de Cro-Magnon. *L'Anthrop*. 69:47-74, 249-272.

Vandermeersch, B. 1981. *Les Hommes Fossiles de Qafzeh (Israël)*. Paris: C.N.R.S.

Villa, P. 1983. Terra Amata and the Middle Pleistocene archaeological record of southern France. *Univ. Calif. Pub. Anthrop*. 13:1-303.

Walker, A. 1981. Dietary hypotheses and human evolution. *Phil. Trans. Roy. Soc. Lond.* B292:57-64.

Walker, A. 1984. Extinction in hominid evolution; pp. 119-152. In: *Extinctions*, ed. M.H. Nitecki. Chicago: Univ. Chicago Press.

Walker, A., and R.E. Leakey. 1986. *Homo erectus* skeleton from West Lake Turkana, Kenya (abstr.). *Am. J. Phys. Anthrop*. 69:275.

Walker, A., M.R. Zimmerman, and R.E.F. Leakey. 1982. A possible case of hypervitaminosis A in *Homo erectus. Nature* 296:248-250.

Weidenreich, F. 1937. The dentition of *Sinanthropus pekinensis. Palaeont. Sin*. 1D:1-180.

Weidenreich, F. 1941. The extremity bones of *Sinanthropus pekinensis*. *Paleont. Sin*. 5D:1-150.

Weidenreich, F. 1943. The skull of *Sinanthropus pekinensis*. *Paleont. Sin*. 10D:1-298.

Weidenreich, F. 1945. Giant early man from Java and south China. *Anthrop. Pap. Am. Mus. Nat. Hist*. 40:1-134.

Whitesides, G.H. 1985. Nutcracking by wild chimpanzees in Sierra Leone, west Africa. *Primates* 26:91-94.

Wolpoff, M.H. 1971. Metric trends in hominid dental evolution. *Case West. Reserv. Univ. Stud. Anthrop*. 2:1-244.

Wolpoff, M.H. 1979. The Krapina dental remains. *Am. J. Phys. Anthrop*. 50:67-114.

Wolpoff, M.H. 1980. *Paleoanthropology*. New York: Knopf.

Wolpoff, M.H. 1984. Evolution in *Homo erectus*: The question of stasis. *Paleobiol*. 10:389-406.

Wolpoff, M.H. 1985. On explaining the supraorbital torus. *Curr. Anthrop*. 26:522.

Wolpoff, M.H., Wu, X.Z., and A.G. Thorne. 1984. Modern *Homo sapiens* origins: A general theory of hominid evolution involving the fossil evidence from East Asia; pp. 485-516. In: *The Origins of Modern Humans*, ed. F.H. Smith and F. Spencer. New York: Liss.

Wood, B.A. 1985. Early *Homo* in Kenya, and its systematic relationships; pp. 206-214. In: *Ancestors: The Hard Evidence*, ed. E. Delson. New York: Liss.

Yi, S., and G.A. Clark. 1983. Observations on the Lower Palaeolithic of northeast Asia. *Curr. Anthrop*. 24:181-202.

Hunting in Late Upper Paleolithic Western Europe

Lawrence Guy Straus

Department of Anthropology
University of New Mexico
Albuquerque, New Mexico

Three papers in this volume (Fisher, Frison, Morlan) deal with the prey and hunting practices of the terminal Pleistocene/initial Holocene Paleoindians and four authors (Behrensmeyer, Trinkaus, Binford, Klein) debate the crucial issue of hunting by early ("proto-human") hominids. The major problems discussed are (1) how to discern hunting from scavenging and hominid agency from natural processes; (2) the possible relationships between hominid physical evolution and hunting; and (3) the increasing importance of hunting and its significance in the radiation of genus *Homo*. However, there is little discussion on hominid hunting that evolved later in the Pleistocene in the Old World. If it is problematical that early hominids and even archaic *Homo sapiens* (Neanderthal et al.) were full-fledged, organized hunters, then who were the first and what were their prey and hunting practices? Two papers mention the most likely answer in passing:

> ...it is primarily for later Upper Paleolithic populations that we have unquestionable archeological evidence of systematic, planned exploitation of larger mammals on a regular basis (Trinkaus, this volume).
> ...there was a quantum advance in human ability to extract animal protein from nature, probably coincident with the adventof the Later Stone Age... (Klein, this volume).

I propose to provide examples of that "quantum advance," that "archeological evidence of systematic, planned exploitation of larger mammals on a regular basis." In western Europe and southern Africa -- subcontinents with high-quality faunal information -- our best evidence for significant changes in hunting, involving the regular, planned, efficient slaughter of large numbers of herd animals and the taking of elusive or dangerous game, appears in the archeological record around the time of the Last Glacial Maximum (20,000 yrs. B. P.), not necessarily at the beginning of the Upper Paleolithic/ Later Stone Age (35,000 - 40,000 yrs. B. P.). There seems to have been a lag of twenty (or more) millennia between the evolution of anatomically modern *Homo sapiens sapiens* and the appearance of large-scale, sophisticated hunting, although many of the technological innovations of the early Upper Paleolithic/ Later Stone Age (e.g., bone projectile points, blades, standardized and hafted compound tools, etc.) may have been "preadaptations."

The purpose of my paper is to characterize briefly the developments in human hunting practices attested to in the archeological records of the Solutrean and Magdalenian (21,000 - 11,000 yrs. B. P.) of western Europe, based on results of recent excavations in Germany, France and Spain. Such intensive extractive systems also characterized the adaptations of late Pleistocene people in other areas such as central and eastern Europe, southern Africa and western North America. They were the evolutionary result of anatomical changes from archaic to modern *Homo sapiens*, technological innovations, developments in social organization and information management, and demographic pressures all leading to a general broadening of the resource base *and* to the specialized, efficient exploitation of a relatively few species of gregarious ungulates.

Which ungulate species archeologically dominates in any individual level, site or region during any given period could be the result of one or several of the following factors: aspects of the physical environment (terrain, vegetation, climate), the nature of ecological relationships with other ungulate species, the season of hunting episodes, aspects of human culture such as decision-making, planning, hunting strategies, tactics and technologies. Shifts in faunal representation within the stratigraphic sequence of a site could reflect changes in the environment or, more directly, changes in the role of the site within a hunter-gatherer settlement subsistence system. Temporal shifts in the fauna of a region could reflect major changes in environment and/or in human subsistence strategies.

Some ungulate species in western Europe do provide limited information on changes in Last Glacial physical environments (e.g., the appearance of saiga, musk oxen, woolly rhino and mammoth at certain periods in southwest France, the reappearance of roe deer and boar and the replacement of reindeer by red deer north of the Pyrenees at the end of the Tardiglacial). However, two facts must be kept in mind: (1) the ungulate remains in Late Upper Paleolithic archeological sites are there as results of human activity -- a "cultural filter" which consisted of purposeful, nonrandom hunting and systematic butchering sequences; they are not the product of unbiased, proportional sampling of the faunas of a catchment area, and (2) most large mammals are quite flexible in their ranges of acceptable habitats, diets, climates and animal neighbors, although the relative abundance of one species versus that of another (e.g., red deer versus reindeer in southwest France and Cantabrian Spain) may be governed by interspecies competition as well as by subtle environmental factors. While our paleoenvironmental reconstructions (based on oxygen isotope analyses, pollen and sediments) still lack precision, as do our "absolute" dates, it seems that the major Late Upper Paleolithic faunas in large regions such as southwest France or Cantabrian Spain did not closely "track" climatic shifts, *at least insofar as they are represented in archeological sites*. Red deer dominate in Iberian sites (including Levante and Portugal as well as Cantabria) under a wide variety of environmental conditions. Reindeer can be the dominant game species under full glacial (cold, dry) conditions or under more temperate, humid ones (e.g., during the Laugerie, Lajcaux, Bolling or Allerod oscillations) in southwest France. But so can horses or bison, possibly as a result of seasonality in site environs, human planning, accidents of animal-hunter convergence, etc. Ibex dominate under a variety of environmental conditions (from Pleniglacial to early Holocene in sites located in or near alpine habitats).

While it is true that different specific tactics would have to be employed by hunters to kill large numbers of each of the key ungulate species, the point is that the need and the means were present in a variety of habitats, under differing environmental conditions and at different seasons to use the basic strategies of specialized herd hunting. The targets varied and obviously only those animals present and accessible were targeted for such hunts. Given the availability of herd species, Late Upper Paleolithic people developed the means of efficiently making large numbers of kills of single species, while also engaging in a wide variety of food gathering activities as part of a broad-based subsistence strategy.

One of the ironies of full glacial conditions in western Europe is the fact that despite substantially lowered temperatures and a dearth of arboreal vegetation, the pervasive, and often now extinct, open vegetation communities (low-latitude periglacial "steppe-tundra," heath, grassland, parkland, etc.) provided vast expanses of lush pasture for a wide variety of medium to large ungulates (most of the surviving are now ecologically separated among tundra, steppe and woodland zones). These Pleniglacial habitats supported ungulate biomasses resembling those of the modern African savannas in terms of size and diversity more than those of Holocene European woodlands (see Butzer 1971). Depending on the specific region, the late Last Glacial ungulate archeofaunas of western Europe included reindeer (*Rangifer tarandus*), red deer (*Cervus elaphus*), horse (*Equus caballus*), bison (*Bison priscus*), ibex (*Capra ibex*), and generally lesser quantities of aurochs (*Bos primigenius*), roe deer (*Capreolus capreolus*), chamois (*Rupicapra rupicapra*), giant deer (*Megoceros giganteus*), boar (*Sus scrofa*), saiga antelope (*Saiga tatarica*), muskox (*Ovibos moschatus*), moose (*Alces alces*), fallow deer (*Dama dama*), wild ass (*Equus hydruntinus*), woolly rhinoceros (*Coelodonta antiquitatis*) and mammoth (*Mammuthus primigenius*). Preying upon these and the smaller game of late Ice Age Europe was a plethora of carnivores, also substantially thinned out since the end of the Pleistocene: four canids, at least four felids, two ursids, and a hyenid, as well as a variety of mustelids (see Kurtén 1968; Delpech 1983; Altuna 1972, 1986; Poplin 1976; Butzer 1971).

Late Upper Paleolithic people ate not only ungulates, but also other land mammals, fish (notably salmonids), marine molluscs, birds, and even seals (all of which constituted small but sometimes significant parts of their diet) along with some plant foods (not attested to by the archeological record, but probably eaten as supplements particularly in the more southerly latitudes of Europe and during the more temperate climatic oscillations). Nonetheless, it is clear that reindeer, red deer, horse, ibex and bison were the mainstays of Late Upper Paleolithic subsistence, and that these species were taken in numbers, and with tactics and weapons which were unprecedented in the history of hominid adaptations (e.g., Straus 1977, 1983a, b). Besides using the meat, marrow and brains of mammals for food, hunters also employed bone for fuel, skins for clothing and shelters, antler for tools and art objects, teeth for ornaments, sinews for cordage, etc.

Earlier characterizations of Upper Paleolithic hunting in Europe have relied on a few old excavations in which stratigraphic subdivisions were naturally coarse at best (and thus represented many individual occupation levels lumped into cultural horizons) and stressed the primordial role of only one species -- the reindeer (whence the term "l'Age du Renne"). In contrast, my discussion relies mostly on data from recent excavations in which faunal remains were systematically collected and analyzed, and in which techniques of stratigraphic dissection were used, allowing us to assume that the faunal (and artifact) assemblages were deposited by, at most, only a few closely spaced occupations (if not a single occupation) of the excavated site area. The new data do not radically change our perceptions of the main bases of Late Upper Paleolithic subsistence in all European regions, but they do provide important details and nuances in our understanding of those now extinct adaptive systems. Approximate ages of the main culture-stratigraphic units used in Western Europe are given in Table 1. Site locations are presented in Figure 1. The review begins in the north with West Germany.

WEST GERMANY

There seem to be two key species upon which Magdalenian hunters in what is today West Germany depended: reindeer and horse. It has been known

Table 1. Upper Pleistocene Chronological Subdivisions in Western Europe.

Paleolithic Stages	Industries	Approximate Dates B.P.	Pollen Zones	Oxygen Isotope Stages
			Preboreal*	1
-------	Azilian	10,000	--------	--------
			Dryas III	
	Upper/Final Magdalenian		Alleröd*	
LATE	-------	13,000	Dryas II	
			Bölling*	
	Lower/Middle Magdalenian		Dryas Ib	
			Prebölling*	
UPPER			Dryas Ia	2
	-------	17,000	Lascaux*	
	Solutrean			
-------	-------	20,500	Laugerie*	
		23,000	Tursac*	
EARLY	Upper Perigordian (Gravettian)			
		29,000	Kesselt*	-------
	Aurignacian		Arcy*	
UPPER		32,000		
	Lower Perigordian (Chatelperronian)			3
-------	-------	35,000	Les Cottés*	
			Hengelo*	
		61,000	Moershoofd*	-------
				4
MIDDLE	Mousterian	75,000		-------
				5
-------	-------	125,000	--------	--------

*: temperate oscillations/interstadials

since the 1930s that large numbers of reindeer were slaughtered on the plains of northern Germany, specifically at the terminal Pleistocene sites of Stellmoor ("Ahrensburgian," ca. 10,500 yrs. B. P.) and Meiendorf ("Hamburgian," ca. 13,500 yrs. B. P.) near Hamburg (Rust 1937, 1943). The Meiendorf Hamburgian stratum had a reindeer MNI (Minimum Number of Individuals) of 127, with traces of other animals, including one horse (Butzer 1971:534). However, we have no idea of the length of time in which such an assemblage may have

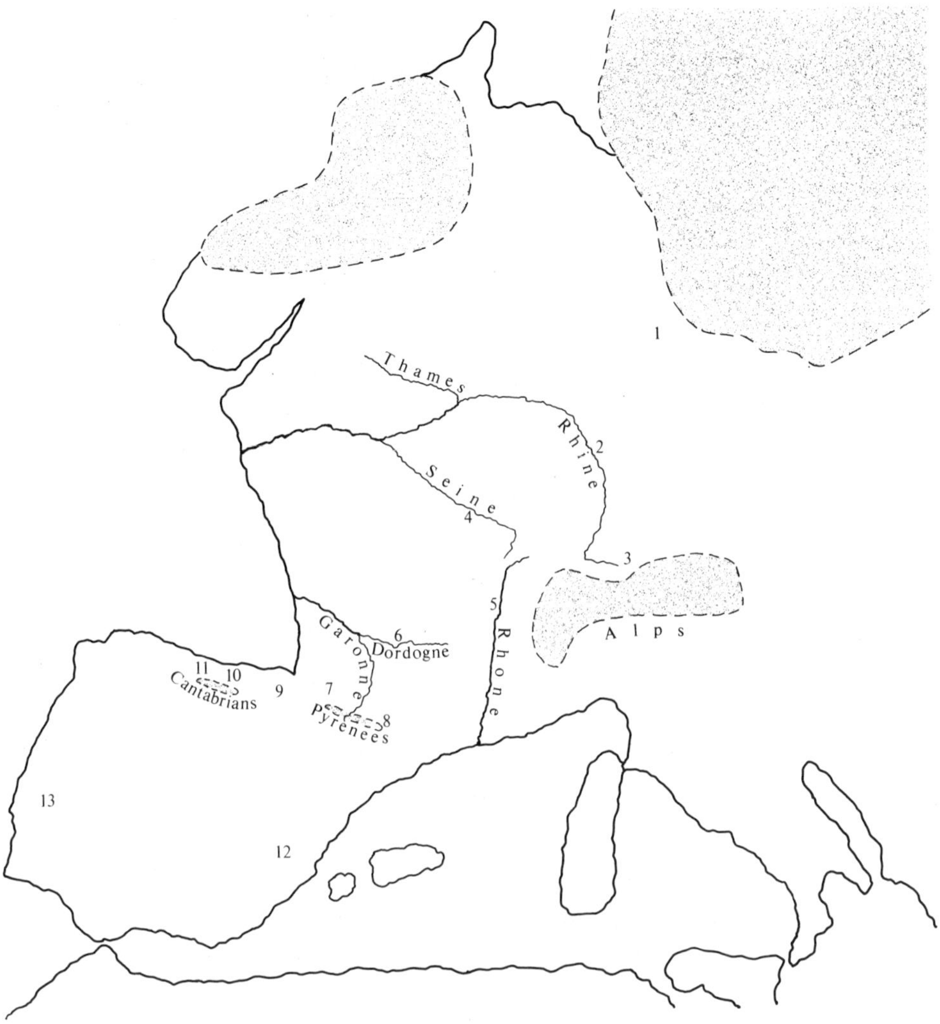

Figure 1. Late Upper Paleolithic Sites of Western Europe Mentioned in the Text. 1. Stellmoor and Meiendorf; 2. Andernach and Gönnersdorf; 3. Petersfels; 4. Pincevent and Verberie; 5. Solutré; 6. Laugerie-Haute, Combe Cullier, Le Flageolet, La Madeleine, Gare de Couze and Le Morin; 7. Duruthy and Dufaure; 8. Enlène and Les Eglises; 9. Ekain and Les Erralla; 10. El Juyo, El Rascaño, Cueva Morín, and El Pendo; 11. La Riera and Tito Bustillo; 12. Parpalló, Les Mallaetes and Volcán del Faro; 13. Caldierao The stippling shows the approximate location of the major ice sheets at ca. 15,000 B.P.; the solid lines show glacial coastlines and major rivers.

accumulated. At these (and other northern German) sites, summer hunting (which Rust argued for on the basis of antler evidence) may have been conducted at the reindeer's natural water-crossings or by driving them into lakes.

Further south, on opposite banks of the middle Rhine near Cologne, two major late Magdalenian open air sites, Andernach and Gönnersdorf, have yielded substantial ungulate faunas almost exclusively dominated by horse remains. Andernach was discovered in 1883 and then excavated by Hermann Schaffhausen (Bosinski 1979a:195; 1979b:58-63), whereas Gönnersdorf was studied in 1968-1976 under the direction of Gerhard Bosinski (1979a, b). Details of the horse-dominated fauna of Andernach are not available; there are small quantities of reindeer, chamois, red deer and, as at Gönnersdorf, several carnivores (notably Arctic fox) and birds (Bosinski 1979a:195). The thin Gönnersdorf occupation level has been dated to 12,660±370 and 12,380±230 yrs. B. P. (late Bolling), slightly older than Andernach. Poplin (1976) analyzed the fauna from the 96 square meter area of the site dug in 1968, listed in Table 2. The figures are all underestimates, as part of this area of the site had already been destroyed before scientific excavation had begun. Ten fish and 80 bird remains were also found. The 1968 faunal assemblage consists of a dense scatter of remains clustered around an area of red ochre-stained sediments and pit features, and seems to be the residue of one winter's occupation, based on seasonality data developed by Poplin from reindeer and fox teeth, reindeer antlers, horse foetal bones, fish and bird remains. The subsequent excavations uncovered several other occupations over an area of half a hectare, but the faunas from those occupations have not yet been published in detail. All are said, however, to be dominated by horse (Poplin 1978:100) but with evidence of some summer use in the areas excavated later. The post-1968 excavations added only three new ungulate species to the overall Gönnersdorf fauna: woolly rhinoceros (Number of Identifiable Specimens [NISP]=1), moose (MNI=1), chamois (NISP=6, MNI=2). As both Andernach and Gönnersdorf are very near the Rhine, it is conceivable that horse herds were hunted by driving them between the river and bluffs, into the river, or off terrace bluffs, though we have no unambiguous evidence of hunting methods (see Levine 1983:38).

In southwestern Germany, new excavations on the terrace in front of the classic Magdalenian cave site of Petersfels (itself entirely dug out between 1927 and 1932) have yielded high quality faunal data. Petersfels is located on a narrow side valley leading up from the upper Danube to a low pass in the Swabian Jura (Alb) over to the upper Rhine and Lake Constance (Bodensee). Several occupation levels were defined and have been dated between 13,000 and 11,500 yrs. B. P. by a large number of radiocarbon determinations (Jaguttis 1983). The principal occupation level (AH3) in sector P1, dates to around 12,500 yrs. B. P. (Bölling), making it contemporary with Gönnersdorf. This occupation was uncovered over an area of 38 square meters; the mammalian faunal assemblages from it are listed in Table 2. Here, as in all Petersfels levels, reindeer is the dominant ungulate, although, as at Gönnersdorf, hare was a major target of Magdalenian hunting (? trapping), probably for pelts as well as for meat (lean and thus low-quality) (Albrecht et al. 1983). Analyses of teeth and antlers suggest that the slaughter of the reindeer took place almost exclusively in the fall. The area of level AH3 in sector P1 is a very small sample of the total site area. Albrecht et al. (1983:122) calculate that in several occupations, the Magdalenians probably killed "at least 1200 reindeer" at Petersfels -- more than 100 during the AH3 occupation alone. They argue for collective hunting during the fall reindeer herd migration between highland (Jura) and lowland (Lake Constance) pastures. The hunters took advantage of the steep-sided valley in front of the cave to drive, channel and slaughter the reindeer. Such a scenario is supported by the good representation of all reindeer carcass parts at the site -- even heads (Albrecht et al. 1983:122-127).

FRANCE

Meticulous, long-term excavations at the open air sites of Pincevent on the bank of the upper Seine (Leroi-Gourhan and Brézillon 1966, 1972; Leroi-Gourhan 1984), and Verberie on the bank of the Oise (Audouze et al. 1981) in north-central France, have produced evidence of heavy dependence on specialized, mass reindeer hunting. Both sites date to the late Magdalenian:

Table 2. Magdalenian Mammalian Faunas at Two Major German Sites

	Gönnersdorf 1968 Unit		Petersfels P1, AH3	
Taxa	NISP	MNI	NISP	MNI
Reindeer	c. 180+	4	1453	25
Red deer	6*	5	6	2
Bovines	2	1	7	1
Chamois	-	-	1	1
Ibex	-	-	2	1
Saiga	2	1	-	-
Mammoth	5	1	-	-
Horse	c. 500	13	185	2
Arctic fox	c. 1400	31	11	"several"
Fox	10	2	-	-
Wolf	7	2	3	1
Wolverine	-	-	1	1
Hare	c. 300	7	258	8

NISP = Number of Identified Specimens
MNI = Minimum Number of Individuals

+ plus antler fragments
* isolated canines

Sources: Poplin 1976; Albrecht et al. 1983

Pincevent's (incoherent) radiocarbon and thermoluminescence dates range between 11,000 and 13,000 yrs. B. P., while the one Verberie date (thermoluminescence) is 13,300±850 yrs. B. P. (Schmider 1984a:231). The huge Pincevent site was repeatedly occupied in summer-fall by reindeer hunters whose occupation lenses are separated by sterile alluvial deposits. From adjacent, contemporaneous "habitations" T112+V105 of Section 36, there are remains of 43 individual reindeer and only traces of horse and hare. From the separate "habitations" 1 (Section 5), E74 (Section 18) and M89 (Section 27), there are reindeer MNIs of 12, 12 and 15 respectively, with essentially no other fauna (Leroi-Gourhan 1984:62; David 1972).

Locus 2 (a hearth-centered unit like the Pincevent "habitations"), making up only part of the large Verberie site, yielded remains of *at least* 10 reindeer killed during a single, brief occupation (season undetermined), with only traces of bovines and horse (Audouze et al. 1981). At both Pincevent and Verberie, the river may have been used in herd hunting, perhaps during migration. Schmider (1984a, b) notes the existence of several other specialized Magdalenian reindeer hunting camps along rivers in Ile-de-France. In contrast, the Magdalenian site of Belloy on the lower Somme is dominated almost entirely by unquantified horse remains (Fagnard 1984:296).

Otherwise, the Upper Paleolithic of France is famous for the large quantities of reindeer remains in the rockshelters of the Périgord area of Aquitaine and for the "horse magma" beneath the rock of Solutré in Burgundy. The Solutré "horse magma" -- a breccia of horse bones as thick as 80 cm in some areas -- dates to the Upper Perigordian (Arcelin and Mayet 1923), with radiocarbon determinations around 23,000 yrs. B. P. (Delibrias and Evin 1974). Recent (unpublished) excavations by Combier are the source of the small sample of 823 horse remains from the "magma" analyzed by Levine, who concludes (1983:39-40) that the "magma" was the result of the repeated slaughter of "probably thousands of horses...within a relatively short period of time." Furthermore, her analyses of the population structure and "the general context of the deposit" suggest "that the main technique employed was herd driving -- possibly into an enclosure against the cliff face." Such behavior seems to have continued into Magdalenian times, judging from the sample Levine analyzed from the uppermost cultural level of this immense open air site.

Less known than Solutré, but also significant, is the site of "Les Trappes," a series of 21 pits dug into the surface of the narrow terrace between the Vézère River and the steep limestone cliff at Les Eyzies. According to Hauser (1911:14-15 and Plan 3) the 1907 excavator of "Les Trappes," the fill of these pits contained "Solutrean flints" (no faunal remains were reported). The holes were placed so as to obstruct the only passage up or down the right bank of the Vézère, downstream of the side valley of La Micoque, right next to the major Solutrean and early Magdalenian living site of Laugerie-Haute.

The faunas from the limited 1956-1959 stratigraphic excavations of Bordes at Laugerie-Haute Est and of Bordes and Smith at Laugerie-Haute Ouest have been described recently by Delpech (1983:68-79 plus his Tables 16 and 17). No individual level (each only 4-12 cm thick) has a very large number of remains, since the excavations cut back only a small width into the deep profiles of the old Peyrony trenches (Smith 1966). Nevertheless, there are several hundred reindeer remains from each Solutrean level and an average of 1,025 from each of four Magdalenian levels. Horse is the second-ranking animal, but it is far less abundant than reindeer. The Solutrean levels span the period of 21,000 - 19,750 yrs. B. P., while the early Magdalenian levels date from 18,500 - 14,000 yrs. B. P., according to a number of (incoherent) radiocarbon determinations (Delibrias and Evin 1974). Delpech's ungulate summary (which lumps some levels and which does not give MNI estimates) is presented in Table 3, showing that reindeer remains make up 93% of the Laugerie-Haute ungulate total.

Across the Vézère River from Laugerie-Haute is the equally famous rockshelter of Pataud, with archeological levels spanning the Aurignacian and Upper Perigordian (*sensu lato*) periods from 34,000 - 22,000 yrs. B. P. (Movius 1975, 1977). The paleontology of the very large Pataud faunal assemblages was studied by Bouchud (1975), who identified 41,736 reindeer remains from the Perigordian levels (which make up well over 90% of the ungulate total) and 3,289 from the Aurignacian levels (almost 100% of the ungulate remains in the basal layers and 62-70% in the principal later Aurignacian horizons). The greater abundance of reindeer in the Perigordian levels is all the more significant because these occupation levels (lenses) were often thinner than the more massive layers of the underlying Aurignacian sequence, although the surface area dug in the Aurignacian levels was smaller than that dug in the uppermost Perigordian due to the need to step the trench sides (Movius 1975, 1977). The Pataud faunas were restudied from a behavioral perspective by Spiess (1979), and the numbers of remains he identified were significantly

Table 3. Ungulate Faunas from Combined Laugerie-Haute Levels (NISP)

	LAUGERIE-HAUTE EST					LAUGERIE-HAUTE OUEST				
								Lower Solutrean		
Taxa	Magdalenian II+, III	Magdalenian I	Upper Solutrean	Middle Solutrean	Lower Solutrean	Upper Solutrean	Middle Solutrean	11A	12A+B	12C+D
Reindeer	2651	1188	172	257	319	1389	437	121	1085	537
Red deer	21	1	--	--	2	3	--	--	1	--
Bovines	18	6	--	1	2	1	--	--	--	1
Chamois	15	6	--	--	2	--	1	--	--	--
Ibex	12	12	1	5	6	--	3	1	2	--
Saiga	47	--	--	--	--	--	--	--	--	--
Mammoth	1	--	--	--	--	21	46	11	6	7
Horse	127	34	36	9	23	12	22	12	27	10

Source: Delpech 1983

fewer than those of Bouchud. Nonetheless, reindeer is still overwhelmingly dominant, especially in the uppermost Perigordian levels (Level 3, lens 2 and Level 4, 0-2 floor, with reindeer MNIs of 40 and 26 respectively). If represented at all, other ungulates have MNIs of 1, 2, 3 or -- in one case -- 4 animals (Spiess 1979: his Table 6.15). Based on tooth eruption, wear and cementum analyses, as well as on foetal bones and salmon remains, Spiess (1979) concludes that the band hunts took place throughout the Pataud sequence exclusively in the cold seasons of the year (mostly late fall-winter), although the length of individual occupations, the size of the cooperating hunting parties and the types of tactics used cannot be determined. One can certainly speculate that the river, cliffs and blind side-canyons of the Les Eyzies area provided excellent natural facilities for trapping substantial numbers of herd animals during individual hunting episodes through use of drive techniques, particularly during fall migrations.

In site after site in the Périgord, Delpech (1983) has documented the numerically overwhelming dominance of reindeer -- a dominance even valid in terms of potentially available meat weight (despite the greater size of individual bovines, horses and even red deer -- none of which are present in any more than small quantities in the Late Upper Paleolithic levels). The largest ungulate assemblages -- most from thin occupation levels -- are summarized (by NISP only) in Table 4. They include Combe Cullier Levels 11 and 13a (15,030±330 yrs. B. P.), Le Flageolet II Level IX (15,000 - 14,000 yrs. B. P.), La Madeleine Levels 7, 9, 12 and 13 (12,600 - 13,000 yrs. B. P.), Gare de Couze Levels C, D, G0, G1 and H (11,750 - 12,450 yrs. B. P.), and Le Morin Levels AI-II, AIII and AIV (10,480±330 yrs. B. P.)(Delpech 1983). All these sites are located in Dordogne or the adjacent areas of Gironde (Le Morin) and Lot (Combe Cullier), and all but Le Morin are primarily composed of reindeer (generally 90% of total NISP). Le Morin, which appears to be the youngest of these Magdalenian sites (though the exact provenience of the single radiocarbon date is unknown), is interesting in that it is quite diverse. Bovine (*Bos/Bison*) remains make up 24% of the total ungulate NISP, horse remains make up 16%, red deer 11%, and boar 5%. (Bison-dominated assemblages like those of the American Folsom or Plano cultures are not found in the Upper

Table 4. Major Ungulate faunas from Other Late Upper Paleolithic Levels of the Perigord.

	Combe Cullier		Flageolet	La Madeleine				Gare de Couze					Le Morin		
Taxa	11	13a	IX	7	9	12	13	C	D	G_0	G_1	H	AI-II	AIII	AIV
Reindeer	458	638	694	686	1292	942	1006	418	1589	820	2832	643	416	293	481
Red deer	--	16	18	--	--	--	--	17	75	47	49	1	125	74	97
Roe deer	--	--	--	--	--	1	--	10	9	10	6	--	8	3	5
Bovines	4	4	4	19	41	5	3	7	17	11	11	7	292	165	210
Chamois	12	20	3	1	--	3	4	--	1	--	1	2	--	--	--
Ibex	3	22	1	--	--	--	--	--	--	--	--	1	--	--	--
Saiga	19	2	9	--	--	--	--	--	--	--	--	1	--	--	--
Boar	--	--	--	--	--	--	--	4	10	3	5	1	49	41	40
Horse	1	8	3	--	7	18	61	5	12	8	9	2	176	128	133
Ass	--	--	--	--	--	--	--	--	--	--	--	--	--	--	1

Source: Delpech 1983

Paleolithic of western Europe. However, bovines, because of their great size, were obviously of dietary importance in Upper Paleolithic Europe, although they may have more often been hunted individually. Along with horse, bison was the favorite animal motif of Franco-Cantabrian Upper Paleolithic cave art -- not the more banal, safer prey, reindeer and red deer.) Otherwise the specialized hunting of reindeer seems to have dominated the subsistence activities of Périgord Magdalenian groups, who nonetheless supplemented their diets with fishing (notably at Gare de Couze) and the occasional (? individual) hunting of other big game (including saiga, especially in the Dryas I period) (Delpech 1983).

The diversification of food resources seen at Le Morin is also apparent in the great, multi-phase Magdalenian site of Duruthy at the southern end of Les Landes 60 km from the Basque Pyrenean crestline, excavated by Arambourou (1978). Like Petersfels (and Dufaure, discussed below), the small cave at Duruthy was dug out by early prehistorians; the modern excavations have been conducted on the terrace in front of the *abri*, at the foot of a prominent cliff which dominates the Gave d'Oloron river at a ford (like so many of the major Upper Paleolithic sites). The excavations, begun in 1957, are still going on. The faunal collections made through 1975 were studied by Delpech (1978, 1983) and are summarized in Table 5. The "Magdalenian III," Level 5 (14,180±200 yrs. B. P.), is dominated by horse with 62% of the total ungulate NISP, followed by bison with 31%, and only 7.5% reindeer remains. The thick, rich "Magdalenian IV," Level 4 (13,840±210 - 13,510±120 yrs. B. P.) shows an increase in bovines (mostly bison - 45%) and reindeer (25%) with a decrease in the relative frequency of horse (28%). Finally, in the vast, cobblestone-paved "Magdalenian VI," Level 3 (top=11,150±200 yrs. B. P.), excavated in whole or part over an area of about 190 square meters, reindeer remains dominate (72%). However, given the late (Allerod) age of this deposit, it is not surprising to find red deer in second place (16% of the total ungulate NISP), with bovines (still mostly bison) and horse making up only 9% and 2% respectively. The reindeer were slaughtered (probably again while crossing the river) in fall-winter, according to Delpech's (1978) analyses of tooth eruption and wear and of antlers (or in fall-winter-spring according to Gordon's [1986] dental

Table 5. Major Ungulate faunas from the French Pyrenees (NISP)

	Duruthy			Enlène	Les Eglises						
Taxa	3	4	5	Magdalenian IV	6	8	8b	8-9	9	9b	Foyer Noir
Reindeer	4306	353	24	1303	2	--	--	--	--	--	--
Red deer	972	26	--	--	1	1	--	--	--	--	--
Roe-deer	2	--	--	--	--	--	--	--	--	--	1
Bovines	564	628	99	239	--	--	--	--	--	--	--
Chamois	3	2	--	5	1	--	--	--	--	--	--
Ibex	13	--	--	2	2040	622	316	460	1600	362	1941
Boar	1	--	--	--	--	--	--	--	--	--	--
Horse	143	396	199	53	--	--	--	--	--	--	--

Sources: Delpech 1978, 1982; Delpech and LeGall 1983

cementum analyses. Fish [mostly *Salmo* spp] remains only appear in the "Magdalenian VI" and are associated with antler harpoons [Delpech 1978, 1983; Arambourou et al. 1986]).

The adjacent site of Abri Dufaure, excavated in 1980-1984 by Straus (1983c, d, 1985a) and still under analysis, shows similar evidence to that of Duruthy. The reindeer (and other large game) were killed between fall and spring according to cementum studies by Spiess. Horse and bovines, as well as reindeer, are well represented in the middle Magdalenian deposit (Stratum 5: 14,000 - 13,000 yrs. B. P.), while reindeer are apparently more abundant (though by far not the exclusive species) in cobblestone-paved Stratum 4 (terminal Magdalenian: 12,000 - 11,000 yrs. B. P.), where there are also some red deer remains. (Detailed faunal analyses are currently being done by Jesus Altuna in San Sebastián.) Harpoons and fish vertebrae are far more frequent at Duruthy than at Dufaure, suggesting some functional differences between the two sites despite the identity of their topographic settings. According to

LeGall (personal communication), one Dufaure pike was fished in spring. A similar pattern of increasing reindeer specialization in the terminal Magdalenian, but with a significant dependence upon horse, is also suggested by the old collections at nearby Isturitz (Bahn 1984). However, heavy dependence on reindeer (albeit together with bovines) has been documented recently in the middle Magdalenian of Enlène in the central Pyrenees (Delpech 1982) under different environmental conditions (see Table 5). The appearance of fish as a supplementary resources does seem to be an exclusively terminal Magdalenian phenomenon in the Pyrenean region (Straus 1983a, b).

The Pyrenees (like their western prolongation, the Cantabrian Cordillera) also have several highly specialized ibex-hunting sites, mostly dating to the Magdalenian (see Straus 1987). The best studied of the French ibex sites is Les Eglises, a Magdalenian (12,900±220 - 11,800±500 yrs. B. P.) site excavated and published by Clottes (1983). Using his analyses of artifact and living floor, lithic raw material sourcing done by Simonnet, and seasonality data developed from ibex teeth, lagopod behavior and salmon vertebrae growth rings by Delpech and LeGall, Clottes shows that hunting parties repeatedly came in late fall-winter to this cave high in the Ariége Pyrenees from lowland bases. At Les Eglises, they procured ibex, as well as willow grouse, on their cold-season pastures. Some or all of these occupations may have occurred during the rut, when male and female ibex were together, since both sexes of adults are well represented in the archeofaunas (Delpech and LeGall 1983). The ungulate faunas (Delpech and LeGall 1983) are summarized as NISP figures in Table 5. The ibex remains pertain to a minimum of 44 individuals, the grouse to 66 and the salmon to 54. It is difficult, due to disturbance of the loose sediments, to tell how many human occupations of the cave may be represented. Ibex hunting (through drives) requires meticulous planning and coordination of several beaters and hunters in order to be successful. Ibex is wary, fast and nimble on its favored steep, rocky habitat, which is dangerous for man. Efficient ibex hunting is testimony to the skill and efficiency of Magdalenian people, as is the mass slaughter of reindeer at water-crossings and the stampeding of horses along narrow defiles.

Specialized ibex sites are also known from the eastern Pyrenees of

Roussillon. However, the Late Upper Paleolithic faunas of Provence are few, poorly preserved and scantily published. Ibex and chamois are found in sites with steep, rocky relief (common in this region). Red deer is often dominant in other sites, and horse, aurochs, boar, roe deer and even giant deer, reindeer and saiga are sometimes present (Bonifay 1979). We have no details on hunting practices or seasonality, despite the abundance of interesting Late Upper Paleolithic sites in Provence.

SPAIN

The faunal analysis in Vasco-Cantabrian Spain is principally the work of zooarcheologist Altuna, with additional studies by Klein and Fuentes. Altuna's early work (1972) involved the analysis of archeofaunas from excavations conducted by J. M. de Barandiarán in the Basque Country between the 1920s and 1960s. Subsequently Altuna has directed his own excavations in Guipúzcoa and analyzed faunas from recent digs in Santander (Cantabria) and Asturias provinces. Aspects of Vasco-Cantabrian faunas have been summarized in English by Freeman (1973, 1981) and Straus (1977, 1983a, b, 1986, 1987). Results of the analyses of the principal Late Upper Paleolithic faunas from the modern excavations at Erralla (Altuna and Mariezkurrena 1985) and Ekain (Altuna and Mariezkurrena 1984) in Guipùzcoa, Cueva Morìn (Altuna 1971), El Rascaño (Altuna 1981), El Juyo (Klein et al. 1981) and El Pendo (Fuentes 1980) in Santander, La Riera (Altuna 1986) and Tito Bustillo (Altuna 1976) in Asturias are presented in Tables 6 and 7. As discussed by Straus (1987), two of these sites, Erralla (Magdalenian 16,270 - 12,300 yrs. B. P.) and El Rascaño (Magdalenian 16,430 - 12,280 yrs. B. P.), and specialized occupations at two other sites, Ekain Level VIb (Magdalenian: 12,050 ± 190 yrs. B. P.) and La Riera Levels 4-6 (Solutrean: ca. 20,000 yrs. B. P.), as well as several other sites either excavated many years ago or not yet fully analyzed, have produced faunas heavily or almost exclusively dominated by ibex remains (Table 6). In addition, the small, early Magdalenian (15,800 ± 350 yrs. B. P.) fauna from the northern Navarra site of Abauntz and the larger (but less precisely dated)

Table 6. Ibex Dominated Faunas from Cantabrian Spain

		Erralla		Ekain	El Rascaño					La Riera (Solutrean)			
Taxa		V	I-III	VIb	5	4b	4a	3	2	4	5	6	4-6*
Reindeer	NISP	1	4	4	--	--	--	--	--	--	--	--	--
	MNI	1	1	1									
Red deer	NISP	198	48	47	113	173	63	26	67	216	512	177	194
	MNI	5	4	5	4	8	4	4	4	3	6	3	3
Roe deer	NISP	2	3	2	--	--	--	--	--	--	--	1	1
	MNI	1	2	2								1	1
Bovines	NISP	1	4	3	2	--	1	1	--	8	--	--	1
	MNI	1	1	2	1		1	1		2			1
Chamois	NISP	83	65	9	--	--	--	2	7	--	--	--	1
	MNI	2	4	1				1	1				1
Ibex	NISP	2375	132	160	1313	1319	673	510	557	425	648	125	362
	MNI	22	8	8	34	36	23	20	15	6	12	4	6
Boar	NISP	--	--	--	--	--	--	--	2	--	--	--	--
	MNI								1				
Horse	NISP	1	5	--	5	3	1	4	7	24	18	7	3
	MNI	1	1		2	1	1	1	1	2	4	1	1

* combined during excavation

Sources: Altuna 1981, 1986; Altuna and Mariezkurrena 1984, 1985

Levels II and III of the Guipùzcoa site of Lezetxiki are dominated by chamois remains (Altuna and Mariezkurrena 1982; Altuna 1972). (*Rupicapra* frequently accompanies *Capra* in Vasco-Cantabrian mountain sites.) Systematic exploitation of montane habitats in this region (as in the Pyrenees and Alps) began in the Late Upper Paleolithic and was characteristic of settlement-subsistence systems at that time.

Reindeer is present only in trace quantities in northern Spain -- mostly in Magdalenian levels of the Basque provinces. It was presumably excluded by its cervid relative, red deer, which is as well adapted to open grasslands and heaths as to woodlands (Straus 1981), and was the principal game species of the Cantabrian Paleolithic (Table 7). As first discussed by Altuna (1972) and Freeman (1973), this species was extensively hunted (even "harvested") in the Late Upper Paleolithic, beginning in the Solutrean and Lower Magdalenian (Straus 1977). Whereas Mousterian (> 100,000 - 35,000 yrs. B. P.) and Aurignaco-Perigordian (35,000 - 21,000 yrs. B. P.) levels always have a few red deer NISP and MNIs, many Solutrean and Magdalenian ones have large numbers of this animal, even when the excavation units in question are vertically and horizontally limited in extent. Indeed the abundance and dominance of red deer continued into the Azilian and Mesolithic of this region, despite post-Pleistocene climatic shifts and the replacement of open vegetation by dense woodlands. One has but to note the minimum numbers of red deer individuals from a few recently excavated levels to judge the importance of mass herd hunting of this large cervid. Such hunting was probably conducted by means of drives and surrounds which took advantage of natural topographic features such as blind side valleys, narrow gorges or passes, rivers, etc., or deep, lightly crusted snow. It is attested to by evidence of "catastrophic" mortality in red deer age profiles and a heavy emphasis on the slaughter of hinds with their young (see Klein et al. 1981; Altuna 1976, 1986; Straus and Clark 1986; Straus 1986). Red deer MNIs, listed in Table 7, often approach or exceed 20, even from thin levels of excavations which only partially sample the respective cave sites. These fauna-rich levels date from the Solutrean of La Riera (20,500 - 17,000 yrs. B. P.) through early-middle Magdalenian levels there and at Ekain and El Juyo (17,000 - 14,000 yrs. B. P.) and the late

Table 7. Red Deer Dominated Faunas from Cantabrian Spain

		Ekain		Cueva Morín	El Juyo		El Pendo	Tito Bustillo		
Taxa		VII	V-III	2	4	6	II	1a	1b	1c
Reindeer	NISP	--	--	2	--	--	--	--	1	--
	MNI			1					1	
Red deer	NISP	650	261	260	1662	1462	427	836	1062	1222
	MNI	24	16	9	17	38	15	9	15	17
Roe deer	NISP	9	16	9	49	27	5	3	--	--
	MNI	3	5	2	3	2	1	1		
Bovines	NISP	9	5	3	44	2	16	19	14	7
	MNI	3	3	1	2	1	1	2	1	2
Chamois	NISP	8	9	8	--	--	5	17	40	19
	MNI	4	3	3			1	2	4	1
Ibex	NISP	81	69	19	23	37	16	126	172	151
	MNI	7	10	3	2	3	2	6	6	3
Boar	NISP	--	1	3	--	--	2	--	--	--
	MNI		1	2			1			
Horse	NISP	6	2	12	42	19	9	62	51	36
	MNI	3	1	3	2	2	1	3	4	3

(continued)

LA RIERA

Taxa		Solutrean									Early Magdalenian			
		7	8	9	10	11	13	14	15	16	17	18	19	20
Red deer	NISP	1912	1268	1783	938	830	603	2494	1160	1797	722	1648	1323	706
	MNI	29	19	25	12	7	7	17	8	18	6	15	18	13
Roe deer	NISP	33	10	48	6	3	3	24	7	2	--	--	2	--
	MNI	3	3	4	2	1	1	2	1	1			1	
Bovines	NISP	5	2	--	--	--	--	2	2	1	1	--	--	--
	MNI	1	1					2	2	1	1			
Chamois	NISP	18	--	13	1	17	18	13	20	7	--	--	1	1
	MNI	2		2	1	1	1	2	2	1			1	1
Ibex	NISP	677	520	464	185	74	178	591	250	458	171	356	372	163
	MNI	10	6	7	5	1	3	7	6	8	4	6	5	6
Horse	NISP	32	9	1	--	--	--	7	9	7	1	6	6	1
	MNI	6	1	1				2	1	3	1	1	1	1

Sources: Altuna 1971, 1976, 1986; Altuna and Mariezkurrena 1984; Klein et al. 1981; Fuentes 1980

Magdalenian levels at Tito Bustillo, El Pendo and Cueva Morín (14,000 - 12,000 yrs. B. P.). Had these sites been dug completely and with modern methods, the red deer NMIs and NISP would have been far larger (as indeed they are from some older excavations, such as Altamira and Urtiaga, which lumped several occupation levels into thick cultural horizons). The point is that prehistoric Cantabrians had shifted from hunting single red deer to hunting herds, probably much the same as prehistoric people in southwestern France had, by Late Upper Paleolithic times, developed herd-hunting techniques for procuring reindeer -- a docile species easily taken *singly* even by Neanderthals.

Nevertheless, red deer generally does not dominate the Cantabrian subsistence base to the same extent that reindeer (or horse) dominates many German or French sites of roughly the same periods. Although there are clearly sites and levels in northern Spain which specialized in the procurement and processing of red deer and, particularly, ibex, the Late Upper Paleolithic subsistence base *as a whole* was quite diversified (Straus 1977), with a wide variety of other ungulate (and other mammalian) species regularly being hunted by means of diverse tactics, including solitary hunting and trapping. In addition, marine molluscs, anadromous and marine fish, and birds were becoming increasingly important dietary supplements from Solutrean through Mesolithic (and even Neolithic) times (Straus et al. 1980; Straus and Clark 1986). Plant foods (e.g., hazels, acorns, berries, seeds, roots, etc.) were probably increasingly incorporated with the growth of woodland and estuarine habitats in the Tardiglacial and early Holocene. Thus, while the "Magdalenians" of northwestern Europe and even much of France took an increasingly specialized approach to subsistence in the broad, open Late Glacial environments of those regions, the prehistoric Cantabrian "Magdalenians," confined to a narrow, high-relief coastal strip, rich in a wide variety of edible species, diversified their resource base while also developing specialized techniques and technologies for the efficient exploitation of the key medium to large ungulates -- red deer and ibex.

Closing this *de facto* north-south transect of western European Late Upper Paleolithic archeofaunas, can be made with the few assemblages quantitatively analyzed by Davidson (1976, 1983) in Levantine Spain (Valencia). The Solutrean and post-Solutrean ("Solutreo-Gravettian" and Magdalenian) levels of Parpalló (an old excavation), Les Mallaetes and Volcán del Faro (both new excavations) have been shown to be contemporary at least in part in the period between about 21,000 - 14,000 yrs. B. P. In this far different Mediterranean environment of the late Last Glacial period (albeit somewhat cooler and drier than today's), diversification was nonetheless also the basis of successful Late Upper Paleolithic subsistence. Davidson (1976, 1983) has shown the numerical importance (in terms of NISP and MNIs) of red deer, ibex and rabbits (*Oryctolagus cuniculus*), and has detailed the evidence for more limited hunting of ass, horse and aurochs, and the taking of birds and molluscs. MNI counts per level can be as high as 71 for red deer and 39 for ibex in Parpalló (with admittedly thick layers) and 26 for rabbit and 212 for ibex at Les Mallaetes. Half of the 18 levels from the limited excavation at Volcán yielded over 200 rabbit bones each. Clearly, as in Cantabrian Spain, low-yield, labor-intensive strategies became part of the repertoire of subsistence activities in Levante, presumably as a result of the need to feed a growing regional human population. An efficient use of the region's resources involved elaborate seasonal scheduling of different subsistence activities -- something probably just as true of the activities of Late Upper Paleolithic hunter-gatherers in Germany and France, as in Cantabria and Levante. Recent excavations at the Gruta do Caldeirao in central Portugal (undated Solutrean), have revealed faunas (currently under analysis) dominated by red deer throughout the seven cultural levels. As at La Riera, however, ibex remains are abundant in some levels. However, rabbit becomes a major supplementary resource in the uppermost deposits (Zilhao 1986).

CONCLUSIONS

In this brief *tour d'horizon* we have seen that Late Upper Paleolithic people in western Europe were efficient big-game hunters, sometimes targeting particular animals (reindeer, horse, red deer or ibex) at specialized sites, and sometimes (particularly in southern latitudes) combining such specialization with the exploitation of very diverse animal food resources. They were armed with thrusting spears, darts (*sagaies*), leisters, projectile weapons (atlatls, whose elaborately decorated "hooks" have been found, possibly bows, judging from the presence of gracile tanged and shouldered points, and harpoons), and probably nets, traps and weirs (judging from the fish, birds, lagomorphs, mustelids and foxes they regularly took). They had mastered the use of the battue, the drive line and the surround. They had learned to take advantage of natural choke points and obstacles to game flight, as well as the vulnerability of reindeer at traditional water-crossing spots along migratory routes. They had learned how to predict the flight pattern of ibex (i.e., straight up rocky cliffs), and so could preposition hidden hunters along crestlines. Similarly, they had undoubtedly learned to take advantage of the "follow-the-leader" behavior of horse and bison herds, and could thus make constructive use of the stampede. In order to be successful in collective hunting, these hunters would have required planning and organizational skills well beyond those needed by their Neanderthal ancestors, suggesting quantum growths in social organizational complexity, symbolic abilities, calendrical knowledge, information acquisition and exchange, transgenerational transmission of knowledge, etc. All of these new developments are probably reflected in ways which we cannot precisely decipher in the rich mobile and cave art of the Upper Paleolithic in western Europe.

This sort of sophisticated hunting and the elaborate organizations it required evolved in Europe and elsewhere after hundreds of thousands of years of something less organized, less efficient, less regular and less massive -- the acquisition of animal products by scavenging and by occasional, opportunistic "hunting" on the part of the ancestors of *Homo sapiens sapiens*.

ACKNOWLEDGEMENTS

Much of my research in northern Spain and southwestern France has been supported by grants from the National Science Foundation, as well as by grants from the L. S. B. Leakey Foundation, the National Geographic Foundation and the University of New Mexico. To all I am very grateful. I wish to thank Matthew Nitecki for suggesting that I write this paper. My thanks also go to Matt Schmader for drafting and Lynn Baca for word processing.

REFERENCES

Albrecht, G., H. Berke and F. Poplin. 1983. Saugetierreste vom Petersfels P1 und Petersfels P3, Grabung 1974-1976. In: *Naturwissenschaftliche Untersuchungen an Magdalénien-Inventaren vom Petersfels, Grabungen 1974-1976*, ed. G. Albrecht, H. Berke and F. Poplin, pp. 63-127. Tübingen: Verlag Archaeologica Venatoria.

Altuna, J. 1971. Los mamíferos del yacimiento prehistórico de Morín. In: *Cueva Morín: Excavaciones 1966-1968*, ed. J. González Echegaray and L. Freeman, pp. 367-399. Santander: Patronato de las Cuevas Prehistóricas.

Altuna, J. 1972. Fauna de mamíferos de los yacimientos prehistíricos de Guipúzcoa. *Munibe* 24:1-464.

Altuna, J. 1976. Los mamíferos del yacimiento prehistórico de Tito Bustillo. In: *Excavaciones en la Cueva de Tito Bustillo*, ed. J.Moure and M. Cano, pp. 151-194. Oviedo: Instituto de Estudios Asturianos.

Altuna, J. 1981. Restosóseos del yacimiento prehistórico del Rascaño. In: *El Paleolitico Superior de la Cueva del Rascaño*, ed. J. González Echegaray and I. Barandiarán, pp. 223-269. Monografías del Centro de Investigación y Museo de Altamira 3. Santander.

Altuna, J. 1984. Bases de subsistencia de origen animal en el yacimiento. In: *El Yacimiento Prehistórico de la Cueva de Ekain*, ed. J. Altuna and J. Merino, pp. 211-280. San Sebastián: Sociedad de Estudios Vascos.

Altuna, J. 1985. Bases de subsistencia de los pobladores de Erralla: macromamíferos. *Munibe* 37:87-117.

Altuna, J. 1986. The mammalian faunas from the prehistoric site of La Riera. In: *La Riera Cave*, ed. L. Straus and G. Clark, pp. 237-274, 421-479. Anthropological Research Papers 36. Tempe, Arizona.

Altuna, J., and K. Mariezkurrena. 1982. Restos óseos del yacimiento prehistórico de Abauntz. *Trabajos de Prehistoria Navarra* 3:347-353.

Arambourou, R. 1978. *Le Gisement Préhistorique de Duruthy*. Mémoires de la Société Préhistorique Française 13. Paris.

Arambourou, R., L. Straus and C. Normand. 1986. Recherches de Préhistoire dans Les Landes en 1985. *Bulletin de la Société de Borda* 111:121-140.

Arcelin, F., and L. Mayet. 1923. Le gisement du Crot-du-Charnier, à Solutré. *Bulletin de l'Association Régional de Paléontologie et de Préhistoire (Lyon)* 1:1-10.

Audouze, F., D. Cahen, B. Schmider and L. Keeley. 1981. Le site magdalénien du Buisson Campin á Verberie. *Gallia Préhistoire* 24(1):99-143.

Bahn, P. 1984. *Pyrenean Prehistory*. Warminster: Aris and Phillips.

Bonifay, M.-F. 1979. Principales caracteristiques des faunes de la fin des temps glaciaires dans la zone périméditérranéenne française. In: *La Fin des Temps Glaciaires en Europe*, ed. D. de Sonneville-Bordes, pp. 277-279. Paris: CNRS.

Bosinski, G. 1979a. *Eiszeitjager im Neuwieder Becken*. Koblenz: Landesdenkmalamt.

Bosinski, G. 1979b. Stratigraphie du Paléolithique supérieur récent et du Paléolithique final dans le bassin de Neuwied. In: *La Fin des Temps Glaciaires en Europe*, ed. D. de Sonneville-Bordes, pp. 193-201. Paris: CNRS.

Bouchud, J. 1975. Etude de la faune de l'Abri Pataud. In: *Excavation of the Abri Pataud*, ed. H. Movius, pp.69-153. Bulletin of the American School of Prehistoric Research 30. Cambridge.

Butzer, K. 1971. *Environment and Archeology* (2nd ed.). Chicago: Aldine.

Clottes, J. 1983. La caverne des Eglises á Ussat. *Bulletin de la Société Préhistorique de l'Ariége* 38:23-81.

David, F. 1972. Témoins osseux. In: *Fouilles de Pincevent*, ed. A. Leroi-Gourhan and M. Brézillon, pp. 295-320. Paris: CNRS.

Davidson, I. 1976. Les Mallaetes and Monduver: the economy of a human group in prehistoric Spain. In: *Problems in Economic and Social Archaeology*, ed. G. Sieveking, I. Longworth and K. Wilson, pp. 483-499. London: Duckworth.

Davidson, I. 1983. Site variability and prehistoric economy in Levante. In: *Hunter-Gatherer Economy in Prehistory*, ed. G. Bailey, pp. 79-95. Cambridge: Cambridge University Press.

Delibrias, G., and J. Evin. 1974. Sommaire des datations 14C concernant la préhistoire en France. *Bulletin de la Société Préhistorique Française* 71:149-156.

Delpech, F. 1978. Les faunes magdaléniennes et azilienne du gisement de Duruthy. In: *Le Gisement Préhistorique de Duruthy*, ed. R. Arambourou, pp. 110-116. Mémoires de la Société Préhistorique Française 13. Paris.

Delpech, F. 1983. *Les Faunes du Paléolithique Supérieur dans le Sud-Ouest de la France*. Paris: CNRS.

Delpech, F., and O. LeGall. 1983. La faune magdalénienne de la grotte des Eglises. *Bulletin de la Société Préhistorique de l'Ariége* 38:91-118.

Fagnard, J.-P. 1984. Le Paléolithique supérieur dans le Nord de la France. *Bulletin de la Société Préhistorique Française* 81:291-301.

Freeman, L. 1973. The significance of mammalian faunas from Paleolithic occupations in Cantabrian Spain. *American Antiquity* 38:3-44.

Freeman, L. 1981. The fat of the land: notes on Paleolithic diet in Iberia. In: *Omnivorous Primates*, ed. R. Harding and G. Telecki, pp. 104-165. New York:Columbia University Press.

Fuentes, C. 1980. Estudio de la fauna de El Pendo. In: *El Yacimiento de la Cueva de El Pendo*, ed. J. González Echegaray, pp. 217-237. Bibliotheca Praehistorica Hispana 17. Madrid.

Gordon, B. 1986. Of men and reindeer herds in French Magdalenian prehistory. In: *The Pleistocene Perspective*, vol. 2, ed. by A. ApSimon, pp. 297-318. London: Allen and Unwin.

Hauser, O. 1911. *Le Périgord Préhistorique*. Le Bugue: Rejou.

Jaguttis, M. 1983. Die radiocarbondatierung des ausgrabung Petersfels. In: *Naturwissenschaftliche Untersuchungen an Magdalénien-Inventaren vom Petersfels, Grabungen 1974-1976*, ed. G. Albrecht, H. Berke and F. Poplin, pp. 47-57. Tubingen: Verlag Archaeologica Venatoria.

Klein, R., C. Wolf, L. Freeman and K. Allwarden. 1981. The use of dental crown heights for constructing age profiles of red deer and similar species in archaeological samples. *Journal of Archaeological Science* 8:1-31.

Kurtén, B. 1968. *Pleistocene Mammals of Europe*. Chicago: Aldine.

Leroi-Gourhan, A. 1984. *Pincevent: Campement Magdalénien de Chasseurs de Rennes*. Paris: Ministére de la Culture.

Leroi-Gourhan, A., and Brézillon, M. 1966. L'habitation magdalénienne no. 1 de Pincevent. *Gallia Préhistoire* 9(2):263-385.

Leroi-Gourhan, A., and Brézillon, M. 1972. *Fouilles de Pincevent: Essai d'Analyse Ethnographique d'un Habitat Magdalénien*. VII Supplément à Gallia Préhistoire. Paris.

Levine, M. 1983. Mortality models and the interpretation of horse population structure. In: *Hunter-Gatherer Economy in Prehistory*, ed. G. Bailey, pp. 23-46. Cambridge: Cambridge University Press.

Movius, H. 1975. *Excavation of the Abri Pataud*. Bulletin of the American School of Prehistoric Research 30. Cambridge.

Movius, H. 1977. *Excavation of the Abri Pataud: Stratigraphy*. Bulletin of the American School of Prehistoric Research 31. Cambridge.

Poplin, F. 1976. *Les Grands Vertébrés de Gönnersdorf: Fouilles 1968*. Wiesbaden: Franz Steiner Verlag.

Poplin, F. 1978. Données de la grande faune sur le climat et l'environnement. In: *Geowissenschaftliche Untersuchungen in Gönnersdorf*, ed. K. Brunnacker, pp. 98-104. Wiesbaden: Franz Steiner Verlag.

Rust, A. 1937. *Das Altsteinzeitliche Rentierlager Meiendorf*. Neumunster: K. Wachholtz.

Rust, A. 1943. *Die Alt- und Mittelsteinzeitliche Funde von Stellmoor*. Neumunster: K. Wachholtz.

Schmider, B. 1984a. Les habitations magdaléniennes de Marsangy. In: *Jungpalaolithische Siedlungsstrukturen in Europa*, ed. H. Berke, J. Hahn and C.-J. Kind, pp. 169-180. Tubingen: Verlag Archaeologica Venatoria.

Schmider. 1984b. *Les Industries Lithiques du Paléolithique Supérieur en Ile-de-France* (2nd ed.). VI Supplément à Gallia Préhistoire. Paris.

Smith, P. 1966. *Le Solutréen en France*. Bordeaux: Delmas.

Spiess, A. 1979. *Reindeer and Caribou Hunters*. New York: Academic Press.

Straus, L. 1977. Of deerslayers and mountain men: Paleolithic faunal exploitation in Cantabrian Spain. In: *For Theory Building in Archaeology*, ed. L. Binford, pp. 41-76. New York: Academic Press.

Straus, L. 1981. On the habitat and diet of *Cervus elaphus. Munibe* 33:175-182.

Straus, L. 1983a. Terminal Pleistocene faunal exploitation in Cantabria and Gascony. In: *Animals and Archaeology: Hunters and their Prey*, ed. J. Clutton-Brock and C. Grigson, pp. 209-225. British Archaeological Reports S163. Oxford.

Straus, L. 1983b. From Mousterian to Magdalenian: cultural evolution viewed from Cantabrian Spain and Pyrenean France. In: *The Mousterian Legacy*, ed. E. Trinkaus, pp. 73-111. British Archaeological Reports S164. Oxford.

Straus, L. 1983c. Paleolithic investigations in Les Landes, France. *Current Anthropology* 24:388-389.

Straus, L. 1983d. The Abri Dufaure Prehistoric Project: preliminary report on the 1980, 1981 and 1982 field seasons. *Journal of Field Archaeology* 10:371-378.

Straus, L. 1985a. Le Magdalénien final de l'Abri Dufaure: un apérçu de la chronologie et de la saison d'habitation humaine. *Bulletin de la Société Préhistorique de l'Ariége* 40:169-184.

Straus, L. 1985b. Stone Age prehistory of northern Spain. *Science* 230:501-507.

Straus, L. 1986. Late Wurm adaptive systems in Cantabrian Spain: the case of eastern Asturias. *Journal of Anthropological Archaeology* 5:330-368.

Straus, L. 1987. Upper Paleolithic ibex hunting in southwest Europe. *Journal of Archaeological Science* 14

Straus, L., and G. Clark. 1986. *La Riera Cave*. Anthropological Research Papers 36. Tempe, Arizona.

Straus, L., G. Clark, J. Altuna and J. Ortea. 1980. Ice-Age subsistence in northern Spain. *Scientific American* 242(6):142-152.

Zilhao, J. 1986. The Portuguese Estremadura at 18 000 B. P. In: *The Pleistocene Perspective*, Vol. 2, ed. A. ApSimon, pp. 107-123. London: Allen and Unwin.

Prehistoric, Plains-Mountain, Large-Mammal, Communal Hunting Strategies

George C. Frison

Anthropology Department
University of Wyoming
Laramie, Wyoming

Prehistoric human animal procurement has become a major area of archaeological interest because current studies in cultural ecology require a better understanding of past human subsistence strategies in hunter-gatherer societies. As with most archaeological research problems, straightforward answers are lacking and are unlikely to emerge from the data base; consequently, the number of different approaches toward solving the problem approximates the number of individual investigators. In reality, however, no single effort will bring about a solution to the problem. The best to be hoped for is that each effort may contribute small increments that, when added together and properly interpreted, will continually lead to more accurate explanations.

This paper deals with hunting from a more pragmatic approach than that used by most investigators. The perspective adopted here is that each animal species demonstrates separate and distinct behavior patterns that should be reflected in individual procurement sites. The investigator in this kind of research effort must develop enough familiarity with the animals in their natural habitat so that he or she can determine the optimum procurement strategy or strategies for each species. The participant observer is in the best position to acquire the necessary information since there are innumerable aspects of animal behavior that cannot be detected or properly understood either by observation alone or gleaned from the published works of others.

Successful prehistoric communal hunting depended largely upon the ability of the hunting group to predict the reactions of animals of a given species to the natural environment. Consequently, landforms, vegetation cover, and time of year of the procurement events are variables that must be known and understood: otherwise, accurate interpretations cannot be realized. The communal animal kill sites referred to in this paper ocurred in a wide variety of landforms. In some cases, especially the more recent ones, the present topography and vegetation is the same or similar to that at the time of the events. In the older sites of Paleoindian age, there have often been major changes in landforms and vegetation: consequently, there is need for continued efforts toward better paleoecological reconstructions. However, if in the final analysis, the procurement feature and the context in which it is recovered will not accommodate the behavior patterns of the species involved, it is time to carefully rethink both the archaeological and the paleoecological data.

Since geological and man-made processes of archaeological site destruction are continuous, it follows that the older the site, the less the chances of preservation. Much of the efforts at prehistoric animal procurement, particularly bison, tended to be strongly associated with arroyo systems making them even more subject to destruction through the continual geologic processes of aggradation and degradation. Significant animal procurement sites almost without exception can be pointed out as accidents of geologic preservation in the sense that small changes in lateral stream cutting, the forces that cause a stream to seek a new course after a period or periods of aggradation or surface degradation would have eliminated the evidence. Related problems of site formation and destruction are discussed by Klein in this symposium.

Related to this is the question of what increment of the original data base remains intact. Even with the intensive archaeological survey and mitigation efforts witnessed in parts of the High Plains in recent years due to energy extraction, the number of significant new animal procurement sites of Paleoindian age (8,000-11,500 years before present) is surprisingly small. Since our expectations have not been met in finding a data base that promises to yield enough new data, this and other concerned investigators have decided that, besides gathering new data, the data from earlier site investigations should be assembled and brought to a level consistent with current analytic and interpretive analysis. In this way, the present data base will see continual usefulness as new methods of recovery, analysis, and interpretation develop.

In terms of methodology, we are witnessing continual improvement in the methods of data recovery, analysis, and interpretation and several papers in this symposium focus directly on this subject. Site formation (and destruction) processes are of ever-increasing concern to archaeologists. This is particularly evident in Klein's paper (this volume) which raises questions as to whether a long-established part of the Old World archaeological record is, in fact, even an archaeological site and may have been erroneously interpreted in the past and consequently provided an incorrect interpretation of past hominid hunting behavior. Determinations of site integrity must be more accurate and will become more so only as the processes of site formation and destruction are better known and understood.

Todd's and Behrensmeyer's studies of taphonomic processes as presented in this volume lead archaeologists further in the direction of more accurate interpretations of faunal remains found in archaeological sites, particularly those where large numbers of animals were killed, processed, and utilized. It is of utmost importance in these kinds of site situations to be able to interpret human versus animal and natural modification of bone. Seasonality determinations of mammoth procurement events may be closer to a reality if Fisher's (this volume) method of time-of-year determinations through proboscidean tusk analysis proves reliable. These four papers in particular focus on problems encountered in prehistoric Plains-Mountain communal animal kills in North America. The other papers in this symposium, although oriented toward many different aspects of hominid hunting and scavenging, have little direct concern with prehistoric communal hunting in North America.

Prehistoric hunters in North America and in particular those on the high plains, intermontane basins and in the adjacent mountain ranges were able to form cooperative hunting groups that allowed them to take larger numbers of animals more efficiently than they could be taken during individual or small-group hunting efforts. Evidence for this is mainly in the form of bonebeds that still remain as evidence of the events; in rare cases, parts of man-made structures used to contain the animals are yet extant. There are other kinds of evidence that may suggest or indicate communal animal procurement. For example, campsites may contain faunal remains for which taphonomic study reveals that a large number of animals were killed during a very short period of time and thereby may have resulted from a communal effort nearby. Another possibility is a structure suggesting shamanistic activity that was always present when supernatural help was enlisted to aid in high-risk economic ventures such as a communal kill.

A PERSPECTIVE ON HUNTING AND PREHISTORIC ANIMAL KILL SITE INTERPRETATION

Each investigator inevitably develops his or her own bias in pursual of research problems. The writer's views toward hunting may be somewhat at odds with those of other investigators but they stem also from a very different background of experience than most other investigators in the same area of research interest. Part of this bias is that, based on long experience, hunting is a one-sided contest that is carried on between man and animals. Each hunting episode is a planned strategy designed to kill an animal or animals whose ultimate chances of eluding the accomplished hunter are indeed slim. Hunting success varies between individuals and depends largely on personal qualities, some of which are endowed genetically, some that are acquired by observation, but mostly those that come from long experience.

The qualities that comprise a good hunter may be briefly categorized as follows: first are personal physical qualities and skills of the hunter; second is the hunter's knowledge of and the ability to predict the behavior of the species being hunted; third is the hunter's knowledge of the hunting territory and how each given species exploits it; and fourth are the limitations imposed by the hunting equipment. A corollary to the first statement is that endurance and skills increase in direct proportion with hunger; a corollary to the second statement is that good hunters go after the most desirable animals. A distinction is usually made also between what are considered dangerous animals and ordinary animals. The hunting of the former may produce situations in which the chances of the hunted animal killing or maiming the hunter are a real and recognized possibility. This kind of hunting usually carries with it greater prestige than hunting the ordinary animals.

There are also two other basic kinds of hunting: trophy and subsistence. Trophy hunting has many culturally imposed limitations as to age, size, and sex of the animals taken, along with other limitations on the strategies of taking and the weaponry that is used. Subsistence hunting (meat hunting) is usually less restrictive but certain institutionalized limitations may be imposed. On the other hand, a certain amount of the trophy hunting instinct is found in about every hunter and can be confirmed by carefully curated and protected sets of antlers, horns, skulls, etc. taken from exceptional animals.

The date of the earliest appearance of *Homo sapiens sapiens* in North America is open to question with little indication that the problem will be settled in the foreseeable future. However, the earliest undisputed evidence of sophisticated hunting groups on the High Plains is found in the Clovis Cultural Complex dated between about 11,200 and 11,500 years ago (see Haynes 1982). Clovis hunters possessed highly-developed stone tools and weaponry and were able to take a wide variety of large ungulates although the full range of their hunting and gathering activities still remains unproven.

The methodology of animal kill site recovery and analysis still has a long way to go. Even in a single component bison kill site such as the Casper Site of Hell Gap age (Frison 1974), there is as yet no convincing way to determine with certainty whether or not we are dealing with a single event, more than one event closely-spaced in time, or more than one event at the same season of the year spaced a year or more apart. We can identify the geomorphic feature involved (in this case a parabolic sand dune), seasonality of the event through tooth eruption schedules, age of the site within broad limits through cultural diagnostics and radiocarbon dates, climate and vegetation through small mammal, pollen, and phytolithic studies, but the finer details of Hell Gap bison procurement escape us partly because we are uncertain whether or not it was a single event. Perhaps the only large Paleoindian bison procurement site that can be safely argued for as a single event is Olsen-Chubbuck (Wheat 1972), where the conditions of the geomorphic feature involved, site preservation, and the inability of the involved hunting group to gain access to all of the carcasses allows a straightforward interpretation of a single event even though the actual events leading up to the kill remain obscure and subject to conjecture.

After studying the full range of complexity encountered in plains/mountain animal procurement sites, the writer feels that a distinction should be made between what on one end of the spectrum is arbitrarily designated as cooperative hunting and that at the opposite end which is designated as communal hunting. Much of what appears from first impressions to have been the result of well-plannned, predictive hunting operations may actually have been the results of cooperating small numbers of hunters in opportunistic situations. Many of the so-called arroyo bison traps, for example, may have been the result of both a group of hunters and a herd of bison being in the right place at the right time. On the other hand, many arroyo traps probably required fences at critical points where the natural configuration of the arroyo was insufficient to hold the animals, drive line construction, and careful planning and organization to bring the animals into proper position. Unfortunately, the evidence for much of this is no longer extant.

Communal hunting is arbitrarily defined as procurement operations that were planned ahead of time and required that some individual was given temporary authority to coordinate the efforts of a number of persons. These designations in some archaeological cases will be difficult to determine and justify particularly in sites such as Jones-Miller (Stanford 1974), Hudson-Meng (Agenbroad 1978), the Agate Basin component at the Agate Basin site (Frison and Stanford 1982), and the Horner site (Frison and Todd 1987) where the numbers of animals are large and the number of separate events cannot, as yet, be satisfactorily determined because of methodological limitations.

The Evidence for Cooperative and/or Communal Hunting on the Central and Northern Plains and the Adjoining Rocky Mountains

A careful look at the archaeological record indicates that North American Plains communal hunting centered almost exclusively on three species that included bison (*Bison bison bison* and several subspecies including *B. bison occidentalis* and *B. bison antiquus*), pronghorn (*Antilocapra americana*) and possibly mammoth (*Mammuthus columbi*). In the foothills and mountains adjacent to the High Plains and in isolated uplifts and other areas within the plains that provided good habitat, a fourth species, mountain sheep (*Ovis canadensis*) was taken communally in Late Prehistoric and Early Historic times (Frison 1985) using brush and stone drive lines, log fences, ramps and catch pens. Net trapping of mountain sheep probably goes well back into the Paleoindian Period (Frison et al. 1986).

The evidence for prehistoric communal taking of either mule deer (*Odocoileus hemionus*) or white-tailed deer (*Odocoileus virginianus*) in traps on the High Plains or the adjacent mountains is unconvincing although mule deer are found in relatively small numbers in archaeological sites from Paleoindian to Historic times. One site in northern Wyoming (Frison and Walker 1984) contained a large quantity of mule deer remains but tooth eruption studies (Simpson 1984) indicated a continuous accumulation of animals over several months rather than a rapid accumulation at one time which could be indicative of a nearby trap. White-tailed deer appear in plains/mountain archaeological sites only during the Late Prehistoric Period with no known evidence that suggests communal efforts.

Camel (*Camelops*) remains in small numbers have appeared in some High Plains Paleoindian sites (see Frison et al. 1978) but convincing evidence of hunting or utilization of camels as food is lacking (see Haynes and Stanford 1984). The same is true of horse (*Equus* sp.) which appears rarely along with mammoth (see Walker and Frison 1986) in archaeological sites but not in the proper numbers or context to suggest they were of any significance as food or an object of systematic hunting.

Elk (*Cervus elaphus*) remains are relatively rare in High Plains archaeological sites until relatively late and well into the Late Prehistoric Period. On the other hand, elk antler tools are found in archaeological contexts as far back as the Folsom period (Frison and Craig 1982:157-172) and in quarry sites during the Early Plains Archaic Period (Frison 1978a:331) but skeletal material is, as yet, lacking in these older contexts. If this is interpreted that elk were not used as food and were present as their antlers would indicate, it is difficult to explain since their size is between that of the bison and deer, their carcass yield of excellent quality meat is high in relation to total weight, and their behavior patterns allow them to be easily taken.

From this extant body of evidence, it can be emphasized once more that only four of the species mentioned above, mammoth, bison, pronghorn, and mountain sheep need to be seriously considered in any discussion of plains/mountain prehistoric communal and/or cooperative ungulate hunting. Remains of larger carnivores and scavengers appear regularly in High Plains archaeological sites dating from Paleoindian through Late Prehistoric. Some of these occurrences may represent food use but there is no evidence to suggest large scale, systematic procurement efforts. A number of archaeological sites are mentioned in the following pages (see Fig. 1) that demonstrate prehistoric communal hunting. These sites are found in open plains, intermontane basins, and on the slopes of the mountains and they by no means exhaust the total number known.

Mammoth Hunting and the Clovis Cultural Complex

There seems little argument that Clovis hunters were involved with mammoth hunting but the data base is not sufficient to indicate how much of their economy depended upon mammoth procurement. The presence of Clovis projectile points in mammoth carcasses is strong evidence for their use in killing the animals. Mammoth hunting has captured the fancy of many, and the general concepts of mammoth hunting have been spread mainly as the result of artistic license and are mostly incorrect. Almost invariably, mammoth hunting in textbooks and museum displays depicts an animal in a bog or being forced over a precipice accompanied by a disorganized group of hunters wildly flinging spears and shouting. A pack of snarling dogs is usually present and one or more of the luckless hunters is being either trampled, gored, or flung through the air or smashed on the ground with the animal's trunk. The animal, with spears protruding, often resembles an oversize pincushion. These kinds of scenes violate all the rules of intelligent, systematic hunting.

In the writer's opinion, the use of bogs to trap large mammals such as the mammoth or late Pleistocene bison is not a concept that can be applied to a true hunting group for several reasons. First, healthy animals are seldom trapped in bogs; second, healthy animals are attracted to bogs and swampy areas to roll in the mud and avoid flies and to take advantage of the vegetation growing there; third, it is difficult and in many cases impossible to extract a bogged animal from the muck; and fourth, butchering a bogged animal is exceedingly difficult and often results in contaminating the meat products to the extent that they are inedible. If healthy animals are encountered in a bog, and from the writer's actual experience, the better strategy is to drive them from the bog and then kill them on open ground.

The question of cooperative/communal mammoth hunting in North America may never be settled but some discussion is appropriate. As a basis for departure into different aspects of mammoth procurement, it could have been that even the taking of a single mammoth represented a cooperative and possibly even a communal effort even though the general character of mammoth hunting may have been largely opportunistic. The suggestion has been made (Saunders 1977) that Clovis hunters were taking entire families at one time. This was based on the fact that the population structure of the animals at the Lehner site suggested a family that was killed at one time. It was assuming also that mammoth social behavior was analogous to that of modern elephants. This is one interpretation that can be extracted from the present data from North American mammoth sites but an alternative explanation should also be explored and one that is also based on the same assumptions of elephant social behavior.

Extensive elephant culling in Zimbabwe for the past several years has been used by the writer as a testing ground for the efficiency of Clovis

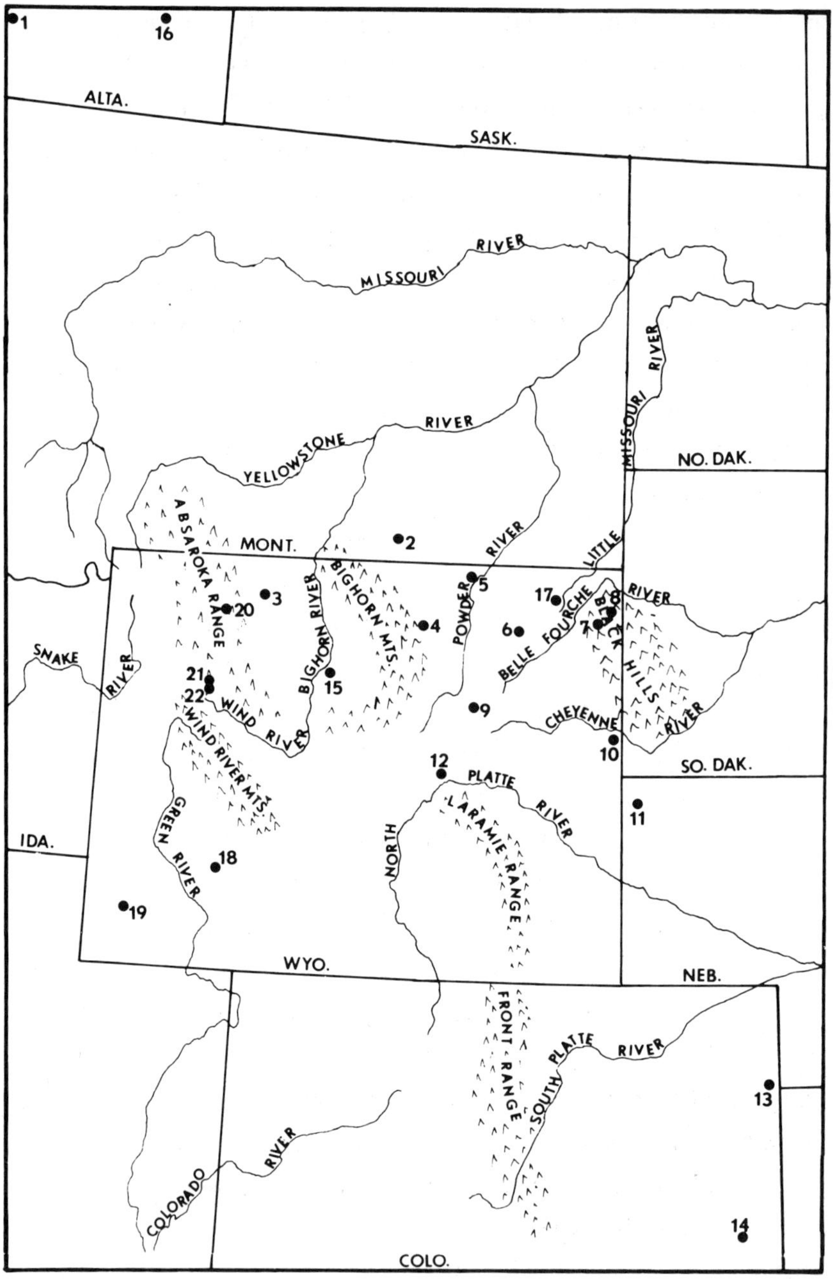

Figure 1. Archaeological sites mentioned in this paper that relate to cooperative and communal

weaponry, tools, and how elephant social structure might affect animal procurement within the limitations imposed by Clovis weaponry (see Frison and Todd 1986). The strategy of present day elephant culling is to take entire families in order to avoid the serious consequences of elephants that have lost their own families and are not accepted into other families. These kinds of animals become distraught and present a danger to visitors in National Parks where culling takes place. By far the most humane and efficient solution is to as quickly and efficiently as possible, kill entire elephant families.

Mature males are forced out of the family group as they reach puberty, forming groups of their own and joining the family only for breeding purposes. Hunting of males is different from hunting nursery herd groups and is not considered here. However, the mammoths found in Clovis sites that contain more than one animal are usually ones that would be found in family groups and which would consist of a matriarch and her offspring except for the mature males.

The strategy of a successful cull is to first kill the matriarch, and the rest of the family will then crowd upon the dead matriarch and they can in turn be dispatched with relative ease (Fig. 2). A charging matriarch elephant is dangerous, so that a firearm is needed that will deliver a bullet of sufficient size with enough velocity to penetrate the brain cavity causing the animal to drop immediately. This requires a high level of skill since it must be done repeatedly and the bullet must be placed carefully in order to reach the brain.

Figure 1 (*continued*) hunting. Bison Sites: (1) Head-Smashed-In (Reeves 1978). (2) Kobold (Frison 1970b). (3) Horner (Frison and Todd 1987). (4) Piney Creek (Frison 1967). (5) Powder River (Frison 1968). (6) Carter/Kerr-McGee (Frison 1984). (7) Hawken (Frison et al. 1976). (8) Vore (Reher and Frison 1980). (9) Ruby (Frison 1971b). (10) Agate Basin (Frison and Stanford 1982). (11) Hudson-Meng (Agenbroad 1978). (12) Casper (Frison 1974). (13) Jones-Miller (Stanford 1974, 1979). (14) Olsen-Chubbuck (Wheat 1972). Mammoth Site: (15) Colby (Frison and Todd 1986). Pronghorn Sites: (16) Laidlaw (Brumley 1984). (17) Bush-Bunger (Frison 1978). (18) Eden-Farson (Frison 1971a). (19) Fort Bridger Pronghorn Trap (Frison 1978). Mountain Sheep Sites: (20) Sheep Mountain (Frison et al. 1986). (21) Bull Elk Pass (Frison 1985). (22) Black Mountain (Frison 1985).

Figure 2. Elephant family members killed as they crowded on the dead matriarch (left of picture) seeking protection.

If the animal is simply wounded and not dropped, the hunter is in danger.

A Clovis projectile point is a well-designed and lethal bit of weaponry but there is little if any possibility of placing a Clovis point in such a way that a matriarch elephant will be immediately killed and/or rendered immobile. A Clovis point using either thrusting spear or atlatl will penetrate the rib cage of an elephant and inflict a wound that will eventually prove lethal, but the time between projectile point penetration and death can vary from minutes to days. The hunter facing a family of elephants is presented only with the head of the matriarch, and penetration into the brain is beyond the capacity of a Clovis projectile point mounted on either a thrusting spear or dart shaft. It is highly unlikely that hunters equipped with Clovis weaponry could face a family of elephants with any amount of success.

A strategy more likely to succeed is to become familiar with the movements of a family group and pick off individuals that stray away from the protective umbrella of the family. An animal that is crippled in this way will eventually fall behind the rest of the family and, once separated, the hunters need only play the waiting game. All of this is done as quietly as possible with the elephant family being unaware of the hunter. Elephants have relatively poor eyesight but their sense of smell is highly developed. With a knowledge of elephant behavior and skill in the use of Clovis weaponry, opportunistic hunting of individual animals in a family would be relatively successful. Cooperation between two or more hunters would further enhance the success of elephant hunting.

Sample sizes of mammoths from sites with unequivocal Clovis associations are too small for meaningful animal population studies and, consequently, studies to determine the seasonality of mammoth hunting has lagged. However, the research presented by Fisher (this volume) on elephant tusk formation and growth may eventually prove significant for determining both age of the animals and the time of year of the events.

The killing of at least one animal at the Colby site (Frison and Todd 1986) is believed to have been a cold weather event based on what is thought to have been a frozen meat cache. The left front quarter of a mature mammoth was surrounded with long bones and a skull of a young (5 year old) male was placed on top of the pile (see Fig. 3). This cache was not opened but one nearby had been opened, the contents removed, and the surrounding bones were scattered. Partial remains of at least seven mammoths were present and it is not believed that they were taken at one time. It is argued, instead, that the site represents a natural *cul de sac* into which crippled animals that had been wounded earlier were driven from time to time. The taking of even a single mature or juvenile mammoth would have created a meat surplus in a Clovis hunting group and justified caching some for later use. Further discussion on the significance of frozen meat caches in Paleoindian times on the High Plains follows later.

Figure 3. Proposed frozen meat cache consisting of the left front quarter of a mammoth with the bones of other mammoths piled around and over it. The Colby site in central Wyoming (Frison and Todd 1986).

Prehistoric Bison Procurement

The extinct subspecies of bison were relatively larger in body size, the horn span was wider and the horns themselves were more massive than in the modern forms. There were probably also subtle behavioral differences but there is nothing that is presently seen in archaeological site assemblages to indicate major differences. Bison were amenable to communal hunting through a number of basic strategies that reflect responses to their general patterns of behavior. The archaeological evidence reflects these responses, and in nearly every case can be interpreted through examination of the observed behavior of modern bison that live in open range conditions that approximate those of prehistoric times. The more that we are able to make systematic observations on bison behavior with the purpose of relating this to pre-horse procurement strategies, the more the archaeological record begins to fall into place.

This is not to suggest that prehistoric communal bison procurement was simple or easy. Bison are large, strong, fleet of foot, perceptive of danger, and react quickly and violently (Fig. 4) when they realize they are being restrained in a corral or natural trap. They can be maneuvered into proper position for driving into corrals, natural landform traps such as arroyos (Figs. 5a), the troughs of parabolic sand dunes (Fig. 5b), over bluffs (Fig. 6 a, b)and/or steep slopes and into depressions such as karsts. There is abundant archaeological evidence to support these statements.

Bison react differently according to internal conditions such as animal age, sex, breeding and birthing season behavior as well as responding to external conditions of weather, terrain, vegetative cover, time of day, time of year, and location of water sources. Successful communal procurement required careful consideration of these factors: however, in the final analysis, it was probably the capricious and unpredictable segment of bison behavior that was the real limiting factor of success. There are days when no known strategy will meet with success and the only alternative is to scrub the effort and try again another day.

Bison corrals with long attached wings were being used on the High Plains at least by A. D. 200 and they were very likely being used in Paleoindian times, but unequivocal archaeological evidence has not yet been recovered. A corral structure in the Powder River Basin of Wyoming at the Ruby site (Frison 1971b) reveals the extent of bison behavior perceived by prehistoric peoples. The final drive lane leading into the corral has a noticeable bend (Fig. 7.2) that most observers attribute to the inability of the group involved to construct a straight fence. However, anyone who has attempted to handle bison in corrals soon learns that a straight drive line leading into a corral does not function as well as a curved one. As long as the animals cannot see the end of the drive lane, they will keep on moving. Once they can see the end, they are apt to turn around unless they are

Figure 4. Female bison charging after being cornered. The author eluded the animal which then crashed through a heavy corral fence.

prodded from behind and kept moving at top speed. Should they be allowed to turn around, they cannot be stopped unless something in the way of a substantial gate blocks the way (see Fig. 4).

It is also interesting and instructive to note that from a point where the animals could first perceive the opening to the entrance of the corral at the Ruby Site, the floor of the drive lane contained large numbers of dart points where the animals were given added encouragement along the way into the corral. Prodding with a sharp instrument (in this case a stone dart

Figure 5. Analog of a natural arroyo trap (a), and a parabolic sand dune trap (b).

Figure 6. (a) The Kobold site in southern Montana; a typical buffalo jump over a perpendicular bluff with several thousand years of use. Arrow indicates direction of the final stampede (Frison 1970b). (b) The Glenrock Buffalo Jump. The space between the arrows is where the animals had to be forced over the jump-off in order to obtain a good kill (Frison 1970a).

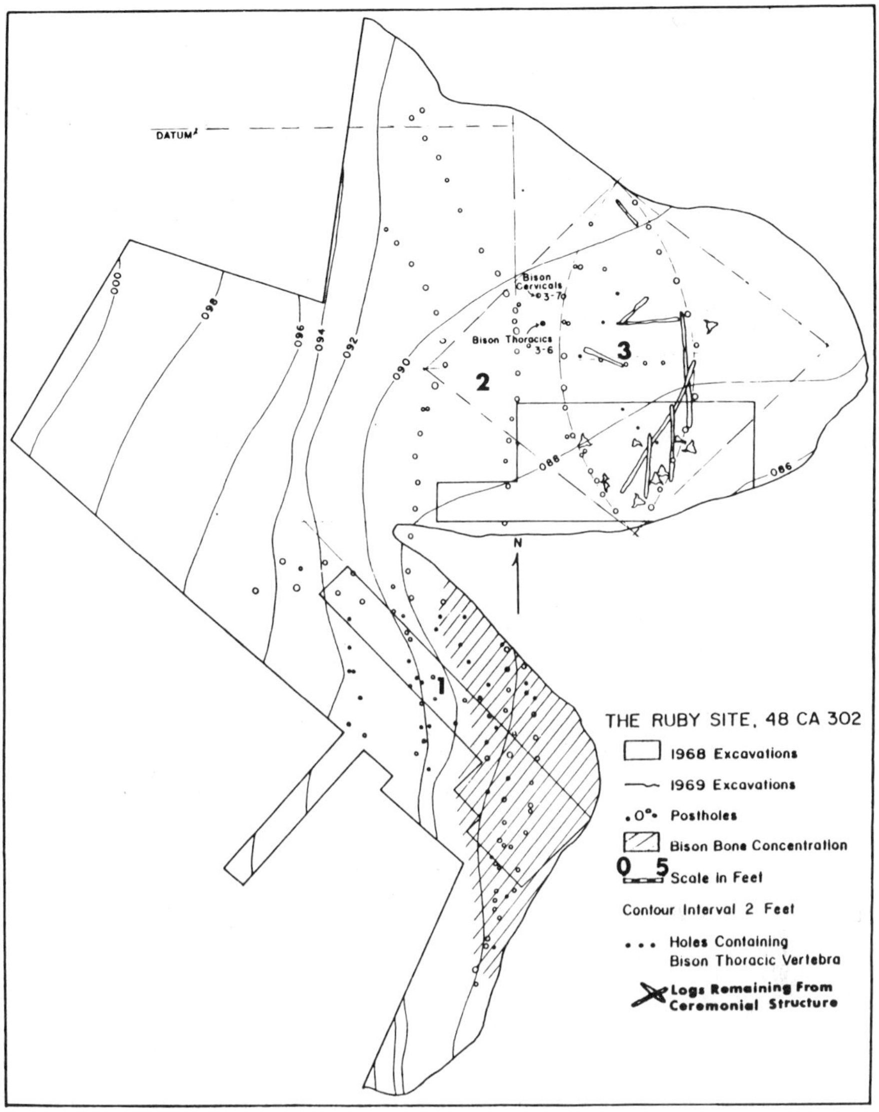

Figure 7. The Ruby site in the Powder River Basin, Wyoming; bison corral (1), drive lane (2), and religious structure (3) (Frison 1971b).

point) from the rear was one way to ensure the forward progress of the animals. Very little improvement could be made on either the design of this corral and drive line complex or the corraling methods that were apparently used.

Bison jumps are the best-known of all New World animal procurement sites. There are many variations but the general idea was to maneuver bison herds into proper position so they could then be stampeded over bluffs, steep slopes, or into deep depressions. Many of these bison jumps required that the animals be brought for long distances (up to several kilometers) and then the final stampede had to be controlled so that the animals piled over a segment of the jump-off (see Fig. 6b) that was high enough or steep enough to kill, cripple, or restrain the required number of animals. The drive line systems associated with Head-Smashed-In bison jump near Lethbridge, Alberta, possibly the largest bison jump in North America and also used over the longest time period, demonstrates the ability of prehistoric hunters without horses to successfully manipulate bison herds (see Reeves 1978).

One of the common fallacies that accompanies discussions of bison jumps is that any number of bison needed only to be headed in the direction of the jump-off and they would simply leap off into space and kill themselves. Nothing is further from the truth. Bison are extremely agile and can easily avoid a jump-off unless there is a herd of sufficient numbers so that the mass of the herd carries those animals in the lead over the precipice. It is virtually impossible to drive a small number of bison (5-10 animals) over a jump-off even with well-trained horses. A large, fast-moving, tightly-packed herd (50-100 animals) is an altogether different situation and one that can produce the desired results.

Arroyo traps were also used extensively and may have been used as extensively as jump-offs. Their use required maneuvering the animals up steep-sided arroys until a natural (knick point) or artificial barrier was reached (Fig. 8). In this case, the smaller herds (probably up to 25 individuals) would have been the more ideal size. Since these sites were in arroyo bottoms, the chances of site destruction through geologic processes were higher than in the case of the jump-off so that the number of arroyo traps that are extant may be only a small number of the total.

Whatever the nature of the bison procurement site, whether it was a

Figure 8. The Hawken site in northeast Wyoming, a 6500 year-old arroyo bison trap: old arroyo bottom (1), bone bed (2), knick point barrier (3), and the present arroyo (4) that cut out around leaving the remnant of the old arroyo intact (Frison et al. 1976).

jump-off, arroyo trap, corral, or sand dune (see Fig. 9), they all fit in with the behavior patterns of the species. Once the true nature of the procurement event is determined through identification of the landform originally involved, which often requires the expertise of a geomorphologist, the archaeologist is in a much better position to reconstruct the cultural aspects of the event.

Figure 9. The Casper site, a Hell Gap period bison kill in the trough of a parabolic sand dune (Frison 1974).

Pronghorn Procurement

The pronghorn *Antilocapra americana* biologically has remained unchanged since Late Pleistocene times on the High Plains. Although skeletal remains have been recovered in Paleoindian sites, their numbers are small in comparison to those of bison (Walker 1982, Walker and Frison 1986, Wilson 1974) and, as yet, there is insufficient evidence to indicate a widespread time depth for communal procurement.

The best known evidence that suggests good time depth to communal pronghorn procurement is the Laidlaw site located in southern Alberta near Medicine Hat. A radiocarbon date on the site is over 3000 years B. P. (Brumley 1984). The procurement situation took advantage of the natural topography and the animals were driven between two converging fences into a small catch pen that was apparently a pit although part may have also been above ground. Similar pronghorn traps are mentioned in the Lewis and Clark Journals, and another (the Bush-Bunger site) at the head of the Little Missouri River in northeast Wyoming (see Fig. 1) still has part of the wooden fences remaining (Frison 1978a: 256-257) which would allow it at most an early historic age. Brumley (1984) has provided an excellent literature review of Plains pronghorn procurement.

The pronghorn and its behavior invites cooperative/communal hunting. It is of relatively small carcass size (an average for all ages and sexes of mature animals is around 20 kg of flesh per animal). They tend to form large herds after the rutting season in late September and maintain herd size until spring. Single animal hunting results in relatively low return in animal weight compared to expenditure of effort. They can be driven and can be restrained within an enclosure of relatively simple construction.

A different method of pronghorn procurement was common to the Shoshoneans of the Great Basin in which the animals were driven into relatively large, round enclosures and were circled within the structure until exhausted and killed with clubs. Eyewitness accounts of these kinds of pronghorn procurement operations are known (Egan 1917, Reagan 1934) and evidence of another is still visible in the southern Green River Basin in southeast Wyoming (Frison 1978a:253-255) (Fig. 10).

Over 200 pronghorn were recovered from a late Late Prehistoric or early Historic Shoshone camp site (The Eden-Farson site; see Fig. 1) also in the Green River Basin area (Frison 1971a) that, based on tooth eruption, were killed within a period of a few days in late October and early November (Nimmo 1971) which strongly suggests that this was the result of communal rather than individual or small group procurement.

The configuration of these kinds of communal pronghorn procurement complexes fall in line with animal behavior. Pronghorn depend largely on their cursorial ability and highly developed vision for survival. However, they do not like to jump or crawl through fences (although they have managed to adapt to fenced areas by learning to both jump and crawl through fences in recent decades) so that a brush fence that deer would simply jump over, will

Figure 10. The Fort Bridger pronghorn trap in southwestern Wyoming; the circular encloslure (1) covers an area of about 4.5 hectares. The driveline (2) was designed to head the animals into the opening of the circular enclosure. Fences were of juniper (Frison 1978:254-255).

constrain the pronghorn. Once inside an enclosure, the hunters could keep the animals running around the inside of the fence while the hunters could take the shorter path near the center. The animals soon became exhausted and were easily killed with clubs.

For all of the cursorial and visual qualities of pronghorn, their behavior patterns demonstrate characteristics that make them easy prey for both communal and individual hunting. Since they prefer open spaces and depend on their vision so they can perceive and run away from danger, they must consequently place themselves in a visible position. As a result, they can be seen for miles, particularly in the early and late sunlight hours when the light rays are nearly parallel to the ground surface. The pronghorn do not hide, as deer are prone to do, and are visible throughout most of the daylight hours.

The pronghorn also demonstrate a unique degree of obstinacy that the hunter can often exploit.

The pronghorn have in addition an almost insatiable curiosity. This character combined with excellent vision is often their downfall. Any object such as a piece of clothing laid on top of a sagebrush will inevitably bring the animals from distances often up to a kilometer or more. In this way it was relatively easy to attract the animals into proper position to stampede them into a trap. Another behavioral trait that is often detrimental to their survival is that once they run away from danger and are out of sight of it, they tend to completely forget about it. Hunters need only to let the animals put on a burst of speed and disappear over the horizon. As soon as they stop, the hunters can usually repeat the same strategy.

Mountain Sheep

Mountain sheep demonstrate behavioral characteristics different from deer, pronghorn, bison and elk, but they are still amenable to communal hunting. They are usually regarded as being confined to the mountains, but there are numerous features within the plains such as minor uplifts, buttes, eroded canyons, and areas of "breaks" along major rivers that are known to have supported mountain sheep populations in historic times.

The mountainous northwest quarter of Wyoming and adjacent areas of Idaho and Montana contain numerous communal sheep traps constructed of wooden fences and cribbed log catch pens. As with bison and pronghorn traps, the operation of sheep traps can be interpreted from an understanding of mountain sheep behavior. They depend largely upon their ability to move in all directions in rough country in order to escape predators. Ewes with young lambs favor open areas that allow unrestricted vision either within or adjacent to rough country in order to detect predators in time to escape.

The locations and designs of the sheep trap complexes imply that they were intended mainly for the taking of nursery herds. Their locations are actually quite predictable, especially in relation to sheep bedgrounds. After the lambs are several weeks old, they form nursery herds that prefer a bare ridge or prominence for a bedground where they can see for relatively long distances. When disturbed, they tend to first go downhill and then turn upslope. The drive lines (see Fig. 11) are usually placed so that the animals will enter the open end of the V and will subsequently be driven either directly into a catch pen (Fig. 11.2) or into a corral where they can later be driven into smaller converging drive lines and then into a catch pen (Fig. 11.1). The principle that works best is to drive them upslope until just before the catch pen is reached and then downslope into the catch pen. This, of course, requires careful selection of the location of the drive complex. Mountain sheep can be maneuvered quite successfully for short distances which made communal trapping relatively easy.

The catch pens reflect a highly developed knowledge of mountain sheep behavior. They are long and narrow with a sloping ramp of logs leading from the terminus of the converging fences to the top of the catch pen itself (Fig. 12). The animals coming down the converging fence lines saw only the ramp with open sky at the top. However, when the top of the ramp was reached, the animals had no choice but to jump off into the catch pen where they were killed with clubs. Differences in terrain and vegetation cover required that each procurement complex be somewhat unique but they all were constructed in response to observable characteristics of mountain sheep behavior.

Another feature of mountain sheep fences is that they lean inwards several degrees from the perpendicular (Fig. 13). The outstanding ability of these animals to scale high, perpendicular obstacles enables them to take a short run at a fence and utilize their feet and legs to propel themselves upwards and, if they can contact any part of the top of the fence with their feet, go on over the top. Mountain sheep will still run at a leaning fence but, in this case, they fall back inside the corral. The sheep fences were constructed with full awareness of this limitation on the ability of the animals to escape over a fence leaning toward them.

Taking mountain sheep in nets may go well back into the Paleoindian period. A net was recovered that had been cached in a dry rockshelter in the Absaroka Mountains of northwestern Wyoming and is almost certainly one that

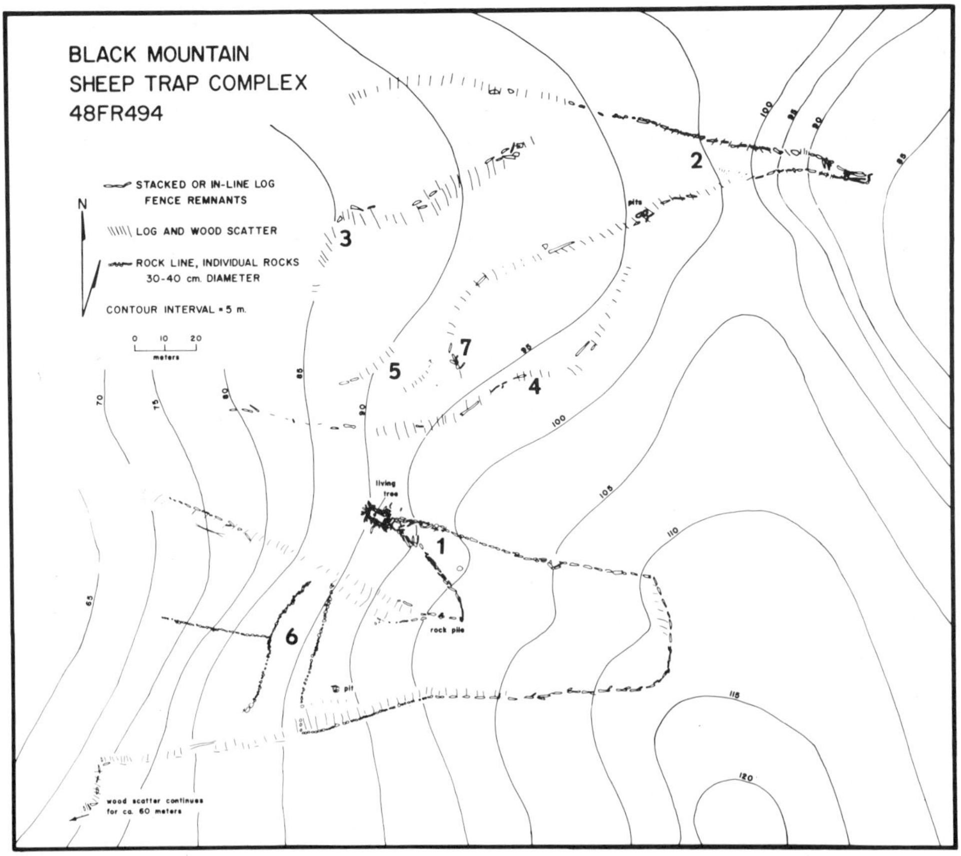

Figure 11. The Black Mountain sheep trapping complex in the southern Absaroka Mountains in northwest Wyoming. Drive lines, circling pen and catch pen (1), trap without circling pen (2), earlier traps that were partially salvaged to build later ones (3-6), and a probable shaman structure built of logs (7) (Frison 1985).

was used to take mountain sheep. It is radiocarbon dated to almost 9000 years ago and, although too badly deteriorated to unfold, it was probably as much as 50 m long and 2m in width. It was probably a net that was stretched across a trail or pass commonly used by mountain sheep so that it would entangle an animal or animals that were driven into it (Frison et al. 1986). One need only to observe the behavior of mountain sheep when tangled in a net to

Figure 12. Ramp (1) and converging fences (2) for trap No. 1 in Figure 10 (see Frison 1985).

understand its use. They will struggle violently for only a short time and give up with relative ease (Fig. 14) compared to a deer or pronghorn that is trapped under the same conditions. In many ways, mountain sheep behavior is reminiscent of that of their distant relatives, the Old World domesticated sheep.

Taking mountain sheep in the large drive line and catch pen complexes was certainly communal in nature. Construction of fences and catch pens along with the collection and driving of the animals into the traps would have required the efforts of a number of persons. Someone with some measure of authority to coordinate efforts would have to have been present also.

The use of the net could have been an entirely different situation. The requirements for construction of the net were considerable: possibly as much as 2 km of two-strand, twisted juniper bark cordage of diameters from 1 mm

Figure 13. Catch pen and converging fences for the Bull Elk Pass mountain sheep trap which is near the Black Mountain complex. Note that the preserved fence remnant leading to the catchpen (arrows) leans inward (Frison 1978:264-270, 1985).

to over 4 mm were used and considerable time would have been needed for construction. On the other hand, its use may have required much less in terms of manpower than a trapping complex. Sucessful operation of the net in mountain sheep procurement could have been done with as few as five persons.

CULTURAL CONSIDERATIONS IN COMMUNAL ANIMAL PROCUREMENT

With the archaeological evidence and considerations of animal behavior as background material, we can now begin to explore some of the cultural parameters that accompanied communal large mammal procurement. This

Figure 14. Docile behavior of mountain sheep ewes and lambs trapped under a drop net.

leads directly into an important aspect of communal large animal procurement activity, that of the supernatural control sought to enhance activities for which uncertainty of the final results is always present. Perhaps it was best stated by Aberle (1966) as activities that "relate to unpredictability and uncontrollability, and to efforts to cope with unpredictability and uncontrollability."

The ethnographic literature is rich in descriptions of ceremonial activity that surrounded communal bison hunting (see Fletcher and LaFlesche 1906:281; Grinnell 1892; Mandelbaum 1940:190-191; Chittenden and Richardson 1905:1028-1029; Gilmore 1924:209) and also communal pronghorn hunting (Steward 1938). The archaeological record contains evidence that can be used to argue for similar activities in bison jumps. Features such as small stone circles that regularly appear in the vicinity of bison jumps (see Frison 1970a:4-5) probably reflect shamanistic activity. Ritualistic treatment of bison skulls in bison jumps is documented (e. g., Fig. 15) and Stanford (1979) recovered both features and artifacts that strongly suggest shamanistic activity in a bison kill that is approximately 10,000 years old.

However, the writer feels that the best prehistoric evidence for ritualistic activity associated with communal bison procurement is associated with a pound or corral in eastern Wyoming that is approximately 1700 years old (Frison 1971b). The corral complex was built on a terrace of an intermittent stream. Paired post holes suggest a corral construction analogous to present-day cattle corrals with poles or logs stacked in rows between the paired posts. A lane formed by two converging fences led into the corral. This part of the complex was covered by alluvium after it was abandoned and was partially preserved, while the drive lines that extended for some distance from the corral were lost through erosion. The corral and the converging fences reflect a high degree of knowledge of bison behavior as was discussed above. On the other hand, there was also a strong reliance on the supernatural to achieve the desired goals as is demonstrated by a religious structure constructed adjacent to the corral. The structure (Fig. 7.3) is lens-shaped in outline and was formed by the intersection of two arcs of circles with radii of about 10.2 m. Post holes outline the structure which is oriented almost north-south. The south half of the structure was apparently roofed over and buffalo skulls were placed around it facing outward. Inside the south half were three of the longest dorsal spines of bison placed in holes with the proximal end upwards. Two holes outside the structure contained bison vertebra. No evidence of use of the structure for domestic purposes was present: it is interpreted instead as evidence for a high level of ritualistic activity in order to increase the possibilities of success in bringing the animals into the corral.

Figure 15. Special treatment of bison skulls in the Vore site (Frison and Reher 1980).

There is archaeological evidence that strongly suggests shamanistic activity at the Green River Basin pronghorn trap mentioned earlier. Incorporated into the circular corral structure was a much smaller structure that had no apparent function in containing the animals. It could very easily have been a shaman's structure. This concept is supported strongly by similar evidence in the better-preserved mountain sheep traps. In these cases, there are also small structures incorporated into the fences (see Fig. 11.7) that suggest shaman activity. These have collapsed in all cases but their original configurations can be reconstructed and they do not fit the overall configurations of the complexes from a stricty utilitarian interpretation. Special treatment of large ram skulls that were placed in trees in the vicinity of the sheep traps with the brain cases opened also suggests ritual behavior.

Leadership in Communal Hunting Activities

Another aspect of the nature of prehistoric hunting behavior can be gleaned from the evidence at the bison pound site described above and also in processing areas associated with bison jumps such as at the Piney Creek site in northern Wyoming (Frison 1967). Communal hunting reflects the need for some kind of temporary authority in order to ensure success since a careful coordination of activities of a number of individuals was necessary in order to locate a buffalo herd of proper size, maneuver it into proper position, stampede it into the corral, and then kill the animals. At any of the stages, except possibly the last, the possibility of failure of the venture was always present. Similar kinds of considerations were to be found in the communal hunting of other species as well as bison.

Until the animals were killed, the procurement effort was a communal effort, but shortly afterwards the spirit of communality broke down. Associated with the Ruby site corral complex but a short distance away is a processing area. Instead of a production line system of processing, there are a number of discrete processing locations indicating that although the procurement effort was communal, the processing of the products of the effort was carried out at the family level and that the authoritarian structure had broken down. The same conclusions can be drawn from the processing area at the Piney Creek buffalo jump site of Late Prehistoric age (Frison 1967).

THE ECONOMICS OF COMMUNAL HUNTING

At this point the question may be asked as to the reason or reasons for communal animal hunting on the High Plains since individual and small group bison hunting would have normally provided sufficient meat at the family level. Consider also that corral and fence building was hard and time consuming work, particularly using stone and bone tools, and the needed wood for structures might have had to be brought from a distance.

Part of the answer probably was the need for surplus meat that had to be stored for insurance against the period or periods of time during the cold weather cycle when hunting could not be performed. Part of the answer also is that the communal kill provided sufficient resources to allow aggregation of the human group with the concomitant social benefits that this provided. For most of the year, the social group was forced to disperse into single and/or multifamily units due to the distribution of food resources.

To postulate further, without these kinds of needs and benefits to the social group, the accumulation of surplus is difficult to explain. Tribal economy is not organized for sustained production (Sahlins 1968:77) and there is no incentive to accumulate surplus that cannot be sold or traded. Short term caching and/or aggregation of the dispersed societal units appear to have been the strongest motives for communal animal procurement in prehistoric times.

Handling of Meat Surplus

Meat products may be either dried or frozen for future use and, in the case of the communal kills, each method can often be indentified in the archaeological record through the presence or absence of certain cultural features and/or by establishing the time of year of the procurement operations using tooth eruption to age the animals. It has been argued (Frison 1982) that frozen meat caches were part of a Paleoindian subsistence strategy on the High Plains. If the two mammoth bone piles at the Colby site (Fig. 3) and a pile of bison carcass units in the Agate Basin level at the Agate Basin site (Fig. 16) represent meat caches, the procurement events would have had to be cold weather manifestations. As soon as warm weather arrived, the contents of the caches would have spoiled. The seasonality of many Paleoindian bison kills indicates that they were also cold weather manifestations as determined by aging of bison through tooth eruption (see Frison 1978b). Some form of

Figure 16. Proposed frozen meat cache of bison carcass units at the Agate Basin site (Frison and Stanford 1982).

caching of surplus meat on the northern High Plains for use during certain parts of the cold weather cycle was apparently part of the yearly subsistence strategy. Freezing of meat from cold weather bison kills does not rule out meat drying in warm periods; however, except for the Olsen-Chubbuck site which was probably a mid or late summer event, the sample of Paleoindian bison kill sites investigated to date on the Northwestern Plains were apparently cold weather events. Meat drying was the strategy adopted in Late Archaic and Late Prehistoric periods along with late summer-early fall animal procurement which is also confirmed through aging of bison teeth. There are also site features that are associated with meat processing (see Frison 1967) in the Late Prehistoric Period that are lacking in Paleoindian sites. Use of cold weather meat caches is well demonstrated in the Agate Basin level at the Agate Basin site (Frison and Stanford 1982) where the animals (bison) were cut into large units, stacked in piles, and extracted for use as needed (Fig. 16).

The frozen meat cache strategy for winter survival places certain restraints on the hunting group involved. The frozen meat cache cannot be moved and it must be protected. Part of the Agate Basin site cache, like the mammoth meat cache at Colby, was not utilized so that it spoiled upon the arrival of warm weather. Perhaps the most descriptive term for these frozen meat caches used by High Plains Paleoindian groups is "insurance cache" (Binford 1981:287). Enough surplus meat products were stored to allow for maximum needs and if not needed, they would spoil upon the arrival of warm weather. In this way they would provide "insurance" against the times that weather conditions would limit or even deny the normal pursuit of hunting. In this kind of situation, an easy winter would not require the use of all of the cached meat and there would be neither a way to salvage it nor remorse in allowing it to spoil. Other restrictions on the activities of the hunting group would have been that the animals had to be killed late enough in the year so that they could be frozen and maintained in that condition for the remainder of the cold weather period. The site for cold weather quarters would preferably have had to be in or within a reasonable distance of a favored wintering area for the animals which could change yearly depending upon the availability of forage. It would also require the presence of a geomorphic feature that could be used or modified to some extent for use as a trap. If such a feature were not present then there would have been the more serious considerations of the construction of a corral structure.

The cold weather bison kills and frozen meat caches on the Northwestern Plains may also reflect a subsistence strategy that is in between the one described for the arctic (see Binford 1978) where hunting is impossible for months at a time and one farther to the south on the Southern Plains where even insurance caching was unnecessary.

PLEISTOCENE EXTINCTIONS

One of the more controversial and puzzling problems in New World prehistory has to do with the causes of Pleistocene extinctions. Pleistocene overkill by human predators has been suggested (see Martin 1967) as a probable cause. However, if we look carefully at the New World archaeological record as it presently stands, the evidence we have is that North America had only a relatively small population of *Homo sapiens sapiens* hunters between 12,000 and 11,000 years ago and these hunters (Clovis) had developed sophisticated tool and weaponry assemblages that were adequate for taking the large ungulates present on the High Plains at that time. However, the archaeological record, at least on the High Plains, does not reflect a massive killing of animals such as would have been required to bring about the extinction of the Pleistocene fauna. The documented animal kill sites of the Paleoindian Period provides evidence of well-planned procurement strategies but the geologic record leaves no clue that would indicate a period of either aggradation or degradation that would have either washed away or covered up the evidence of a human population and/or the lithic component of the weaponry needed. Equally strong and equally significant evidence against the concept of Pleistocene overkill is the evidence from human behavioral studies. The production of economic goods breaks down once the immediate production goals are reached when incentives to further production such as trade or profit are lacking (see e.g., Sahlins 1968). Human behavior as observed in studies of primitive human groups at the societal level of the late Pleistocene hunters argues against the overkill hypothesis. Any overkill model devised to account for Pleistocene extinctions that fails to consider human behavioral characteristics must remain suspect. The introduction of the horse into Plains Indian societies in historic times did result in a classic example of overkill but in this case, the incentives of trade for European goods, ritualized warfare, and acquisition of horses across tribal boundaries were added quite suddenly to simple tribal societies.

It is highly unlikely that the lack of faunal evidence for mass killing of animals during the late Pleistocene can be interpreted as a geological phenomenon. The methodology for identifying in situ, nonculture bearing, Late Pleistocene deposits through geological, taphonomic, and soils studies is continually improving. If Pleistocene extinctions resulted from human overkill, it is highly unlikely that the evidence for this amount and kind of human activity could continue to escape detection.

DISCUSSION AND CONCLUSIONS

When all of the disparate bits of data on prehistoric, cooperative/communal, large mammal hunting on the North American High Plains are assembled, analyzed, and integrated, a rational picture of High Plains hunters and hunting begins to emerge. The first known North American group that could lay claim to having reached a stage of true hunters were groups of the Clovis Cultural Complex that were present between 11,000 and 12,000 years ago. Groups ancestral to Clovis have not yet been satisfactorily identified in either the Old or New World. On the other hand, from its general character, Clovis hunting groups as suggested by the New World archaeological record are a reflection, however dim, of Upper Paleolithic Old World Hunting groups.

Clovis hunters were taking mammoths. If mammoth behavior resembled that of African elephants, even the taking of a single animal might have been regarded as a communal effort unless it was an animal or animals that were sick or had met with accidents. Human predation on mammoth families could have hastened the extinction of the species, considering that Clovis hunters were taking mostly animals from the nursery herds of a species that did not produce offspring until about 16 years of age and with a gestation period of 22 months. Clovis weaponry was adequate for killing mammoths, but not by frontal assault on a family with the expectations of dropping a matriarch determined to protect her family. The matriarch herself would have been a poor choice to single out as the first priority: a less wary individual would have presented better chances for success. A better strategy would have been a more opportunistic approach, and wound animals that strayed for short periods from the protective umbrella provided by the family. Without the matriarch being aware of the source of the danger, the wounded animal would gradually drift away from the herd. Depending upon the seriousness of the wound, the hunters could either wait or haze it into a natural barrier if one were nearby, and administer the final *coup de grace*.

The actual time of the beginning of communal bison hunting is not known but it was well established between 10,000 and 11,000 years ago and continued into historic times. Bison were taken in arroyo and sand dune traps and in artificial corrals. The latter probably fits the model of true communal hunting more so than the former two. Small hunting groups could take advantage of natural features in arroyos that required little or no modification to form adequate restraints for bison. Corral building was different and required the cooperative efforts of relatively large numbers of individuals. The extant body of data suggests that there was a continual gradient in complexity of procurement operations and concomitant personnel needs from the unmodified arroyo or sand dune trap to situations that required the construction of extensive, sophisticated artificial features capable of withstanding the violent responses of large, wild animals that suddenly found their activities restrained.

Driving bison over bluffs and/or sharp inclines into corrals, sinkholes, or steep-sided arroyos was a well established procurement strategy by 5000 to 6000 years ago, and probably began much earlier, considering the evidence at Bonfire Shelter at the southern end of the Great Plains (Dibble and Lorrain 1968). The operations involving the movement of large herds of bison may have been a response to the smaller bison that appeared in post-Altithermal times and tended to form in larger herds. The gradual physiological changes toward bison of smaller size is thought to be a response to selection for and adaptation to post-Pleistocene conditions. All of these techniques of handling bison can be understood and accepted in the context of their observable behavior.

Identification of the original landform used in any bison procurement site is necessary in order to be able to interpret the strategy used. A buried bone bed (e.g., Fig. 17), may have resulted from driving animals into an arroyo trap, over a steep incline, or into a corral and subsequent geological activity may have removed the means of identifying the original feature involved. Until the

Figure 17. In situ bone bed (a) at the 3000 year-old Powder River site (Frison 1968) in north-central Wyoming and the present arroyo (b) that downcut through it. A detailed geologic study revealed that the bone bed lies in what was the arroyo bottom 3000 years ago and that a knick-point formed a barrier that held the animals.

original landform is identified, the procurement stategy remains in doubt.

Prehistoric bison hunting was done on foot which limited particularly the efficiency of communal hunting when compared to hunting the same animals with the aid of horses as occurred in historic times. Lacking the advantage of anything comparable to horses, the prehistoric hunting group had to follow the bison rather than being able to move the animals about at will. Under these conditions, there was a critical number of animals that determined at what point large communal efforts had to be replaced by single and small group hunting efforts requiring alternative procurement strategies. The human effort required for the successful operation of a large communal kill could not be justified unless there were sufficient animals in the general area.

Bison populations fluctuated continually in response to winter survival conditions and available spring and summer feed which determined animal condition. Females will not breed while in poor body condition and this directly affects the annual calf crop. Feed conditions change from one area to another so that an area attractive to bison one year may not be so the next year. Prehistoric hunters were forced to move in response to animal movements so that a trap used one year might not be used another year.

Bison may move several miles when disturbed while pronghorn and mountain sheep are more apt to remain in a more restricted home territory. These kinds of factors added still another element of uncertainty to communal bison hunts. The evidence from a large, Late Prehistoric bison jump in northeast Wyoming showed that the jumping events were a response to increased animal populations resulting from periods of annual rainfall conditions that either positively or negatively affected forage (see Reher and Frison 1980). In this case, there were periods of several years between procurement operations. It is not known if the jumps were a yearly event occuring in alternative locations. It may have been that the large communal hunts were not annual events and other, less spectacular methods of procurement prevailed only during the years when bison populations were down.

Communal hunting provided more food than could be consumed on the spot and surpluses were prepared for short term storage. There is good evidence for frozen meat caches in Paleoindian times but there is no reason to rule out drying of meat also when animals were killed in warm weather. The present body of evidence indicates that the communal bison kills were usually restricted to cold weather in Paleoindian times. On the other hand, warm weather killing may have been done using different strategies for which the evidence has not preserved as well as the evidence we have in the form of bonebeds from cold weather kills. We do know, however, that drying of meat products from large, communal, warm weather kills was the favored method in the later time periods and the time of year of these kills can be confirmed through taphonomic analysis of bonebed deposits.

Communal animal procurement was shrouded in a good deal of religious activity that is always present in economic ventures in primitive societies when the chances of failure are high. Ethnohistorical and ethnological evidence as well as the archaeological record provide evidence for these kinds of ritualistic activities. Religious structures built for shamanistic purposes can usually be found associated with communal animal procurement complexes. The shamans job was to exercise spirit control over the animals and thereby enhance the chances of success in bringing them into the traps. Religious practices probably also limit a strictly functional interpretation of communal hunting even though these practices were designed to further ensure what was expected to occur.

Communal hunting spawned nebulous forms of leadership which lasted only until the immediate goals in terms of animal products were met. As soon as the animals were killed, they were divided between the members of the group according to predetermined rules and were then processed for future use at the family level. There were no incentives to kill more animals once these immediate goals were met. Some of the Late Prehistoric bison kills were very large and involved large numbers of animals but even in sites such as the Vore Bison Jump in northeast Wyoming (Reher and Frison 1980) where a single season's kill may have involved several hundred or even more than a thousand animals (Fig. 18), there is no evidence that any of the carcasses were not utilized.

These facets of human economic behavior at the lower levels of social organization do not fit with the concept of human overkill that has been used to explain Pleistocene extinctions. However, human predation may have hastened the extinction of the mammoth which may already have been in serious trouble due to unfavorable climatic conditions. On the other hand, the bison, the pronghorn and the mountain sheep which were the most intensely hunted species throughout the known time period of human occupation on the High Plains were able to survive. The differing and often directly opposed explanations for Pleistocene extinctions indicate that there are no simple, straightforward explanations.

Prehistoric High Plains hunters were accomplished predators. They were totally familiar with, and able at a moment's notice to draw upon, a

Figure 18. Bison bone bed at the Vore site (Reher and Frison 1980). The thick bone level represents a single season's kill of at least several hundred animals. Both earlier and later kills are seen in profile. The kill was made by driving the animals into a karst.

comprehensive body of knowledge concerning the habits of the numerous species and their responses to different internal and external conditions. The typical prehistoric hunter needed only to look out of his lodge on any given morning, and carefully and quickly analyze all of the observable conditions of weather, time of year, hunting area terrain, vegetative cover and condition, moonlight or dark night conditions, and many others too numerous to mention. He would have then taken this body of data and integrated it with the behavior patterns of the animal or animals he planned to pursue on that day. He could then depart from camp confident that the probabilities of meat for the family before dark were strongly in his favor. This would be true if he were going hunting by himself, with one or two other companions, or in the company of several other hunters to engage in a communal effort. The reasons for success were as stated at the beginning of this paper: the hunter was aware of his own physical abilities and limitations, he was totally knowledgeable of the behavior of the species he hunted, he was familiar with the hunting territory and how the animals exploited it, and he realized the limits imposed by the weaponry and/or procurement method used.

ACKNOWLEDGEMENTS

This paper is based on data collected over more than a quarter of a century. Monetary support came from several sources, including in alphabetical order: Carter Mining Co.; Kerr-McGee Mining Co.; L. S. B. Leakey Foundation; National Science Foundation; University of Wyoming; Wyoming Archaeological Foundation; Wyoming Archaeological Society; Wyoming Council for the Humanities; and the Wyoming Recreation Commission. It is impossible to name all the individuals that, down through the past decades, have contributed to the thoughts and ideas expressed here. However, as a group they have my sincere gratitude.

REFERENCES

Aberle, D. F. 1966. Religio-magical phenomena and power, prediction and control. *Southwestern Journal of Anthropology* 22:221-229.

Agenbroad, L. D. 1978. *The Hudson-Meng Site: An Alberta Bison Kill in the Nebraska High Plains*. Washington, D.C.: University Press of America.

Binford, L. R. 1978. *Nunamiut Ethnoarchaeology*. New York: Academic Press.

Binford, L. R. 1981. *Bones: Ancient Men and Modern Myths*. New York: Academic Press.

Brumley, J. H. 1984. The Laidlaw Site: An Aboriginal Antelope Trap from Southeastern Alberta. *Archaeological Survey of Alberta, Occasional Paper* No. 23, ed. D. Burley. Edmonton, Alberta.

Chittenden, H. M., and A. T. Richardson. 1905. *Life, Letters and Travels of Father Pierre Jean DeSmet, S. J. 1801-1873*, Vol. 3. Francis P. Harper.

Dibble, D. S., and D. Lorrain. 1968. Bonfire Shelter: A Stratified Bison Kill Site, Val Verde County, Texas. *Texas Memorial Museum, Miscellaneous Papers* No. 1.

Egan, H. 1917. *Pioneering the West, 1846-1848*. W. M. Egan, Ed., Salt Lake City.

Fletcher, A. C., and F. La Flesche. 1906. The Omaha Tribe. *Bureau of American Ethnology, Annual Report* No. 27.

Frison, G. C. 1967. The Piney Creek Sites, Wyoming. *University of Wyoming Publications* 3:1-92.

Frison, G. C. 1968. Site 48SH312: An early Middle Period bison kill in the Powder River Basin of Wyoming. *Plains Anthropologist* 13:31-39.

Frison, G. C. 1970a. The Glenrock Buffalo Jump, 48CO304: Late Prehistoric Period Buffalo Procurement and Butchering. *Plains Anthropologist Memoir* No. 7.

Frison, G. C. 1970b. The Kobold Site, 24BH406: A post-altithermal record of buffalo-jumping for the Northwestern Plains. *Plains Anthropologist* 15:1-35.

Frison, G. C. 1971a. Shoshonean antelope procurement in the Upper Green River Basin, Wyoming. *Plains Anthropologist* 16:258-284.

Frison, G. C. 1971b. The buffalo pound in Northwestern Plains prehistory: Site 48CA302, Wyoming. *American Antiquity* 36:77-91.

Frison, G. C. 1974. *The Casper Site: A Hell Gap Bison Kill on the High Plains*. New York: Academic Press.

Frison, G. C. 1978a. *Prehistoric Hunters of the High Plains*. New York: Academic Press.

Frison, G. C. 1978b. Animal population studies and cultural inference. In: *Bison Procurement and Utilization: A Symposium*, ed. M. Wilson and L. B. Davis. *Plains Anthropologist Memoir* 14.

Frison, G. C. 1982. Paleo-Indian winter subsistence strategies on the High Plains. In: *Plains Indian Studies: A Collection of Essays in Honor of John C. Ewers and Waldo R. Wedel, Smithsonian Contributions to Anthropology* No. 30.

Frison, G. C. 1984. The Carter/Kerr-McGee Paleoindian site: Cultural resource management and archaeological research. *American Antiquity* 49(2):288-314.

Frison, G. C. 1985. Prehistoric and early historic mountain sheep procurement in the central Rocky Mountains. *Archaeological Survey of Alberta: Occasional Paper* No. 26:267-276.

Frison, G. C., R. L. Andrews, J. M. Adovasio, R.C. Carlisle, and Robert Edgar. 1986. A late paleoindian animal trapping net from Northern Wyoming. *American Antiquity* 51:352-361.

Frison, G. C., and C. Craig. 1982. Bone, antler, and ivory artifacts and manufacture technology. In: *The Agate Basin Site*, ed. G. C. Frison and D. J. Stanford. 157-173. New York: Academic Press.

Frison, G. C., and D. J. Stanford, eds. 1982. *The Agate Basin Site: A Record of the Paleoindian Occupation of the Northwestern High Plains*. New York: Academic Press.

Frison, G. C., and L. C. Todd. 1986. *The Colby Mammoth Site: Taphonomy and Archaeology of a Clovis Kill in Northern Wyoming*. University of New Mexico Press.

Frison, G. C., and D. N. Walker, eds. 1984. The Dead Indian Creek site: An archaic occupation in the Absaroka Mountains of northeastern Wyoming. *Wyoming Archaeologist* 27(1-2).

Frison, G. C., D. N. Walker, S. D. Webb, and G. M. Zeimens. 1978. Paleo-Indian Procurement of *Camelops* on the Northwestern Plains. *Quaternary Research* 10:385-400.

Frison, G. C., M. Wilson, and D. J. Wilson. 1976. Fossil bison and artifacts from an early altithermal period arroyo trap in Wyoming. *American Antiquity* 4:28-57.

Gilmore, M. R. 1924. Old Assiniboine buffalo-drive in North Dakota. *Indian Notes* 1:204-211.

Grinnell, G. B. 1892. *Blackfoot Lodge Tales. New York*, 1892.

Haynes, C. V. 1982. Were Clovis progenitors in Beringia? In: *Paleoecology of Beringia*, ed. D. M. Hopkins, J. V. Matthews, Jr., C. E. Schweger, and S. B. Young. New York: Academic Press.

Haynes, G., and D. J. Stanford. 1984. On the possible utilization of *Camelops* by early man in North America. *Quaternary Research* 22:216-230.

Mandelbaum, D. G. 1940. The Plains Cree. *Anthropological Papers, American Museum of Natural History* 37:155-316.

Martin, P. S. 1967. Pleistocene overkill. *Natural History* 76:32-38.

Nimmo, B. W. 1971. Population dynamics of a Wyoming pronghorn cohort from the Eden-Farson Site, 48SW304. *Plains Anthropologist* 16:285-288.

Reagan, A. B. 1934. Some notes on the history of the Uintah Basin. In: *Northeastern Utah, to 1850. Proceedings Utah Academy of Science, Arts and Letters* 11:55-64.

Reeves, B. O. K. 1978. Head-Smashed-In: 5500 Years of Bison Jumping in the Alberta Plains. *Plains Anthropologist Memoir* 14:151-174. ed. L. B. Davis and M. Wilson.

Reher, C. A., and G. C. Frison. 1980. The Vore Site, 48CK302, A Stratified Buffalo Jump in the Wyoming Black Hills. *Plains Anthropologist Memoir* No. 16.

Sahlins, M. D. 1968. *Tribesman*. Prentice-Hall.

Saunders, J. J. 1977. Lehner Ranch Revisited. In: *Paleoindian Lifeways*, ed. E. Johnson. *The Museum Journal* XVII:48-64. West Texas Museum Association, Texas Tech University.

Simpson, T. 1984. Population Dynamics of Mule Deer. In: *The Dead Indian Creek Site: An Archaic Occupation in the Absaroka Mountains of Northwestern Wyoming*, ed. G. C. Frison and D. N. Walker. *The Wyoming Archaeologist* 27:83-96.

Stanford, D. J. 1974. Preliminary report on the excavation of the Jones-Miller Hell Gap site, Yuma County, Colorado. *Southwestern Lore* 40:30-36.

Stanford, D. J. 1979. Bison kill by Ice Age hunters. *National Geographic* 155:114-119.

Steward, J. H. 1938. Basin-Plateau Aboriginal Sociopolitical Groups. *Bureau of American Ethnology, Bulletin* 120.

Walker, D. N. 1982. Cultural Modification of Bone from Pronghorn (*Antilocapra americana*) and Other Small Mammals. In: *The Agate Basin Site*, ed. G. C. Frison and D. J. Stanford. 270-274. New York: Academic Press.

Walker, D. N., and G. C. Frison. 1986. The Late Pleistocene Mammalian Fauna from the Colby Mammoth Kill Site, Wyoming. In: *The Colby Mammoth Site: Taphonomy and Archaeology of a Clovis Kill in Northern Wyoming*. 191-205. University of New Mexico Press.

Wheat, J. B. 1972. The Olsen-Chubbuck Site: A Paleo-Indian Bison Kill. *Society for American Archaeology, Memoir* No. 26.

Wilson, M. 1974. The Casper Local Fauna and its Fossil Bison. In: *The Casper Site: A Hell Gap Bison Kill on the High Plains*. ed. G. C. Frison. 125-171. New York: Academic Press.

Analysis of Kill-Butchery Bonebeds and Interpretation of Paleoindian Hunting

Lawrence C. Todd

Department of Anthropology
University of Denver
Denver, Colorado

Unlike remote time periods where the degree of large animal predation is a key research question (Binford 1985; Bunn and Kroll 1986; Isaac and Crader 1981), hunting has long been considered a key component in the adaptation of many anatomically modern *Homo sapiens* and, in particular, the earliest Paleoindian inhabitants of the Western Hemisphere. While there is little question that hunting of large body-sized animals occurred, the organizational role of predation to Paleoindian survival has received limited attention. Use of large herd mammals by Paleoindians on the North American Plains has often led to characterization of these groups as specialized "Big Game Hunters." Since the first well-documented and accepted association between human artifacts and fossils of bison at the Folsom site in the late 1920s (Cook 1927; Figgins 1927), interpretations of Paleoindian subsistence strategies have frequently been based on evidence from these multianimal bonebeds.

The question posed here is whether evaluation of the role of these sites and of hunting in Paleoindian adaptation to late Pleistocene-early Holocene environments can be accomplished with current analytic approaches. Given the quantity of information from late Pleistocene-early Holocene bonebeds in western North America (Figure 1), the overall high quality of data, and the increasing body of taphonomic research (see Behrensmeyer, this volume), interpretations of hunting practices represented by these relatively recent sites

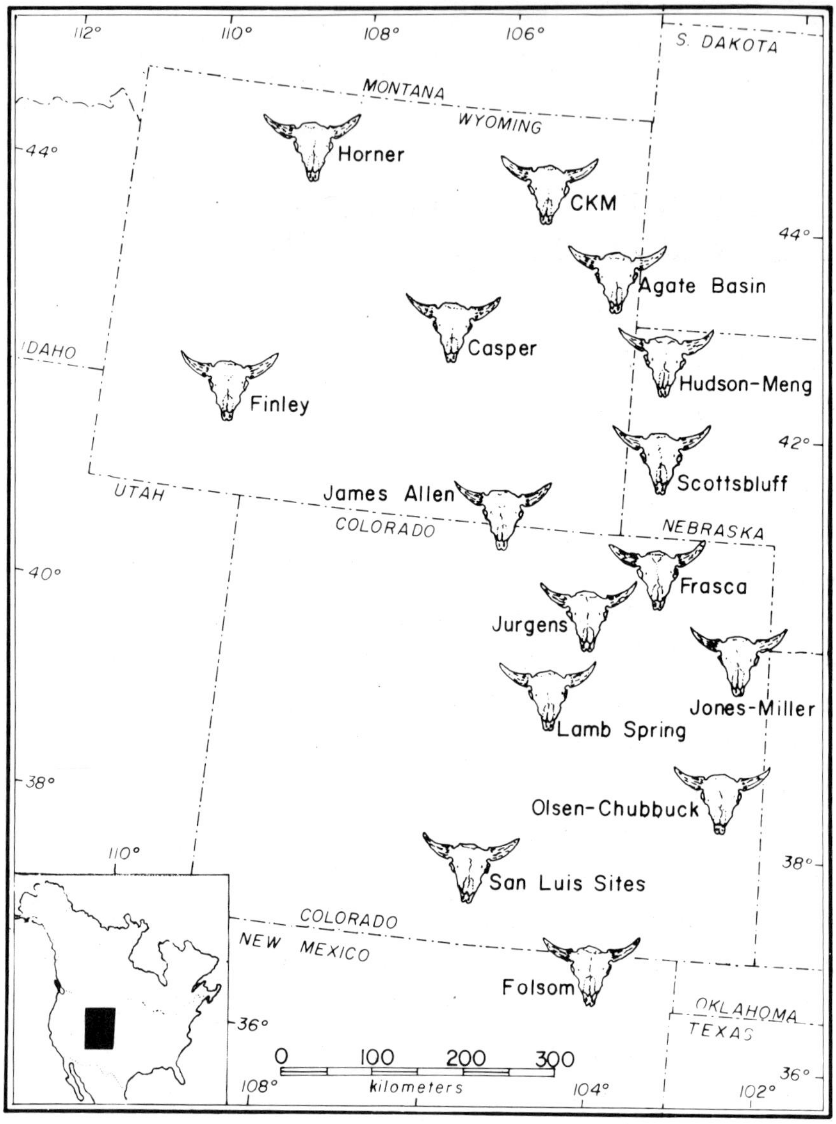

Figure 1. Selected Paleoindian kill-butchery sites on the North American northern Plains: Horner (Frison and Todd 1987); CKM (Frison 1984); Agate Basin (Frison and Stanford 1982); Casper (Frison 1974); Hudson-Meng (Agenbroad 1978); Finley (Moss et al. 1951); Scottsbluff (Barbour and Schultz 1932); James Allen (Mulloy 1959); Frasca (Fulgham and Stanford 1982); Jurgens (Wheat 1979); Jones Miller (Stanford 1974, 1978, 1984); Lamb Spring (McCartney 1984;

should be among the most accessible in the archaeological literature.

However, the difficulties of making behavioral interpretations based on archaeological faunal assemblages are a common theme to the investigation of the role of hunting in human adaptations (Behrensmeyer this volume; Binford 1981, 1984, 1985, this volume; Bunn pers. comm. ; Frison this volume; Hill 1984; Klein this volume).

While many of the studies dealing with the difficulties of using animal bones, as in the interpretation of past behaviors, have focused on the distant past, the need for the development of methods to provide a more secure understanding of the use of animals is of the utmost importance in *all* studies of archaeological faunal assemblages. There has been a tendency for questions of formational histories to have received the most attention in the context of the relative contribution of scavenging or hunting in the diet of our hominid ancestors while, at the same time, assuming that the archaeological record of more recent, anatomical modern humans can be used as the basis for interpreting earlier assemblages (Isaac and Crader 1981; Klein, this volume).

While there is no question that anatomically modern humans have been, and continue to be, highly effective predators, there should also be little question of the range of variability in the role of hunting as evidenced in the ethnographic or ethnohistorical record and that the faunal remains from virtually every archaeological site have been modified to some degree by post-occupational process. The variability and the degree of behavioral flexibility which stands behind modern human predation imply that broad empirical generalizations about the role of hunting in even recent archaeological sites are untenable. Although the presence of meat-eating in late Pleistocene or early Holocene sites may be unambiguous, assumptions about the organization of hunting behaviors is no more appropriate for the recent sites than it is in the earlier time ranges. Before such materials can be meaningfully employed in the interpretation of older sites, investigation of the

---

Figure 1 (*continued*) Rancier et al. 1982; Stanford et al. 1981); Olsen-Chubbuck (Wheat 1972); San Luis Sites -- Linger (Dawson and Stanford 1975; Hurst 1941, 1943), Zapata (Wormington 1957), Stewart's Cattle Guard (Jodry 1986); Folsom (Figgins 1927).

parameters of organizational variability within anatomically modern human hunting may be crucial in the understanding of differences between recent predation and earlier forms of meat-eating (Binford 1981, 1985). The organizational role of hunting in recent hunter-gatherer settlement-subsistence systems, as evidenced by the archaeological record, is not unambiguous.

An example of this point is developed here using several recent formational analyses which suggest that the use of game by Paleoindians may have been significantly different than that which would be expected from ideas developed from records of Late Prehistoric northern Plains hunter-gatherers, and that widespread conclusions about the organization of activities drawn from direct interpretation of the early North American mass-kill sites may miss the mark. Faunal remains at these sites are frequently recovered from concentrated bone middens (Figure 2) which may contain the remains of several hundred animals. These bonebeds are deposits that often have complex formational histories -- the development of patterns within them shares many attributes with the formation of bonebeds at natural mass mortality sites (Behrensmeyer, this volume; Todd 1983, 1987). Therefore, the first problem in the analysis of suspected kill sites is analysis to isolate, at the assemblage level, patterns of variability that can differentiate general properties of mass death sites from those that might be informative of human action.

A second point is that many current interpretations are based on an idea that the role of Paleoindian kill sites was functionally identical to the later, historically documented North American Plains bison hunting groups. However, the growing body of data from the late Pleistocene/early Holocene bonebeds, indicates that this may not be the case -- Paleoindian use of mass-kills does not conform to Late Prehistoric models. For example, Frison (1982:200) has noted that bison kill sites in the Late Prehistoric usually have associated processing areas with "stone-heating pits, stone boiling pits, and piles of bone reduced to varying sizes for boiling out the bone grease," and that products from mass-kills in the form of "both dried meat and pemmican provided the necessary surpluses to insure winter survival." However, "attempts to apply the same model to Paleo-Indian bison kills have not been successful" (Frison 1982:200). Before turning to questions of the differences between Paleoindian

Figure 2. Portion of the Horner II bonebed.

and later uses of bison, the nature of the archaeological record of Paleoindian bonebeds will be examined.

APPROACHES TO THE INVESTIGATION OF KILL-BUTCHERY BONEBEDS

In the initial investigations of early man in the New World, the demonstration of association between artifacts and groups of large animals was considered sufficient documentation to begin discussion of hunting and subsistence practices. The animal remains were often not considered in detail and the bonebeds were frequently treated as mildly interesting matrices from which artifacts were extracted (Wheat 1972:159).

During the past fifteen years, as the general approaches to archaeological analysis have changed, there has been an increasing awareness of the need for detailed study of the bonebeds as a basis for reliable statements about past hunting tactics, butchery practices, and ecology of prey. A primary difference between earlier and more recent excavations of bonebeds has been the recognition that the structure and composition of the bonebeds themselves can be a source of information that is not available from even the best analyses of the associated stone tools. Faunal remains at kill sites are now seen as requiring archaeological analysis complimentary to that of the stone tools and have increasingly been incorporated as an important component of reports rather than as specialist appendices.

However, by the late 1970s and early 1980s this important perspective had begun to be uncritically interpreted to mean that kill-butchery bonebeds could be analyzed as if they were equivalent to humanly produced artifacts. Several sets of conventions or definitional statements derived from frequently observed patterns in kill sites were developed and began to be widely applied. Wheat (1971, 1978, 1979) has presented the most explicit list of conventions for recognizing "what constitutes a kill site as opposed to a butchering site and, in turn, of the nature of camp sites with the associated bone beds and middens" (Wheat 1978:84). These conventions are summarized in Table 1.

The primary variables used as criteria in Wheat's definition include: contextual properties, species composition, skeletal element frequencies, fragmentation, articulation, and properties of the distributional structure of bones within the deposit. All these attributes can indeed provide information about the formational history of a deposit and have been well reported for several North American Paleoindian bison kills (Frison 1974; Wheat 1972, 1979). However, in some cases the assumption has been made that variability along any of these dimensions is a direct indication of differences in human action.

Table 1. Conventions for classifying bonebeds as developed by Wheat (1978).

SITE TYPE	CHARACTERISTICS
KILL SITE	(1) always closely associated with a natural or contrived trap (2) bonebed normally consists entirely, or almost entirely, of bones of a single species (3) should contain high percentages of high weight and low meat value bones (4) individual bones usually unbroken (5) a high percentage of bones should be found in articulation, and the average number of bones per articulated unit should be high often including complete or nearly complete skeletal units (6) the bonebed will be more concentrated than at other types of sites (7) a high percentage of butchering tools should be present
BUTCHERING SITE	(1) no associated trap, but the number and nature of bonesindicate that a kill may have been present nearby (2) absence or near absence of those elements usually left at the kill site, such as complete skulls (3) preponderance of easily transported meat-bearing articulated and disarticulated bones (4) no complete or nearly complete articulated skeletons, and the percentage of articulated bones and the average size of articulations is lower than at kill sites (5) preponderance of butchering tools
LONG-TERM CAMP	(1) no indication of natural trap (2) overwhelming presence of easily transported articulated units such as front and rear quarters and vertebral segments (3) low number of articulated elements, no large articulated units are expected (4) many disarticulated bones are reduced to fragments (5) a wide range of species (6) wide range of artifact types
SHORT-TERM CAMP	(1) no associated trap (2) very sparse scattering of bones (3) few or no articulated bones (4) limited range of skeletal elements (5) bones often extremely fragmented (6) several species possible (7) moderate range of artifact types

As the examples discussed here illustrate, there are a number of factors besides human action that can create Table 1 patterning within these variables.

As the number of documented bonebeds and the sophistication of study methods have increased, it has become clear that the information content of kill-butchery sites has barely been tapped, and that our level of understanding of these relatively recent, usually well-preserved sites is actually quite limited. Many of the basic assumptions, such as those equating archaeological skeletal disarticulation with intentional dismemberment by humans, or those implying that all Paleoindian mass-kills were sources of products for long-term food storage, can no longer be taken for granted.

For example, the traits listed in Table 1 can be demonstrated to be ambiguous by examining noncultural bonebeds and comparing the results with the expectations given in the table. Figure 3 illustrates a portion of the Miocene-age *Stenomylus* Quarry (Sioux County, Nebraska) excavated by the Carnegie Museum in the early 1900s (Peterson 1911). The *Stenomylus* bonebed is composed primarily of a single species and is relatively concentrated. Based on the portion of Peterson's (1911) map illustrated in Figure 3, fifteen crania -- commonly interpreted as "high weight and low meat value bones" -- are prevalent; most of the bones are unbroken and many are articulated including a number of nearly complete skeletons. These are all attributes that meet the criteria for recognition of a kill site as listed in Table 1. The only criteria not met here and in a number of other paleontological cases (Todd 1983:25-29) are the lack of either a natural "trap" or of associated artifacts. The presence of an obvious "trap" is *not* a characteristic of many of the bonebeds accepted as being of cultural origin -- neither of the kill-butchery bonebeds described below (Horner II and Jones-Miller) is associated with natural traps. Given the current interpretative framework for kill-butchery bonebeds, aside from the lack of artifacts and the age of the deposit, then there would be little indication that this Miocene deposit was not the result of human predation. In other words, current conventions of what constitutes evidence of human action based on patterns within Paleoindian period bonebeds, are ambiguous.

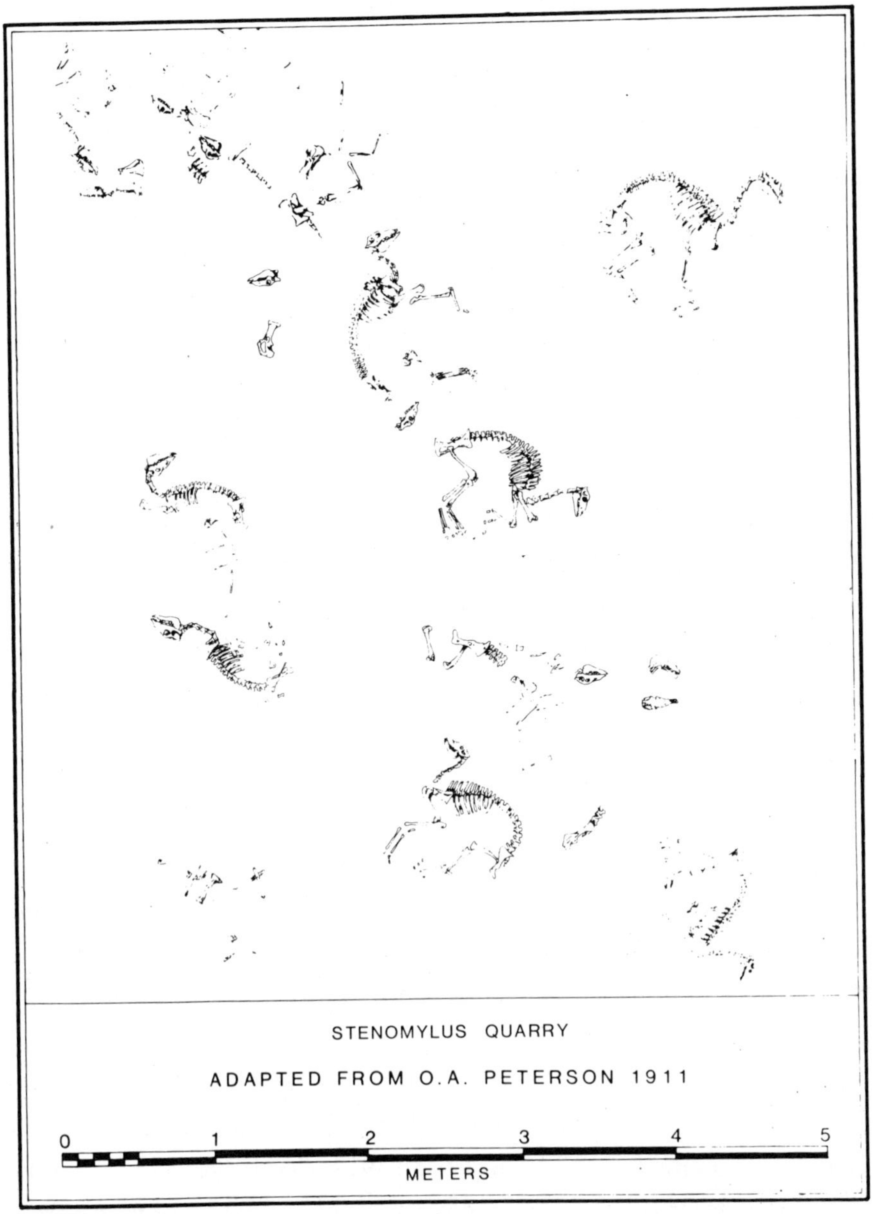

Figure 3. Paleontological bonebed exhibiting characteristics similar to those suggested to be diagnostic of humanly produced kill-butchery sites.

FORMATION OF KILL-BUTCHERY BONEBEDS

My discussion begins with a review of recent studies of two bison bonebeds: Horner (Frison and Todd 1987; Todd 1983) and Jones-Miller (Todd 1984; Todd and Stanford 1986); these have been directed toward the identification and resolution of this interpretative ambiguity (see Frison and Todd [1986] for description of a similar analysis of the mammoth bone assemblage from the Colby site). Locations of Horner, Jones-Miller, and several other Paleoindian bonebed sites on the northern Plains are shown in Figure 1.

Investigation of the role of hunting and processing of large game animals in Paleoindian settlement-subsistence systems requires a better understanding of the formation of these bonebeds. The Jones-Miller and Horner sites are used to demonstrate several approaches to the examination of patterns within kill-butchery bonebeds.

The Horner Site Bonebeds

The Horner site is on a Pleistocene river terrace 40 m above the present channel of the Shoshone River in northwestern Wyoming (Albanese 1987). Although excavations were begun in 1949, a final report has only recently been completed (Frison and Todd 1987). At least three distinct bonebeds are represented at the Horner site, two of which were excavated by Princeton University and the Smithsonian Institution in 1949-1952: Horner I which includes two bonebeds (Frison 1978; Jepsen 1953; Todd, Witter and Frison 1987; Wedel 1961; Wormington 1957), and the Horner II bonebed excavated by the University of Wyoming in 1977-1978 (Frison 1978; Todd 1983, 1987). Element counts for the Horner II bonebed are given in Table 2. No topographic trap is evidenced, all the bonebeds are deposited on an essentially flat terrace surface (Albanese 1987).

The 1949 excavations, of over 360 m^2 uncovered a concentrated multi-animal bison bonebed and associated stone tools (see Bradley and Frison

Table 2. Minimum animal unit values for bison kill-butchery site assemblages.

SKELETAL ELEMENT	CODE	MINIMUM ANIMAL UNITS (MAU) HNR	BS1	BS2	GNR	JSM	CSR	GRT	GRB	WDL	OSB	WCP	LSP
cranium	CR	28.00	7.00	13.00	35.00	102.00	66.00	27.00	98.00	55.00	143.00	17.00	8.00
mandible	MR	55.50	4.50	35.00	11.00	245.00	77.00	26.50	79.50	137.00	112.00	14.00	8.00
atlas	AT	57.00	9.00	39.00	29.00	141.00	45.00	12.50	110.00	89.00	101.00	16.00	9.00
axis	AX	41.00	12.00	36.00	21.00	152.00	38.00	13.50	111.00	90.00	93.00	17.00	6.00
cervical (3-7)	CE	30.00	3.00	7.20	17.20	158.00	33.60	17.60	76.00	33.00	84.20	16.00	7.00
thoracic	TH	45.71	1.00	2.14	10.86	129.29	34.00	7.07	36.00	25.89	114.07	16.60	3.80
lumbar	LM	50.40	2.80	3.00	9.00	166.40	37.00	13.80	61.40	29.00	99.20	16.00	5.20
sacrum	SA	59.00	6.00	27.00	13.00	88.00	50.00	17.00	57.00	50.00	111.00	15.00	3.00
scapula	SC	62.50	5.00	22.00	11.00	157.50	45.00	25.00	68.00	111.00	100.50	13.00	6.00
humerus, proximal	HMPR	38.50	2.00	6.50	16.00	44.00	20.00	5.00	15.00	12.50	86.50	13.00	2.00
humerus, distal	HMDS	61.50	2.00	7.50	14.50	153.00	47.50	9.50	22.50	35.00	89.00	15.50	7.50
radius, proximal	RDPR	61.00	3.00	9.50	11.50	157.50	33.50	13.50	43.00	30.00	81.00	15.00	6.00
radius, distal	RDDS	58.50	3.00	13.50	12.00	155.00	34.50	17.50	56.00	31.00	90.50	15.50	5.00
metacarpal, proximal	MCPR	57.50	2.00	21.00	15.50	131.50	41.00	34.50	83.00	48.50	82.00	13.50	4.00
metacarpal, distal	MCDS	57.50	2.00	22.50	12.50	139.50	41.50	31.00	71.00	50.50	88.00	13.50	4.00
innominate	IM	58.00	1.50	6.50	24.00	124.50	42.50	15.50	63.50	54.00	94.00	15.00	7.50
femur, proximal	FMPR	46.50	2.50	16.00	8.50	94.00	26.00	9.00	17.00	30.50	87.50	14.00	4.50
femur, distal	FMDS	52.00	1.50	5.00	13.50	72.00	27.00	9.00	14.00	29.00	81.50	13.00	3.00
tibia, proximal	TAPR	47.50	2.50	12.00	15.00	91.00	31.50	8.50	14.50	40.00	84.00	14.00	5.50
tibia, distal	TADS	55.00	5.00	15.00	11.00	141.00	24.50	13.00	25.00	73.00	87.50	14.50	8.50
metatarsal, proximal	MTPR	63.00	4.50	13.00	12.00	136.50	29.50	30.50	88.00	52.20	82.50	12.00	6.50
metatarsal, distal	MTDS	65.00	3.50	18.00	12.00	138.50	28.50	28.00	80.50	49.00	89.00	12.00	4.50

Assemblage codes: HNR, Horner II (Todd 1987:Table 5.6); BS1, Boarding School bone layer 1 (Kehoe 1967: Table 8); BS2, Boarding School bone layer 2 (Kehoe 1967:Table 8); GNR, Garnsey (Speth 1983:Figure 28); JSM, Jones-Miller (Fulgham n.d.); CSR, Casper (Frison 1974); GRT, Glenrock upper level (Frison 1970); GRB, Glenrock lower level (Frison 1970); WDL, Wardell (Frison 1973); OSB, Olsen-Chubbuck (Wheat 1972); WCP, Wind Cave mortality sample (Todd, Burgett, and Klukas 1986); LSP, Lamb Spring (McCartney 1984: Table 4).

1987; Frison 1987). Several large articulated skeletal units were recovered. Unfortunately, records on this faunal assemblage are limited to two areas, comprising less than 5% of the excavations. The remains of at least 22 bison are tabulated in field notes with the frequency of identifiable bones approaching 70 elements/m^2. Although the full extent and composition of this bonebed cannot be determined, it is estimated that at least 200 animals were present (Jepsen 1953) and it is clear that the records represent a small portion of the 1949 bonebed. Only a few pieces of the bison bones from the 1949-1952 excavations were saved, but a count of right astragali which, with mandibles and teeth, seem to have been the only systematically collected bones, gives a minimum number of individual (MNI) count of 158.

The second bonebed uncovered at Horner I contained scattered, disarticulated, often fragmented, bison bones associated with an assemblage of stone tools, several hearth areas, and a number of shallow pits which extended into the underlying cobble substrate. The remains of other species (including pronghorn and deer [Walker 1987]), some with indications of butchery were also recovered. Excavation records are more complete for this area and an MNI of 36 is based on mandibles. The frequency of bones seldom exceeds 5 elements/m^2, although in some areas nearly 25 elements/m^2 were reported. These concentrations correspond to the areas of the site with the highest frequencies of hearths, pits, and stone tools.

The preservation of the bone in both Horner I bonebeds was poor. This portion of the site was buried approximately 30 cm below the present surface and the bones exhibit post-depositional deterioration. Most bones also displayed rather severe weathering (Behrensmeyer's [1978] stages 3-4), indicating that they were exposed on the surface for years prior to burial. Given the condition of collections and records, it is difficult to reconstruct past activities in this area.

The third bonebed (Horner II, located 40 m south of the 1949-1952 excavation) was buried under two meters of sediments, and contains the remains of 65 to 70 bison and associated stone tools (Figure 2). Both radiocarbon dates and projectile points indicate that this bonebed is at least a thousand years older than the Horner I bonebeds. The density of identifiable bones often exceeds 70 elements/m^2; many bones are articulated, and several nearly complete skeletons occur along the central axis of the bonebed. Analysis of eruption and wear patterns of the mandibular dentitions indicates that the mortality occurred in the late fall or early winter (Todd and Hofman 1978, 1987). Few bones are broken and most surfaces are unweathered and have suffered little post-depositional deterioration, although near the margins of the bonebed some are more severely damaged. The edges of the bone concentration are sharply defined by a marked drop in density. The bones are located in a very shallow swale, but no natural containment is present. Bone condition indicates that the site was buried soon after the death of the bison.

Disarticulation of the Horner II bison has been investigated in several ways. The frequency of articulated units at recent bison death sites (Hill 1979 a, b; Hill and Behrensmeyer 1974, 1975; Todd 1983; Todd, Burgett, and Klukas 1986) can be compared with the disarticulation within the bonebed. However, an important component in distinguishing causes of disarticulation is the degree that the bones, after separation, are dispersed. In situ decay and disarticulation may leave few joints articulated after several years although components of once-articulated joints may remain spatially associated. On the other hand, if disarticulation is the result of dismemberment by humans or removal by scavengers, less spatial association of once-articulated bones would be expected.

Based on this premise, part of the Horner II analysis has focused on the reassembly of dispersed bison carcasses. Metric and nonmetric attributes (i.e., muscle attachment size and shape, and placement of fovea and foramen) that have been demonstrated to be individually distinctive in the skeletons of modern bison are used in the recognition of dispersed anatomical element pairs. The recognition of "mated elements," or left and right bones from the same animal, is the easiest and most reliable form of this "anatomical reassembly" (Todd 1983, 1987; Todd and Frison 1986). Within the Horner II bonebed, the average distance between disarticulated upper limb bones from opposite sides of a skeleton is low, usually less than 2 m. Compared to average distances between paired elements at recent bison death sites (Table 3: WCNP) without cultural dismemberment, dispersal of disarticulated elements within the Horner

Table 3. Average distances (meters) between dispersed element pairs.

SKELETAL ELEMENT	HORNER II N	HORNER II MEAN	WCNP N	WCNP MEAN	PLUMBAGO N	PLUMBAGO MEAN
mandible	11	0.1	5	1.3	--	---
scapula	--	---	10	11.0	7	0.8
humerus	9	1.0	13	22.8	19	0.4
radius	10	1.1	14	25.4	9	0.7
metacarpal	--	---	10	27.6	5	0.4
femur	8	2.1	12	9.8	19	0.4
tibia	11	0.9	12	8.9	10	0.4
metatarsal	--	---	8	10.1	--	---

REFERENCES: Horner II (Todd 1983:Table 21); WCNP (Todd, Burgett, and Klukas 1986); Plumbago (Todd 1983: Table 2).

II bonebed is extremely low. The average distances between pairs at Horner is also low when compared to those from a group of domestic cows that have suffered little scavenging, but have lain exposed on the ground surface for nearly 50 years (Table 3: Plumbago). Even though a number of the bones within the bonebed were disarticulated, most are probably very near their original anatomical position. This limited dispersal argues against extensive dismemberment during butchery as the primary cause of disarticulation and suggests that decay may be one of the major factors influencing variability in articulation at Horner II.

Bone counts (Table 2) also point to limited removal of large portions of the carcasses. Almost all anatomical segments are present in nearly equal proportions, with all MNI estimates close to the maximum MNI value of 65. While only a few articulated skeletons were recorded from near the center of the bonebed, it seems that these are a small subset of essentially complete carcass remnants present when the site was abandoned by the Paleoindians. In any case, the assumption is highly suspect that the articulated bones in this well-preserved, apparently rapidly buried site are equivalent to prehistoric butchering units. This assumption is even more suspect for other sites which were apparently exposed longer prior to burial.

The Jones-Miller Site

Excavations at the Jones-Miller site in east-central Colorado were conducted by the Smithsonian Institution in the mid 1970s (Stanford 1974, 1978, 1979, 1984). The margins of the site were located and the entire bonebed was excavated. Skeletal element counts are summarized in Table 2. Nearly 300 bison and a limited number of smaller vertebrates (Stanford 1978:92; 1984:622-624) are represented. The bonebed is in a shallow -- 45 cm deep by 20 m wide (Stanford 1978:90) -- swale cut into the third terrace of the Arikaree River. A second swale, also containing bison bone is about 80 m west (Stanford 1984:94). One of Wheat's criteria for a kill site, a topographic trap or containment, is clearly absent at the Jones-Miller site and Stanford (1984:622) notes that:

> no major geophysical features in the immediate area could have been used to trap a large herd of bison. Neither drainage is sufficiently deep to serve as a jump, and there are no sand dunes that could have been used as a trap. An extensive examination has failed to yield any post molds that would represent a corral type of impoundment. It may, in fact, be that the kill did not take place at the Jones-Miller site, but at some point nearby, and that the animals were butchered into quarters and brought to the site for secondary butchering.

Stanford suggested a pound or corral made of perishable materials (1978:95-97), perhaps snow, and ice-covered or incorporating snowdrifts in the swale, may have provided a containment for the bison.

Analysis of eruption and wear of dentition (Reher n.d.) indicates that at least two seasonal modes of mortality are represented, perhaps indicating two episodes of use, either during a single year or over a short period of several years. The first seasonal mode indicated a late fall to winter mortality; the second took place slightly later in the year. The horizontal (Reher n.d.) and vertical (Fulgham n.d.) distributions of the mandibles by these seasonal modes have indicated that some spatial differences in the locations of the different mortality events may be represented. Analysis of a wide range of skeletal elements (Graham n.d.; Todd 1986) indicates that females (about 85-90% of the adults) and juveniles predominate. Portions of the bonebed contain extremely high densities, with over 1000 identified bones recorded for some of the two-by-two meter excavation units.

Many bones were fragmented and very few were articulated. Stanford (1978:91) observes that "most articulations consist of small vertebral sections and lower leg units, such as metapodials and phalanges. Only a small number of skulls have been found, and most were thoroughly butchered. The butchering, then, seems to be nearly complete, with no articulated animals wasted." The initial reports on the site considered degree of disarticulation and fragmentation as direct indicators of intensity of butchery. Recently, Stanford (1984:626) has noted the possible role of other agencies, such as carnivore scavenging, as factors making "determination of butchering activities...extremely complicated."

Indeed, there are indications that the following processes have modified the faunal assemblage subsequent to human activity: (1) feeding by carnivores resulting in destruction, deletion, and rearrangement of bones; (2) weathering that may have influenced later, post-occupational fracture, and eventually bone destruction; (3) trampling by large animals; (4) extensive gnawing by rodents in some areas of the site; and (5) post-depositional chemical deterioration and root etching. These occurred unequally across the site. Spatial differences in the intensity of these processes create intrasite variability in the faunal assemblage that can diminish the potential for recognition of humanly produced modifications, such as cutmarks. Such nonuniform modifications can be misidentified as differences in "activity structure." Investigations of two such factors at the Jones-Miller site, carnivore gnawing and post-occupational breakage and dispersal (Todd and Stanford 1986), are summarized here.

Haynes (1981:357) has previously reported that "a cursory examination of femora, tibiae, and humeri revealed a great deal of carnivore gnaw damage" and suggested that small- to medium-sized canids, perhaps domestic, accounted for most of the damage (1981:357, 360). One of the most common observations in studies of carnivore action is the consistent patterns of differential destruction of articular ends of long bones (Binford 1981; Binford and Bertram 1977; Brain 1981; Haynes 1980a, 1980b, 1982; Richardson 1980). Binford (1981:217) has suggested that differences in survival of proximal and distal ends of the humerus and tibia are particularly important "for purposes of *diagnosing* the relative role bone destruction plays in modifying an assemblage." There are clear differences in the frequencies of damage to the proximal and distal ends of these long bones at the Jones-Miller site, which, based on Binford's (1981) assemblage data, falls within the zone of carnivore-ravaged assemblages (Todd 1987: Figure 5.25).

The observations on carnivore damage presented here are not comprehensive. Not all element classes have been examined for evidence of gnawing, and within those classes described below, only a sample of the bones has been studied. The numbers of specimens examined and the frequencies of carnivore modification are listed in Table 4. In this paper, only damage to the humeri will be described.

In recording the carnivore-damaged bones at Jones-Miller, a series of two digit, bone specific codes were used (see Figure 4 for illustration of humerus damage codes). The codes were developed after examination of the entire assemblage to determine the types of damage. The gnawing codes were then established to capture that variability. The gnawing referred to here is that produced by carnivores only -- gnawing by rodents was considered separately. Two subsets of each code were used. The first, with values from 30-39 was for bones with definite evidence of carnivore damage such as tooth pitting, gouging or furrowing (Figure 5). A second subset of codes (40-49) was for

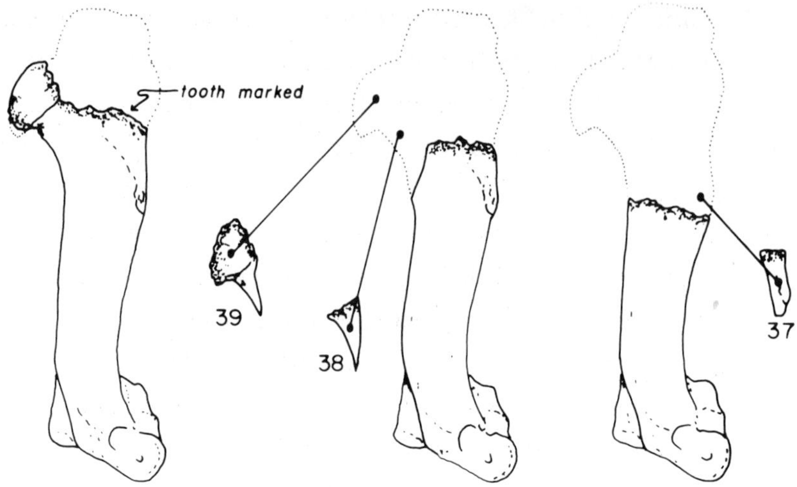

Figure 4. Patterns of carnivore damage to bison humeri at the Jones-Miller site. Numbers are codes for each of the commonly encountered fragments.

bones of similar general morphology but having deteriorated surfaces with lowered potential for observation of diagnostic gnawing marks. These are described as erose (i.e., appearing as if gnawed). Codes for erose bones have the same second digit as the clearly gnawed similar bones. For example, a humerus without the lateral tuberosity with gnaw marks is coded as "32." A humerus with similar morphology, but without clear tooth marks would be coded as "42."

A total of 159 humeri in the Jones-Miller collection have been examined for evidence of gnawing. These included both individual segments and components of conjoined bone groups. Conjoined bone groups are composed of reassembled fragments of a single bone. The pattern of destruction with the Jones-Miller assemblage begins with the gnawing on, and removal of, the major (lateral) tuberosity, followed by loss of the minor (medial) tuberosity, and

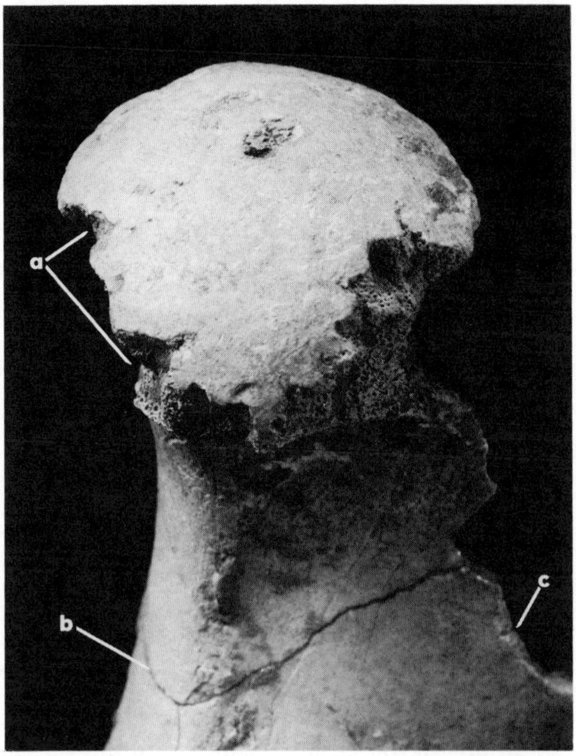

Figure 5. Carnivore-damaged humerus from the Jones-Miller site: (a) tooth furrows, (b) typical "V"-shaped fracture as gnawed head is broken from shaft, and (c) beveled, tooth marked area.

finally by isolation of the head (Figure 4:code 32). Further gnawing removes the head, leaving distal articular ends with attached segments of the shaft (Figure 4:codes 33, 34).

Although destruction of the proximal ends is common, a series of gnawed detached fragments (Figure 4: code 39; Figure 6:a-c) have been identified. Binford reported similar specimens as having been produced by wolves gnawing on caribou humeri (Binford 1981:Figure 3.43) and they have also been observed in other bison bonebeds (Frison 1974: Figure 1.12). These fragmented humeral heads often have distinct tooth furrowing (Figure 6:a) and pitting (Figure 6:b,c). After removal of the tuberosities, the humeral heads often are isolated and supported by only small segments of shafts. Continued gnawing, and pulling on the humeral head often removed it leaving a V-shaped notch on the caudal surface of the proximal shaft (Figure 5b) with a corresponding pointed projection on the distal shaft fragment below the humeral head (Figure 6:b,c).

Possibly in cases of younger animals with incomplete ossification, heads are fully consumed, leaving tooth marks on the proximal margins of the posterior shafts (Figure 6:d-f). The curvature and internal thickening of this portion (Figure 4:code 38) make it slightly more resistant to gnawing than surrounding, thinner sections of the shaft and it is often isolated and snapped off. Gnawed portions of the shaft with the deltoid tuberosity (Figure 4:code 37) are similarly produced.

Several bones, including the proximal humerus, both ends of the femur, the proximal tibia, and the patella have been rather extensively modified by carnivores while metapodials and radii exhibit few gnawing marks (Table 4). This modification may have altered not only elements counts, but also seems to have caused repositioning of the bones. Examination of the carnivore action can determine whether modification occurred equally across the site, or whether there were localizations of gnawed bones.

Figure 7 shows the distribution of the gnawed and erose humeri. There are no obvious clusters and the gnawed specimens are not concentrated near the margins of the bonebed as reported at Horner II (Todd 1983, 1987). More of the carnivore damaged bones appear to be in the southern half of the site. However, since the frequency of gnawed bones is highly correlated with the total number of bones (Todd 1984), the apparent increase of carnivore damage from north to south is largely a function of an increase in frequency of total bones along that axis (Figure 8).

In general, much damage to bones at Jones-Miller seems to result from carnivore action rather than modifications by humans. The percentages of complete limbs plotted against the percentage of carnivore damaged bones indicate an inverse relationship (Figure 9). The usually complete metapodials and radii exhibit little gnawing while those most often fragmentary -- tibiae,

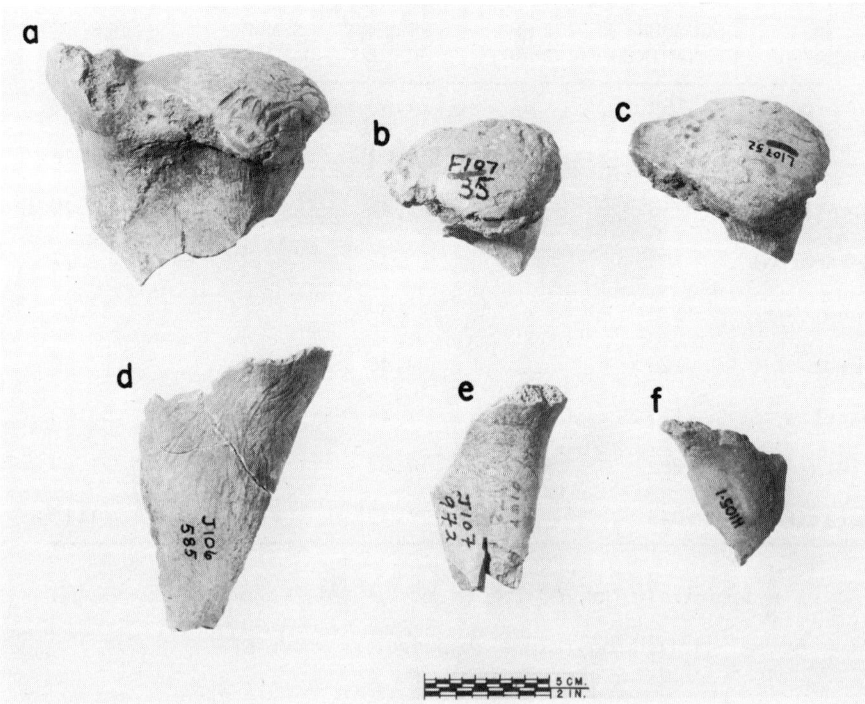

Figure 6. Proximal fragments resulting from carnivore modification to the Jones-Miller humeri: (a-c) heads (Figure 4, code 39), and (d-e) posterior shafts (Figure 4, code 38).

humeri, and femora -- displayed the greatest percentage of carnivore modification. This indicates that degree of fragmentation is not a definitive criterion for interpretation of the intensity of human action in a kill-butchery bonebed that has been subjected to carnivore modification.

A second modification which can alter or create patterns within a bonebed occurs after human and carnivore modification, and includes processes such as treadage which may continue to fracture and scatter bones long after they have been defleshed and dried. Post-occupational breakage and dispersal at Jones-Miller is evidenced by conjoined bone fragments classified as either

Table 4. Summary of carnivore damage on examined bones from the Jones-Miller bonebed.

BONE	NUMBER EXAMINED	NUMBER GNAWED	PERCENT GNAWED	NUMBER DAMAGED(a)	PERCENT DAMAGED
humerus	159	44	27.67	103	64.78
radius	324	10	3.09	11	3.40
ulna	236	60	25.42	82	34.75
metacarpal	259	2	.77	2	.77
innominate	178	63	35.39	71	39.89
femur	125	43	34.40	84	67.20
patella	115	38	33.04	62	53.93
tibia	202	36	17.82	43	21.29
metatarsal	343	4	1.17	4	1.17

a - definitely gnawed (tooth marks and fragment morphology) plus erose (morphology alone) bones.

"green," intermediate, or dry-bone fractures (Todd and Stanford 1986). Breakage has been classified as being dry-bone when fractures occur along well developed longitudinal weathering cracks. These fragments are characterized by rectangular, interlocking fracture surfaces, indicating that the bone had been defleshed and dried prior to breakage. While causes of the dry-bone breaks are unknown, direct human or carnivore action is unlikely. The distance between these conjoined dry-bone fragments is used to indicate degree of post-occupational movement within the bonebed.

The conjoined bone fragment groups are summarized in Table 5. Many of the identified conjoined bone fragments (41.5%) are from dry-bone breakage and only a few (19.8%) indicate fresh bone fractures. The average distance between these dry-bone conjoins is 1.6 m with a maximum distance separating conjoined dry-bone fractured pieces of 14.4 m.

The average distance between the green-bone fractured conjoined pieces

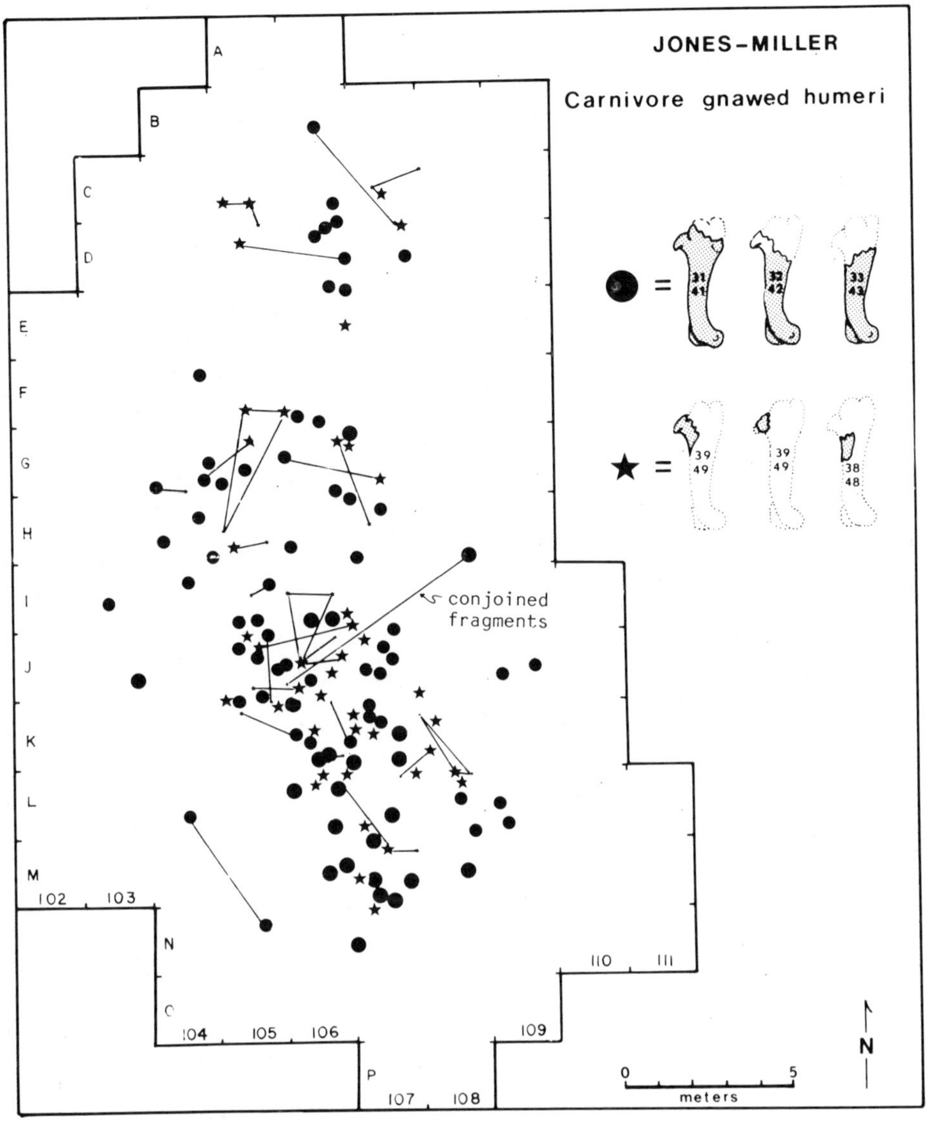

Figure 7. Distribution of carnivore damaged (gnawed and erose) bison humeri from the Jones-Miller bonebed.

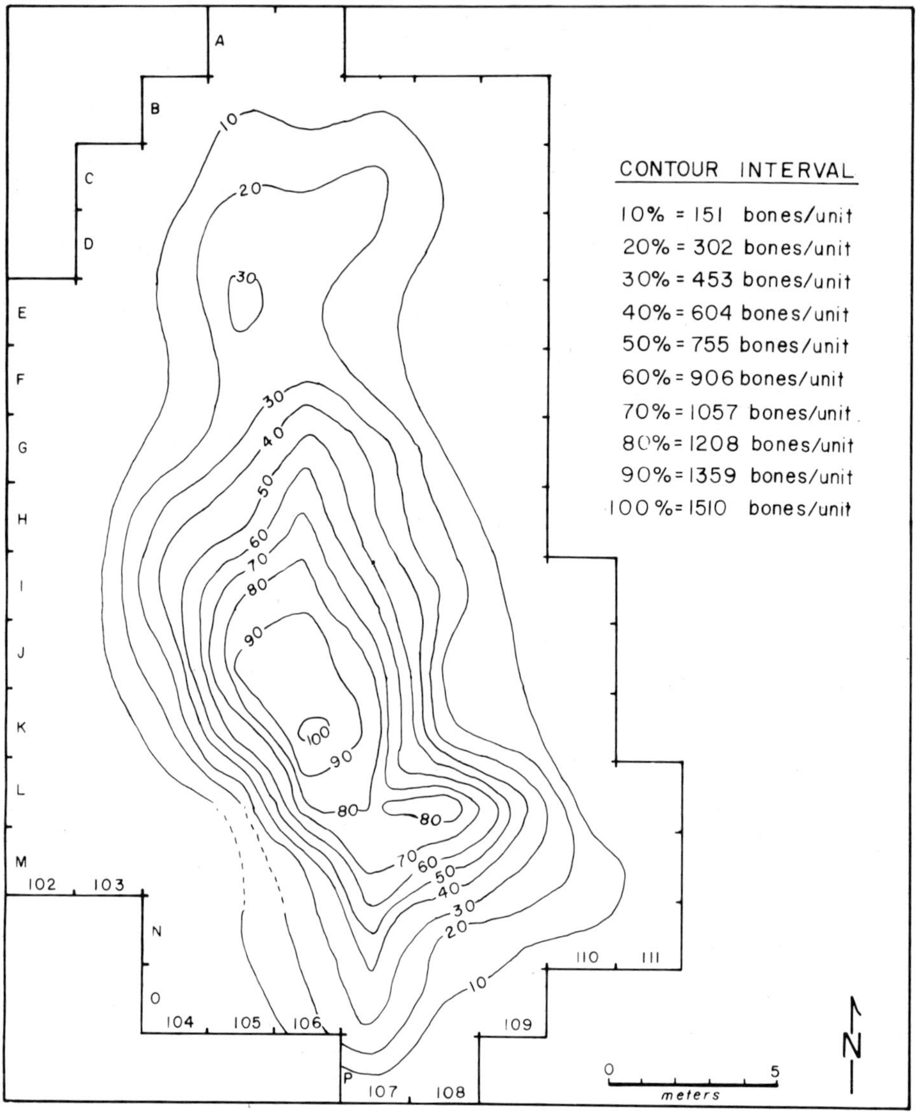

Figure 8. Frequency of bones per 2 X 2 m excavation unit at the Jones-Miller site.

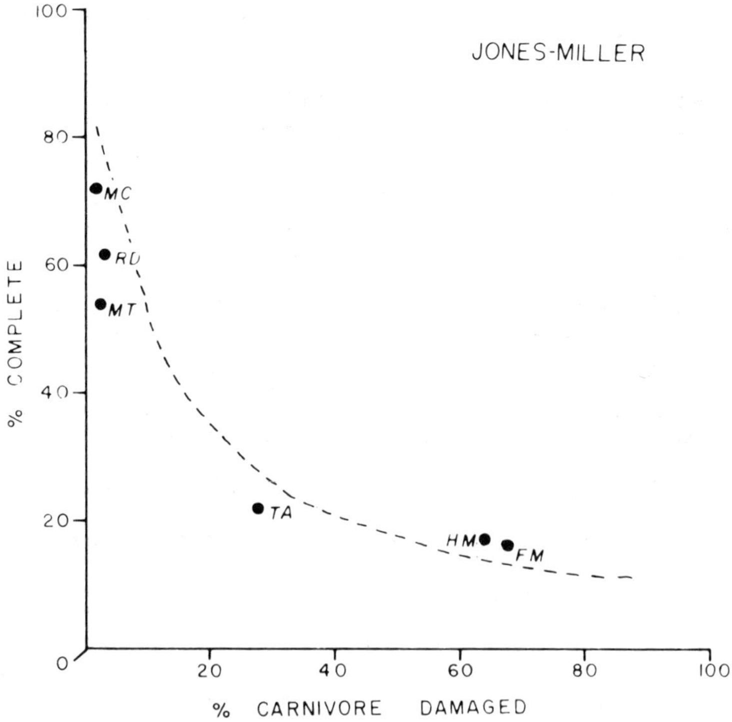

Figure 9. Relationship between carnivore damage and completeness for limb bones from the Jones-Miller bonebed: FM, femur; HM, humerus; MC, metacarpal; MT, metatarsal; RD, radius; and TA, tibia.

is slightly greater than that of the dry-bone fractured conjoins (Table 5). The greater dispersion of green-bone fractures may result from the same noncultural processes as those which displaced the dry-bone conjoin groups. Since the green-bone fractures, regardless of cause, generally would have occurred prior to the dry-bone breaks, the time for scattering of green-bone fractured pieces was longer. The mean distance plus one standard deviation for dry-bone fractures gives a conservative estimate of dispersion of 3.5 m, indicating that *post-occupational horizontal movement of up to 3.5 meters is common within the Jones-Miller bonebed*. Conjoined long-bone dispersal

Table 5. Distances (m) between conjoined long bone fragments from the Jones-Miller site by breakage type.

FRACTURE TYPE	N	MEAN	SD	MIN	MAX
GREEN	155	2.07	1.79	0.10	9.03
INTERMEDIATE	127	1.59	1.37	0.04	8.58
DRY	325	1.55	1.97	0.02	14.37
INT+DRY+EPI[a]	458	1.57	1.82	0.02	14.37

NOTE: N, number of conjoined distances; MEAN, average distance between conjoined fragments in meters; SD, standard deviation; MIN, minimum distance between conjoined pieces; MAX, maximum distance between conjoined fragments.

a - breaks interpreted as being of "non-cultural" origin; INT, intermediate break; DRY, dry bone break; EPI, epiphysis to diaphysis refit, immature specimens (N = 6).

patterns, particularly dry-bone breaks, indicate a significant amount of scattering took place after the utility of the bonebed to both humans and carnivores was exhausted. Given this scattering, distributional patterns of human butchery probably have been significantly blurred or obliterated.

It is not surprising that because of carnivore action, and the extreme post-occupational breakage and dispersal, few articulated units are present. The limited amount of articulation at this site probably bears little relationship to the pattern of dismemberment caused by human abandonment of carcasses. Few of the conjoined bone fragment groups are green bone-breaks (Table 5). Since there are many causes of green-bone breakage, including gnawing, and since there is evidence of carnivore-produced fragmentation, only a small portion of these green bone fractured elements may be the result of human activity[1].

The analysis of bison bones from the Jones-Miller and Horner sites indicates that human activity was a limited component in the creation of patterns within the bonebeds and that identification of human activity is not a direct undertaking. At both sites, which may represent almost polar extremes in terms of length of time between mortality and burial, the nature of hunting and butchering appears much different when a range of formational processes is considered than when the bonebed is identified a priori as direct evidence of human action.

Other northern Plains Paleoindian bonebeds (Figure 1) provide a more complete picture of the range of variability in these types of sites. The counts of skeletal elements in the sites which have basic frequencies of proximal and distal ends of the long bones are included in Table 2. These sites form the basis for many current interpretations of hunting by Paleoindian groups. However, the problems of unraveling the formational histories are still far from answered and interpretations should not be based on the assumption that their variability is entirely the result of human activity. The number of sites serving as bases for an eventual understanding of past hunting practices is relatively large, but their interpretation, while analytic approaches continue to be refined, requires reexamination of previously described assemblages.

PALEOINDIAN BISON KILLS: PATTERN RECOGNITION

The preceding summaries have pointed out some of the variability in Paleoindian bison bonebeds. With the exception of McCartney's (1984) analysis of Lamb Spring, interpretations of kill-butchery sites have usually been undertaken as if human action were the predominant, if not exclusive, cause of patterning within the assemblages. The Horner II bonebed was buried quickly and not subjected to the range of post-occupational modifications indicated for the Jones-Miller site. In terms of comparative analysis, this observation provides a useful baseline. If post-occupational processes alone are responsible for assemblage differences in Plains bonebeds, then comparative analysis of Horner and Jones-Miller should point to a marked difference between the sites. However, as discussed below, this is not the case. While the severity of post-occupational modifications varies considerably and the sites are clearly not fossilized hunting events, assemblage level patterns indicate that all Paleoindian period bison bonebeds are more similar to one another than they are to Late Prehistoric period bonebeds from the same region.

Bone Counts and Late Prehistoric Models

Many interpretations of bison hunting by Paleoindians have relied heavily on ethnographic or ethnohistorically cases and on interpretations developed from Late Prehistoric kill sites, particularly those from the northern Plains and from semisedentary Plains villagers (White 1953, 1954; Wilson 1924) for whom bulk food processing and storage were important components of the settlement-subsistence systems. In fact, Wheat has suggested that

> If the archaeologist is dealing with prehistoric kills, he must assume that the habits of prehistoric and historic bison were essentially the same, *and that the methods of hunting and utilizing them were similar*. There can be no doubt that acculturation of the Indian to the use of the gun and horse changed, in some cases drastically, the hunting methods of the historic Indians. *Yet, the techniques observed by the earliest explorers of the interior North America must have remained essentially unchanged over thousands of years. If these assumptions are valid, then we are in a position to comprehend and interpret the evidence recovered in our excavations of bison kills* (1972:85; emphasis added).

One way to examine the degree of similarity between prehorse Late Prehistoric and Paleoindian sites is based on skeletal element frequencies converted to standardized minimum animal unit values, MAU values (Binford 1984:50). Other data such as cutmark locations and frequencies, nature and degree of bone breakage, distributional patterns of elements, and nature of post-occupational modification (as discussed for the Jones-Miller and Horner sites), are often not reported completely and therefore cannot be used for comparative analysis. Skeletal element frequencies provide one of the most easily attainable, and potentially one of the more powerful tools for comparisons of faunal assemblages (e.g., Binford 1978, 1981, 1984; Binford and Stone 1986).

Information on skeletal element frequencies (including counts of proximal and distal frequencies of long bones) is, however, available from five large northern Plains Paleoindian bison bonebeds (Casper, Jones-Miller, Horner II, Olsen-Chubbuck, and Lamb Spring). Both Casper (Frison 1974) and Jones-Miller contain Hell Gap projectile points. Horner II, Lamb Spring (McCartney 1984) and Olsen-Chubbuck (Wheat 1972) are Cody Complex (or Alberta/Cody) sites. Four Late Prehistoric period sites (Boarding School in northwest Montana, Glenrock in central Wyoming, Wardell in west-central Wyoming, Garnsey in southwest New Mexico) are also included in Table 2. The Boarding School (Kehoe 1967) and the Glenrock sites (Frison 1970) contain multiple bone layers, here treated as separate assemblages. Wardell (Frison 1973) and Garnsey (Speth 1983; Speth and Parry 1978, 1980) are tabulated as single assemblages. Although there are two levels recorded at Wardell, these are combined to allow inclusion of all 274 mandibles reported by Reher (1973). The final assemblage is based on element counts from 17 bison from Wind Cave National Park (WCNP), South Dakota which suffered light scavenging by coyotes and had a relatively low percentage of elements destroyed or removed (Todd, Burgett, and Klukas 1986).

For the present these assemblages can be conceived of as forming three groups. The first group is the large northern Plains Late Prehistoric autumn kills (Boarding School, Glenrock, and Wardell) which represent procurement and processing of bison as part of a settlement-subsistence strategy incorporating seasonal food storage. The second group includes the Garnsey site and the Wind Cave sample, both expected to exhibit patterns different from the northern Plains sites. At Garnsey, unlike the other Late Prehistoric sites, the season of use (spring), the repeated use of the locality for small kills (averaging 6 to 8 animals, [Speth 1983:25]), the possibility of the opportunistic nature of the kills (Speth 1983:170), and the location (eastern New Mexico), all suggest that the site played a different role within a settlement-subsistence system than the northern Plains sites. Processing of large quantities of food for over-wintering may not have been a primary consideration at Garnsey but rather, as argued by Speth (1983; Speth and Spielmann 1983), processing decisions may have been based on more immediate requirements such as consumption at a near-by occupation site. It is possible that the Garnsey site was used by pueblo farmers (Speth, personal communication 1986), who would have faced a different set of processing/transport decisions than more mobile hunter-gatherers. The Wind Cave sample has not been processed by humans; all elements were removed by small vertebrate scavengers (coyotes) in relation to their meat utility. The third group includes the five Paleoindian sites for which the nature of processing and use of bison as food is examined here.

In order to evaluate the degree of similarity between these assemblages, the matrix of skeletal element frequency data has been normalized in relation to the overall chi-square values. These normalized values are then used to calculate squared Euclidean distances between the variables which in turn are the basis for a maximum-linkage clustering (see Binford and Stone 1986; Harpending and Rogers 1985). The results are illustrated in Figure 10.

Based on element frequencies, two distinct clusters are indicated. The first contains all Paleoindian sites (Casper, Olsen-Chubbuck, Horner II, Lamb Spring, and Jones-Miller) as well as the Garnsey and the Wind Cave assemblages. The second cluster includes all the northern Plains Late Prehistoric sites (Glenrock Lower, Glenrock Top, Wardell [both levels combined], Boarding School 1, and Boarding School 2). While such clusterings have no inherent meaning, several points are clear. First, given the wide range of formational processes represented, all of the Paleoindian bonebeds are more similar to one another than to any of the northern Plains Late Prehistoric sites which have often served as models for interpreting the Paleoindian sites. This marked difference indicates that application of models of Late Prehistoric subsistence to Paleoindian cases is inappropriate. Second, the Garnsey site which may represent processing to meet different processing

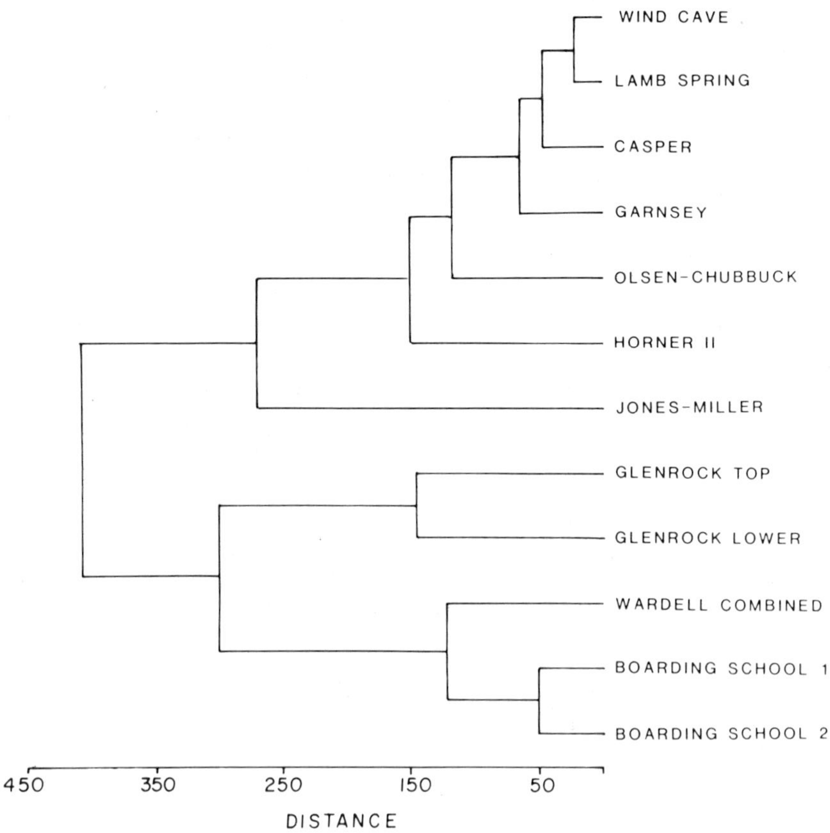

Figure 10. Similarities among Plain bison bonebeds based on skeletal element frequencies (Table 2, MAU).

and transport problems than the northern Plains Late Prehistoric sites and the Wind Cave mortality sample where no butchery has occurred, are included within the Paleoindian cluster.

The relative frequencies of forelimb bones from sites within the Paleoindian and northern Plains Late Prehistoric clusters are compared in Figure 11 (of standardized MAU values derived from Table 2). The Late Prehistoric group (Figure 11b) is characterized by high frequencies of scapulae

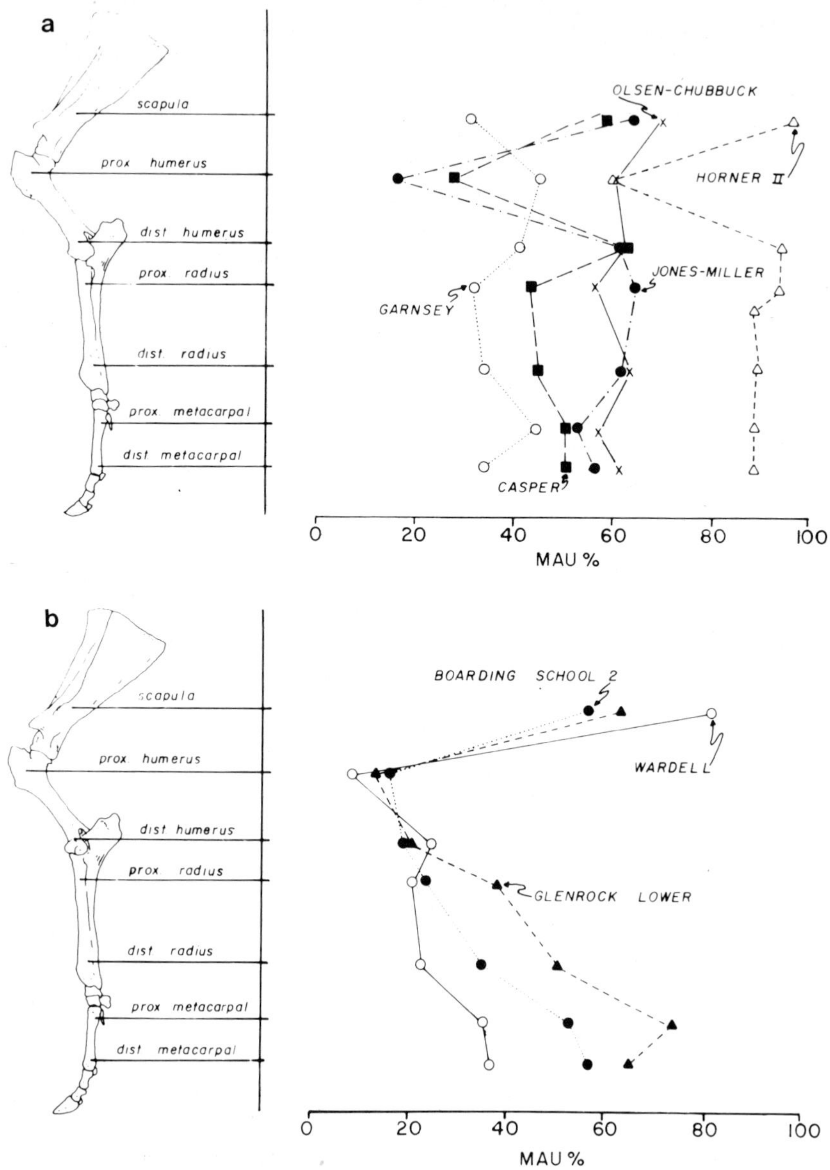

Figure 11. Relative frequencies of forelimb bones from (a) Paleoindian sites and the Garnsey Late Prehistoric site and (b) northern Plains Late Prehistoric sites.

and metacarpals and low frequencies of humeri and radii. The percentage values for all forelimb bones in the Paleoindian-Garnsey cluster (Figure 11a) are nearly equal. The exception to this equitable frequency distribution in the Paleoindian cases is the proximal humerus which is underrepresented in the Casper, Jones-Miller, and Horner II samples. As noted above, this is interpreted as the result of destruction by carnivores. The limited destruction at Olsen-Chubbuck may be a function of the narrow arroyo that contains the bonebed, which may have restricted carnivore access to the bones.

The patterns of element representations for the rearlimb also indicate that the sites in the Paleoindian-Garnsey cluster (Figure 12a) tend to have similar values for all bones of the leg. Garnsey exhibits a slight difference with a relative high frequency of innominates with lower values for all of the leg, which may again indicate disarticulation and removal of entire limb units. The Glenrock assemblage in the northern Plains Late Prehistoric group (Figure 12b) illustrates a classic pattern of segmental selection with high frequencies of innominates and lower legs and low frequencies of femora and tibiae. While the Boarding School 2 and Wardell samples do not exhibit as much segmentation as Glenrock, there are less uniform patterns of representation than in the Paleoindian cluster.

The counts for all bones from the sites included in Table 1 indicate that, in general, the Paleoindian cluster exhibits a high degree of similarity between the frequencies of most elements. In the case of Garnsey, a few element groups such as the crania and atlas and axis vertebrae clearly predominate, while most other element groups have very similar frequencies. At the Twilla site (Hughes 1977), another post-Paleoindian bison kill from the southern Plains, with anatomical part frequencies reported by maximum values per element rather than as proximal and distal counts (barring inclusion in this analysis), the pattern of equability for limb bones is high (as at Garnsey) and contrasts with the northern Plains Late Prehistoric sites. At Twilla, the counts for scapulae, humeri, radii, and metacarpals are 25, 28, 24, and 29 respectively (Hughes 1977:Table 3). Similar rearlimb values are also reported (innominates = 26; femora = 28; tibiae = 21; and metatarsals = 31). While the data are limited, it would thus appear that the bone frequencies at Garnsey

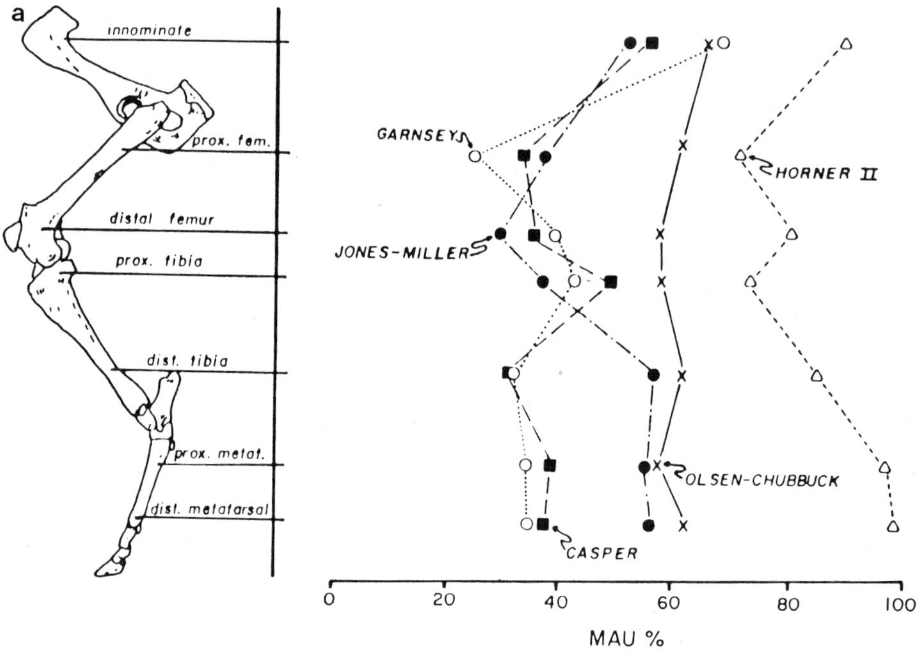

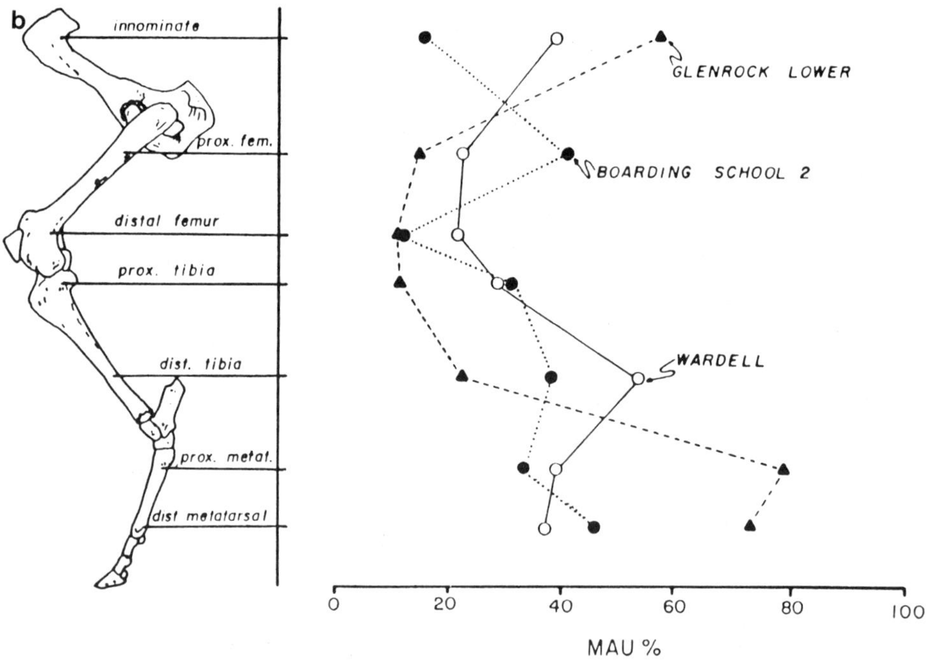

Figure 12. Relative frequencies of rearlimb bones from (a) Paleoindian sites and the Garnsey

may be indicative of a broader regional pattern of processing on the southern Plains that results in treatment of limbs as single units with little segmentation at the kill. On the other hand, all sites in the northern Late Prehistoric Plains cluster display marked differences of element frequencies of adjacent bone groups, often with more lower (metapodial and phalanges) than upper legs.

Thus, it seems that a major difference between the northern Plains cluster and the Paleoindian cluster is in the deletion of bones from the Paleoindian group (including Garnsey and Wind Cave) by complete limb units rather than by dismembered anatomical subsets. In the northern Plains Late Prehistoric group, there is a greater degree of segmentation of anatomical units apparently due to differences in element utility. Such segmentation may indicate a selection of specific resources associated with certain element groups. This selective segmentation may reflect the requirements of transport, scheduling, and food utility decisions. Complete limb removal could be controlled by transport or product-specificity constraints. Lessening of such constraints is possible when (1) a small number of animals is taken or a large labor force was available, (2) consumer groups move into the area of the kill or reside close to the kill, (3) products for consumption are not removed in terms of anticipated future needs (as with carnivore consumption or immediate consumption by humans), or (4) temporal (scheduling conflicts) or physical conditions (weather, carcass condition, etc.) precluded selection at the kill site.

It has been suggested (Kelly and Todd 1986) that the emerging pattern of faunal exploitation by Paleoindians is more consistent with a framework in which mass-kills served to insure the short-term predictability of one "resource patch" for immediate consumption, rather than as sources of food products for long-term storage. It is possible that locating a mobile resource may be more difficult during certain seasons when either bison locations are unpredictable or human mobility is restricted. The pattern of late fall to early winter kills during the Paleoindian Period (Frison 1982) may be a function of this seasonal difference (see also Fisher, this volume).

Figure 12 (*continued*) Late Prehistoric site and (b) northern Plains Late Prehistoric sites.

In summary, several points about the nature of the role of hunting based on data from bison bonebeds of western North America can be made.

(1) These sites are not fossilized "events;" even when buried shortly after abandonment by humans, a number of post-occupational processes may have modified the faunal remains. As the Horner II and Jones-Miller sites indicate, patterning within kill-butchery bonebeds is usually the result of interactions between a number of agents. While the lithic assemblages associated with such sites may have a high degree of "event/agent specificity," the bones need not. The nature of integrity and resolution (Binford 1981) of faunal and lithic assemblages within a single site may not be identical and investigation of these properties for both data sets is required. Analytical refitting of bifacial resharpening flakes and identification of spatially separated sequential episodes of individual tool rejuvenation, as conducted by Frison for the Casper bonebed (1974:95-101), give the impression of being able to develop event level interpretations of activities at the site. However, even at Casper, where fine-grained reconstructions of stone tool use are appropriate, the associated faunal remains may have been modified by processes which have not acted on the lithics. Additional formational studies of kill sites require a more detailed examination of the interrelationships between stone tool and faunal assemblages.

(2) Simply because Paleoindian sites are relatively recent, the role of human action in their formation cannot be assumed. As with the investigation of all archaeological faunal assemblages, whether from Plio-Pleistocene hominid sites or from modern ethnoarchaeological or taphonomic studies, interpretations must be analytically derived rather than observationally inferred. Variability in element fragmentation, disarticulation, dispersal, and deletion/accumulation or anatomical parts are attributes that can carry information on the formational history of a deposit. However, they can not be taken as direct indicators of intensity of human modification.

(3) Formational analyses can be used to expand our knowledge of human hunting behaviors beyond those represented in the ethnographic/ethnohistoric record. Hunting strategies are problem solving components of a subsistence system, and are variable through time and across space. Predation by fully modern humans, even of a single species such as bison at mass-kills, is not a unidimensional phenomena but is extremely flexible and responsive to changes in ecological community structure (Foley 1982, 1984; Reher 1977). Understanding dimensions influencing differences in faunal use decisions such as economic utility (Binford 1978, 1981, 1984) or seasonal variation in such utility (Speth 1983; Speth and Spielmann 1983) allows recognition of past strategies that are different from, as well as similar to, those documented through observations of living hunter-gatherer groups.

END NOTES

[1] There are a number of bones, particularly metatarsals, from the Jones-Miller bonebed that exhibit definite indications of impact fracturing. However, when viewed in the context of breakage of the assemblage as a whole, the metatarsals are among the least fragmented elements at the site. This may indicate that although breakage of bone by humans did occur, it represents only a limited source of bone modification.

[2] In cases where bone grease manufacture is a component of the processing, human modification can also result in nearly complete destruction (or at least reduction to unrecognizable small fragments) of articular ends of long bones; however, the common association of this type of activity with features and characteristic distributional patterns (Binford 1983) and fracture patterns (Todd, Fogarty, and Rapson 1986) makes the distinction between such deletion of articular ends and carnivore consumption possible.

ACKNOWLEDGEMENTS

John Speth, D. J. Rapson, and M. H. Nitecki commented on drafts of this paper. The analysis of the Jones-Miller materials was conducted under a post-doctoral fellowship at the Smithsonian. I would like to thank Dennis Stanford for allowing me to work with the Jones-Miller collection; the techniques that he developed for recording the Jones-Miller site make the collection the best documented kill-butchery assemblage, and I am very glad to have had the chance to work with it. George Frison allowed me to work with the Horner bison bones and has provided a series of comments on the interpretation of hunting practices throughout the development of the thoughts presented here.

REFERENCES

Albanese, J. 1987. Geological investigations. In: *The Horner Site: Type Site of the Cody Cultural Complex*, ed. G.C. Frison and L.C. Todd, in press. Orlando: Academic Press.

Agenbroad, L. D. 1978. *The Hudson-Meng Site: An Alberta Bison Kill in the Nebraska High Plains*. Washington, D.C.: University Press of America.

Barbour, E. H., and C. B. Schultz. 1932. The Scottsbluff Bison Quarry and its artifacts. *Nebraska State Museum Bulletin* 34(1).

Behrensmeyer, A. K. 1978. Taphonomic and ecological information from bone weathering. *Paleobiology* 4(2):150-162.

Binford, L. R. 1978. *Nunamiut Ethnoarchaeology*. New York: Academic Press.

Binford, L. R. 1981. *Bones: Ancient Men and Modern Myths*. New York: Academic Press.

Binford, L. R. 1983. *In Pursuit of the Past: Decoding the Archaeological Record*. New York: Thames and Hudson.

Binford, L. R. 1984. *Faunal Remains from Klasies River Mouth*. New York: Academic Press.

Binford, L. R. 1985. Human ancestors: Changing view of their behavior. *Journal of Anthropological Archaeology* 4:292-327.

Binford, L. R., and J. B. Bertram. 1977. Bone frequencies -- and attritional processes. In: *For Theory Building in Archaeology*, ed. L. R. Binford: 77-153. New York: Academic Press.

Binford, L. R., and N. M. Stone. 1986. Zhoukoudian: A closer look. *Current Anthropology* 27(5):453-468.

Bradley, B., and G. C. Frison. 1987. Projectile points and specialized bifaces from the Horner site. In: *The Horner Site: The Type Site of the Cody Cultural Complex*, ed. G. C. Frison and L. C. Todd, in press. Orlando: Academic Press.

Brain, C. K. 1981. *The Hunter or the Hunted? An Introduction to African Cave Taphonomy*. Chicago: University of Chicago Press.

Bunn, H. T., and E. Kroll 1986. Systematic butchery by Plio/Pleistocene hominids at Olduvai Gorge, Tanzania. *Current Anthropology* 25(5):431-442.

Cook, H. J. 1927. New geological and paleontological evidence bearing on the antiquity of mankind in America. *Natural History* 27(3):240-247.

Dawson, J., and D. Stanford. 1975. The Linger site: A re-investigation. *Southwestern Lore* 41(4):11-16.

Figgins, J. D. 1927. The antiquity of man in America. *Natural History* 27(3):229-239.

Foley, R. 1982. A reconsideration of the role of predation on large mammals in tropic hunter-gatherer adaptations. *Man* 17(3):393-402.

Foley, R. 1984. Early man and the Red Queen: Tropical african community evolution and hominid adaptation. In: *Hominid Evolution and Community Ecology: Prehistoric Human Adaptation in Biological Perspective*, ed. R. Foley: 85-110. London: Academic Press.

Frison, G. C. 1970. The Glenrock Buffalo Jump, 48CO304: Late Prehistoric Period Buffalo Procurement and Butchering. *Plains Anthropologist* Memoir 7.

Frison, G. C. 1973. The Wardell Buffalo Trap, 48SU301: Communal Procurement in the Upper Green River Basin, Wyoming. *Anthropological Papers of the Museum of Anthropology, University of Michigan* 48.

Frison, G. C., ed. 1974. *The Casper Site: A Hell Gap Bison Kill on the High Plains*. New York: Academic Press.

Frison, G. C. 1978. *Prehistoric Hunters of the High Plains*. New York: Academic Press.

Frison, G. C. 1982. Paleo-Indian winter subsistence strategies on the High Plains. In: *Plains Indian Studies: A Collection of Essays in Honor of John C. Ewers and Waldo R. Wedel*, ed. D. H. Ubelaker and H. J. Viola: 193-201. *Smithsonian Contributions to Anthropology* No. 30.

Frison, G. C. 1984. The Carter/Kerr-McGee Paleoindian site: Cultural resource management and archaeological research. *American Antiquity* 49:288-314.

Frison, G. C. 1987. The tool assemblage, unspecialized bifaces, and stone flaking material sources for the Horner site. In: *The Horner Site: The Type Site of the Cody Cultural Complex*, ed. G. C. Frison and L. C. Todd, in press. Orlando: Academic Press.

Frison, G. C., and D. Stanford. 1982. *The Agate Basin Site: A Record of Paleoindian Occupation of the Northwestern Plains*. New York: Academic Press.

Frison, G. C., and L. C. Todd. 1986. *The Colby Site: Taphonomy and Archaeology of a Clovis Kill in Northern Wyoming*. Albuquerque: University of New Mexico Press.

Frison, G. C., and L. C. Todd, eds. 1987. *The Horner Site: Type Site for the Cody Cultural Complex*. Orlando: Academic Press. In press.

Fulgham, T. n.d. Bison bone distributions at the Jones-Miller site. Unpublished manuscript on file at the Smithsonian Institution.

Fulgham T., and D. Stanford. 1982. The Frasca site: A preliminary report. *Southwestern Lore* 48(1):1-9.

Graham, R. n.d. Biometrics of the Jones-Miller *Bison* population and Holocene evolution of the genus *Bison*. Unpublished manuscript on file at the Smithsonian Institution.

Harpending, H., and A. Rogers. 1985. Antana: A package for multivariate data analysis. Unpublished manuscript in possession of the senior author.

Haynes, G. 1980a. Evidence of carnivore gnawing on Pleistocene and Recent mammalian bones. *Paleobiology* 6(3):341-351.

Haynes, G. 1980b. Prey bones and predators: Potential ecological information from analysis of bone sites. *Ossa* 7:75-97.

Haynes, G. 1981. *Bone modification and skeletal disturbances by natural agencies: Studies in North America*. Unpublished Ph.D. dissertation, Department of Anthropology, Catholic University of America, Washington, D.C.

Haynes, G. 1982. Utilization and skeletal disturbances of North American prey carcasses. *Arctic* 35(2):266-281.

Hill, A. P. 1979a. Butchery and natural disarticulation: An investigatory technique. *American Antiquity* 44(4):739-744.

Hill, A. P. 1979b. Disarticulation and scattering of mammal skeletons. *Paleobiology* 5(3):261-274.

Hill, A. P. 1984. Hyaenas and hominids: Taphonomy and hypothesis testing. In: *Hominid Evolution and Community Ecology: Prehistoric Human Adaptation in Biological Perspective*, ed. R. Foley, 111-128. London: Academic Press.

Hill, A. P. 1985 Natural disarticulation and *Bison* butchery. *American Antiquity* 50(1):141-145.

Hill, A. P., and A. K. Behrensmeyer. 1984. Disarticulation patterns of some modern East African mammals. *Paleobiology* 10(3):366-376.

Hughes, D. T. 1977. *Analysis of certain prehistoric bison kills in the Texas Panhandle and adjacent areas*. Unpublished Master's thesis, University of Arkansas.

Hurst, C. T. 1941. A Folsom location in the San Luis Valley, Colorado. *Southwestern Lore* 7(2):31-33.

Hurst, C. T. 1943. A Folsom site in a mountain valley of Colorado. *American Antiquity* 8(3):250-253.

Isaac, G. L., and D. C. Crader. 1981. To what extent were early hominids carnivorous? An archaeological perspective. In: *Omnivorous Primates: Gathering and Hunting in Human Evolution*, ed. R. S. O. Harding and G. Telecki, 37-103. New York: Columbia University Press.

Jepsen, G. L. 1953. Ancient buffalo hunters of northwestern Wyoming. *Southwestern Lore* 19(2):19-25.

Jodry, M. A. 1986. Stewart's Cattle Guard site: A Folsom campsite and butchering locality in south central Colorado. Paper presented at the 44th Plains Conference, Denver, Colorado.

Kehoe, T. F. 1967. The Boarding School Bison Drive Site. *Plains Anthropologist* Memoir 4.

Kelly, R. L., and L. C. Todd. 1986. Coming into the country: Early Paleoindian hunting and mobility. Manuscript submitted for publication.

McCartney, P. H. 1984. Taphonomy and archaeology of a Cody Complex bison kill in Colorado. Unpublished manuscript on file at the Smithsonian Institution.

Moss, J., in collaboration with: K. Bryan, G. W. Holmes, L. Satterthwaite, Jr., H. P. Hanson, C. B. Schultz, and W. D. Frankforter. 1951. Early Man in the Eden Valley. *University of Pennsylvania Museum Monographs* 6.

Mulloy, W. T. 1959. The James Allen site near Laramie, Wyoming. *American Antiquity* 25(1):112-116.

Peterson, O.A. 1911. A mounted skeleton of *Stenomylus hitchcocki*, the Stenomylus Quarry, and remarks upon the affinities of the genus. *Annals of the Carnegie Museum* 7(2):267-273.

Rancier, J., G. Haynes, and D. Stanford. 1982. 1981 investigations at Lamb Spring. *Southwestern Lore* 48(2):1-17.

Reher, C. A. n.d. Archaeological and paleoecological analysis of bison mandibles from the Jones-Miller site. Unpublished manuscript on file at the Smithsonian Institution.

Reher, C. A. 1973. Appendix II: The Wardell *Bison bison* sample: Population dynamics and archaeological interpretation. In: *The Wardell Buffalo Trap, 48SH301: Communal Procurement in the Upper Green River Basin, Wyoming*, G. C. Frison. *Anthropological Papers of the Museum of Anthropology, University of Michigan* 48:89-105.

Reher, C. A. 1977. Adaptive process on the shortgrass Plains. In: *For Theory Building in Archaeology: Essays on Faunal Remains, Aquatic Resources, Spatial Analysis, and Systemic Modeling*, ed. L. R. Binford: 13-40. New York: Academic Press.

Richardson, P. R. K. 1980. Carnivore damage on antelope bones and its archaeological implications. *Paleontologia Africana* 23:109-125.

Speth, J. D. 1980. Late Prehistoric bison procurement in southeastern New Mexico: The 1978 season at the Garnsey site (LA-18399). *Museum of Anthropology, University of Michigan Technical Report* 12.

Speth, J. D. 1983. *Bison Kills and Bone Counts: Decision Making by Ancient Hunters*. Chicago: University of Chicago Press.

Speth, J. D., and W. J. Parry. 1978. Late Prehistoric bison procurement in southeastern New Mexico: The 1977 season at the Garnsey Site. *Museum of Anthropology, University of Michigan Technical Report* 8.

Speth, J. D., and K. A. Spielmann. 1983. Energy sources, protein metabolism, and hunter-gatherer subsistence strategies. *Journal of Anthropological Archaeology* 2:1-31.

Stanford, D. 1974. Preliminary report on the excavation of the Jones-Miller Hell Gap site, Yuma County, Colorado. *Southwestern Lore* 40:30-36.

Stanford, D. 1978. The Jones-Miller site: An example of Hell Gap bison procurement strategy. In: *Bison Procurement and Utilization: A Symposium*, ed. L. B. Davis and M. Wilson: 90-97. *Plains Anthropologist Memoir* 14.

Stanford, D. 1979. Bison kill by ice age hunters. *National Geographic* 155(1):114-121.

Stanford, D. 1984. The Jones-Miller site: A study of Hell Gap bison procurement and processing. *National Geographic Research Reports: 1975 Projects* Vol. 16:615-635.

Stanford, D., W. R. Wedel, and G. R. Scott. 1981. Archaeological investigations of the Lamb Spring site. *Southwestern Lore* 47(1):14-27.

Todd, L. C. 1983. The Horner site: Taphonomy of an early Holocene bison bonebed. Unpublished Ph.D. Dissertation, Department of Anthropology, University of New Mexico, Albuquerque.

Todd, L. C. 1984. Investigation of activity structure in kill-butchery sites: The Jones-Miller site. Paper presented at the 42nd Plains Conference, Lincoln, Nebraska.

Todd, L. C. 1986. Determination of sex of bison upper forelimb bones: The humerus and radius. *The Wyoming Archaeologist* 29(1-3):109-124.

Todd, L. C. 1987. Taphonomy of the Horner II bonebed. In: *The Horner Site: The Type Site of the Cody Cultural Complex*, ed. G. C. Frison and L. C. Todd, in press. Orlando: Academic Press.

Todd, L. C., G. R. Burgett, and R. W. Klukas. 1986. Analysis of large mammal skeletal disarticulation and dispersal in temperate settings. Unpublished manuscript in possession of the authors.

Todd, L. C., and G. C. Frison. 1986. Reassembly of bison skeletons from the Horner site: A study in anatomical refitting. In: *Piecing Together the Past: Application of Refitting Studies in Archaeology*, ed. J. L. Hofman, in press. London: B.A.R. International Series.

Todd, L. C., M. E. Fogarty, and D. J. Rapson. 1986. Long bone fragmentation and interpretation of faunal assemblages: Approaches to Comparative analysis. Paper presented at the 44th Plains Conference, Denver Colorado.

Todd, L. C., and J. L. Hofman. 1978. A study of bison mandibles from the Horner and Finley sites: Two Paleoindian bison kills in Wyoming. In: *Wyoming Contributions to Anthropology, Volume 1. University of Wyoming Publications* 42(1):67-104.

Todd, L. C., and J. L. Hofman. 1987. Appendix 8: Bison mandibles from the Horner and Finley sites. In: *The Horner Site: Type Site of the Cody Cultural Complex*, ed. G. C. Frison and L. C. Todd, in press. Orlando: Academic Press.

Todd, L. C., and D. J. Stanford. 1986. Application of conjoined bone data to site structural studies. In: *Piecing Together the Past: Application of Refitting Studies in Archaeology*, ed. J. L. Hofman, in press. London: B.A.R. International Series.

Todd, L. C., R. V. Witter, and G. C. Frison. 1987. Excavation and documentation of the Princeton and Smithsonian Horner site assemblages. In: *The Horner Site: Type Site of the Cody Cultural Complex*, ed. G. C. Frison and L. C. Todd, in press. Orlando: Academic Press.

Walker, D. N. 1987. Horner site mammalian local fauna. In: *The Horner site: Type site of the Cody Cultural Complex*, ed. G. C. Frison and L. C. Todd, in press. Orlando: Academic Press.

Wedel, W. R. 1961. *Prehistoric man on the Great Plains*. Norman: University of Oklahoma Press.

Wheat, J. B. 1971. Lifeways of early man in North America. *Arctic Anthropology*. 8(2):22-31.

Wheat, J. B. 1972 The Olsen-Chubbuck site: a Paleo-Indian bison kill. *Society for American Archaeology Memoir* 26.

Wheat, J. B. 1978. Olsen-Chubbuck and Jurgens sites: four aspects of Paleo-Indian economy. In: *Bison Procurement and Utilization: A Symposium*, ed. L. B. Davis and M. Wilson. *Plains Anthropologist Memoir* 14:84-89.

Wheat, J. B. 1979. The Jurgens site. *Plains Anthropologist Memoir* 15.

White, T. E. 1953. Observations on the butchering techniques of some aboriginal peoples, No. 2. *American Antiquity* 19:160-164.

White, T. E. 1954. Observations on the butchering techniques of some aboriginal peoples, No. 3, 4, and 5. *American Antiquity* 19:254-262.

Wilson, G. L. 1924. The horse and dog in Hidatsa culture. *Anthropological Papers of the American Museum of Natural History* 15 (pt. 2).

Wormington, H. M. 1957. Ancient Man in North America. *Denver Museum of Natural History, Popular Series* 4.

The Pleistocene Archaeology of Beringia

Richard E. Morlan

Archaeological Survey of Canada
Canadian Museum of Civilization
Ottawa, Ontario

After more than a century of discovery and debate and nearly two centuries after Thomas Jefferson's speculation that North American Indians and northern Asiatics must have had a common origin (see Wilmsen 1965), we still do not know when man entered the New World. Two hundred thousand years or more are claims for greatest antiquity (Budinger 1983; Irving 1985; Steen-McIntyre et al. 1981), but neither these nor the more modest claims of others (Bryan 1978; MacNeish 1976, 1978, 1979; Morlan 1980) have been accepted in critical reviews (Dincauze 1984; Owen 1984; Waters 1985). There is widespread evidence for human occupation after 12,000 years ago, and new South American data imply that people reached North America at an earlier time (Bryan 1986).

Although the time of human colonization of the New World is not known, the most likely place of entry can be specified. Large and small mammals of many species have repeatedly moved through the region around Bering Strait, between Alaska and Siberia, during the past several million years (Harington 1980a; Repenning 1980), and people probably used this route while pursuing the game on which their lives depended. The arrival of people in northwestern North America was probably not a result of an arrow-like purposeful migration. Giddings (1954: 88) described a more likely pace and manner of colonization when he wrote: "...a very sparse Arctic population slowly spread over the belt of northern climate, neither pursuing nor evading pursuit, but simply existing and adjusting at random to the environment, the sons sometimes hunting beyond the range of their fathers but never really leaving home."

During Pleistocene glaciations, lowered sea levels exposed the floors of the Bering and Chukchi Seas and created a broad link between Siberia and Alaska, a vast lowland area commonly known as the Bering Land Bridge. The climate was too dry to permit glacierization except in major mountain ranges, so an ice-free area stretched from the Yukon Territory westward across most of northern Siberia. This region is known as Beringia. For phytogeographic purposes, Yurtsev (1984a) defines Beringia as extending from the Kolyma River on the west to the Mackenzie River on the east, but the distribution of known archaeological sites necessitates coverage of a larger zone known as "Megaberingia" (Yurtsev 1984b:265). This zone extends to the Lena River on the west, includes Kamchatka on the south, and is defined by the late Wisconsinan limits of Laurentide and Cordilleran ice on the east (see Prest 1969). I will refer to the ice-free areas of Alaska and the Yukon as eastern Beringia and to the Siberian areas as western Beringia (see Fig. 1).

Judging from paleontological data (Guthrie 1982, 1984a, 1985a; Harington 1977, 1978, 1980b; Matthews 1982), Pleistocene Beringia was a land of opportunity for hunting societies, but the opportunity could not be seized until certain technological accomplishments had been achieved. There is no evidence for extensive fishing or maritime hunting and gathering prior to the Holocene, and relatively few edible plants are found in Beringia. Human residents of Pleistocene Beringia must have been expert hunters with specialized weaponry (see Workman 1980). Other requirements must have included tailored skin clothing, adequate winter shelter, expert production and control of fire, and skills in food storage for times of scarcity (Irving 1978a: 94; Fladmark 1983: 39). Many northward flowing, extremely cold rivers had to be crossed to reach Bering Strait, either on winter ice that could have been locally treacherous or in watercraft. In long hours of winter twilight and darkness, it is hard to imagine that northern people, if only for psychological reasons, would live for long without a means of lighting as well as heating their dwellings.

Some paleontologists and archaeologists have viewed Beringia merely as a

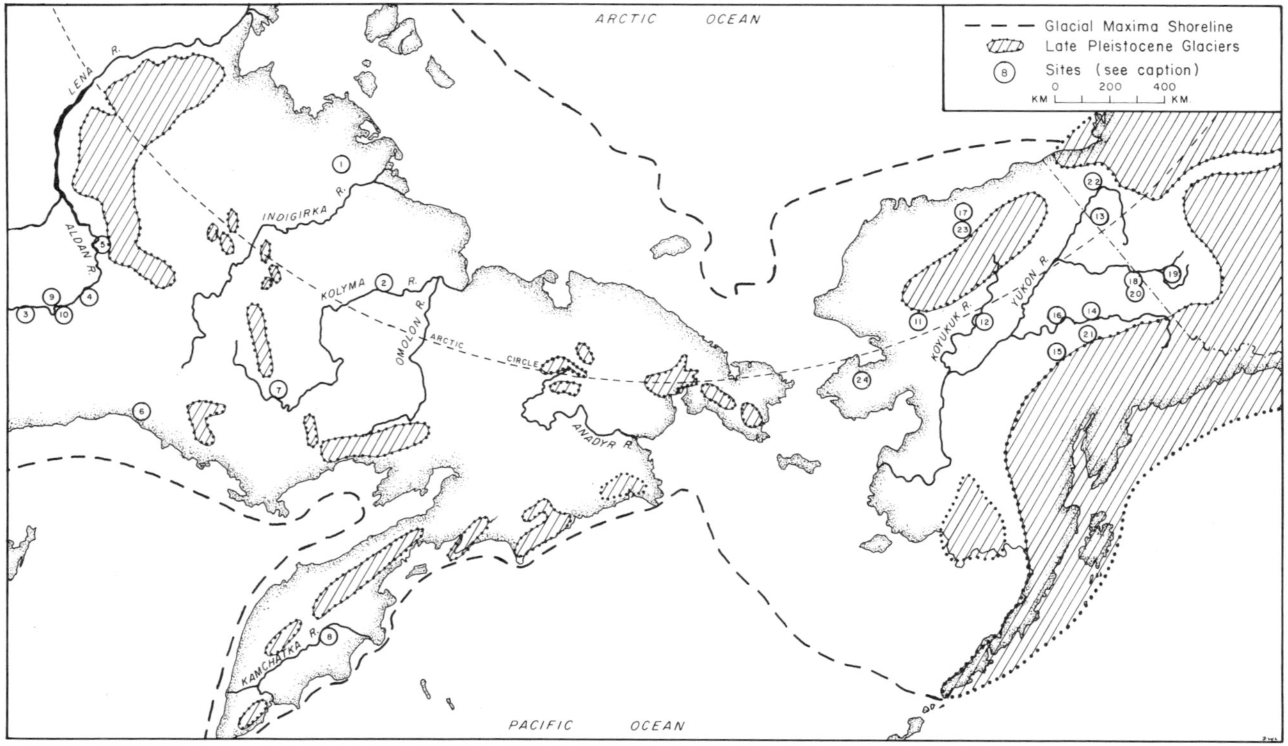

Figure 1. Beringia at the height of the last glaciation. Shoreline locations and glacial ice distribution after Hamilton and Thorson (1983), Hopkins (1972), Prest (1969), and Velichko et al. (1984). Western Beringian sites: 1. Berelekh; 2. Bochanut; 3. Dyuktai Cave; 4. Ezhantsy; 5. Ikhine; 6. Kukhtui; 7. Maiorych; 8. Ushki; 9. Ust'-Mil'; 10. Verkhne Troitskaya. Eastern Beringian sites: 11. Akmak; 12. Batza Tena; 13. Bluefish caves; 14. Chindadn; 15. Dry Creek: 16. Fairbanks; 17. Gallagher; 18. Jack Wade; 19. Klondike; 20. Lost Chicken; 21. Mt. Hayes; 22. Old Crow Basin; 23. Putu; 24. Trail Creek.

highway linking the Old World with the New and have generally ignored its role as a center of biological and cultural evolution. This is partially because until recently there was little evidence concerning the ancient environments of Beringia and none concerning its Pleistocene archaeology. In the absence of direct evidence, intercontinental migrations of mammals, including humans, were inferred from sites outside Beringia (e.g., various papers in Hopkins 1967). Field work during the past twenty years in both eastern and western Beringia has advanced the knowledge of Pleistocene geology, paleontology and archaeology (Hopkins et al. 1982; Kontrimavichus 1984; West 1981a), but our understanding is still in its infancy.

A chrono-stratigraphic framework proposed by Hopkins (1982) integrates various local sequences into a series of named intervals that permit Beringian researchers to describe events and attempt correlations without using nomenclature devised for distant regions (see also Schweger et al. 1982; Hamilton and Thorson 1983; Tseitlin 1979; Giterman et al. 1982). This framework spans the past 80,000 years during which three major Pleistocene intervals are recognized (Fig. 2). The Happy Interval, more than 50,000 years old, represents the earlier, and by far the larger, of two glaciations that occurred in Beringia since the Sangamonian or Kazantzev interglaciation. It is the Beringian manifestation of isotope stage 4 (Shackleton and Opdyke 1973) and corresponds approximately to the Early Wisconsinan and Zyriankan glaciations. Warmer temperatures interspersed with several cool oscillations characterize the Boutellier Interval which is equivalent to isotope stage 3, the Karginsk Interstadial complex of western Beringia and the mid-Wisconsinan of central North America. Isotope stage 2 is represented by the Duvanny Yar Interval, a period of cold climate beginning around 25,000 years ago, corresponding to the Sartan glaciation in western Beringia and the late Wisconsinan in mid-continental North America. Climatic amelioration and gradual inundation of the Bering Land Bridge brought about a relatively rapid change in vegetation as seen in the pollen record by a sudden rise in birch around 14,000 years ago.

Archaeological data will be summarized with reference to this framework, beginning in western Beringia and moving eastward. Each of the sites will be

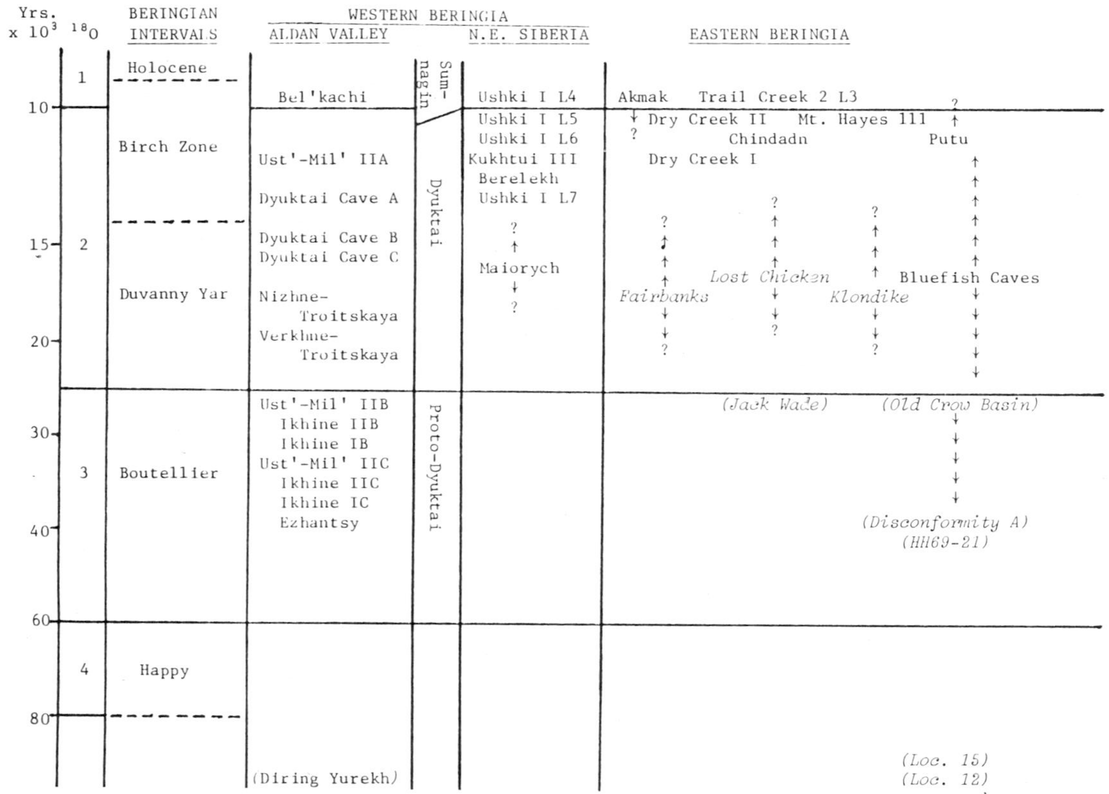

Figure 2. Principal Beringian sites and localities in relation to thousands of years B.P., oxygen isotope stages identified by Shackleton and Opdyke (1973), and chrono-stratigraphic intervals defined by Hopkins (1982). Cultural traditions (Sumnagin, Dyuktai, and Proto-Dyuktai) and chronology of western Beringian after Mochanov (1977: Table D); for eastern Beringian sites, see text and Table 2. Locations yeilding only redeposited materials are entered in italics. Parentheses enclose locations of uncertain artifact identification.

presented according to the views of the principal investigator, followed, when necessary, by a discussion of conflicting views and/or problems that remain to be solved. In addition to relatively secure archaeological data there are claims that human occupations in both western and eastern Beringia are of great antiquity. Should these claims be substantiated, they would indicate that specialized hunting and the other skills that are prerequisite to human life in Beringia were developed much earlier than now perceived.

THE EARLIEST BERINGIAN CLAIMS

The date of the earliest western Beringian claim of between one and two million years ago has appeared in newspapers (*The Arizona Republic* 26 Oct. 1984; *The Guardian* 26 Oct. 1984; *The Mammoth Trumpet* 1(2), 1984). Over 1000 artifacts, comprising a chopper tool industry, have been recovered by Y.A. Mochanov from the Diring Yurekh site on the 100 m terrace of the Lena River system. Considerably more information is needed to determine the age and possible technological status of this find.

The earliest claimed eastern Beringian date results from work by W.N. Irving and his colleagues, along Old Crow River, northern Yukon Territory. Fossil bone and tusk fragments from Old Crow River Locality 12, and other nearby sections, are interpreted as artifacts on the basis of fracture patterns and polishing (Irving 1982; Jopling et al. 1981; Julig et al. 1983). Some of the bones have been dated by the uranium series and are believed to be older than 350,000 years (Irving 1985: 549). I have previously reported altered bones as artifacts and possible artifacts from the so-called midsection deposit and from an erosional contact known as Disconformity A at Old Crow River Locality 15 (Morlan 1980, 1983a; Morlan and Cinq-Mars 1982), but subsequent studies suggested alternate explanations (Morlan 1983b, 1984a, 1986a).

When first reported (Morlan 1979, 1980; Morlan and Matthews 1978), Disconformity A was thought to be perhaps 60,000 years old based on interpolation between overlying radiocarbon dates of about 41,000 years B.P. and underlying Old Crow tephra with a preliminary fission track date of less than 80,000 years B.P. (Briggs and Westgate 1978). New radiocarbon dates on bones from Disconformity A show that the erosional event may be somewhat younger than 40,000 years B.P. (Morlan 1986b), while additional analysis of Old Crow tephra and associated fossils suggests the age of the ash to be between 87,000 and 105,000 years old (Schweger and Matthews 1984; Westgate et al. 1985). If some of the bones from Disconformity A are indeed artifacts, they may be less than 40,000 years old, but I do not believe that any of them can be proven to represent human activity (Morlan 1986a). Even a cutmarked bison rib from a deposit closely associated with Disconformity A at Locality HH69-21 (*Mammoth Trumpet* 1(1), 1984) cannot be unconditionally accepted as a butchered bone in view of recent discoveries that such marks are closely mimicked by animal trampling (Fiorillo 1984; Andrews and Cook 1985).

WESTERN BERINGIA: THE BOUTELLIER INTERVAL

According to Mochanov (1977, 1978a, 1980, 1984), human populations were well established in the Aldan valley in western Beringia about 35,000 years ago. The Ezhantsy site and various components of Ikhine I, Ikhine II, and Ust'-Mil' II together comprise the proto-Dyuktai industry. Several recent reviews of these sites in English (Michael 1984; Dolitsky 1985; Yi and Clark 1985) have focused on stratigraphy and chronology and have provided little information on the artifacts themselves. I will enumerate, from published illustrations, some of the artifact types; the associated faunal remains are listed in Table 1.

Although not directly dated, Ezhantsy is regarded as the oldest component because its artifacts were found at the bottom of the floodplain alluvium on Terrace III with some tools partly embedded in the underlying gravel. Most other components are underlain by floodplain alluvium and by gravel. Among the 622 stone tools recovered at Ezhantsy (Mochanov 1977: 49-58) are diabase chopper-like pebble cores, a variety of chert cores, numerous burins on flakes (70% of the assemblage), microblades, burin spalls, core rejuvenation flakes, and a broken oval biface. The last tool may have been designed as a wedge-shaped core by a procedure similar to the Yubetsu

Table 1. Faunal remains from components of the Dyuktai Tradition in northeastern Siberia (after Mochanov 1977: Table G). Frequencies refer to the number of identified specimens. * interpreted as a tool; ** not listed in table G, but mentioned on page 48. Ezh., Ezhantzy; Ber., Berelekh; VT, Verkhne Troitskaya.

	Proto-Dyuktai									Dyuktai							
		Ikhine I		Ikhine II				Ust'-Mil'			Dyuktai Cave					Ust'-Mil'	
	Ezh.	IC	IB	IIG	IIV	IIB	IIA	IIC	IIB	VT	IX	VIII	VIIv	VIIb	VIIa	IIA	Ber.
Large Herbivores																	
Mammoth	8	1*	3	9		4	7	2	7	8	132	621	7	4	24	4	78
Horse	20	2	4	43	14	50	71	7		9	2		9	4	1	9	
Bison	8	3	8	29	14	58	50	5		23	6	3	3	1	2	3	3?
Reindeer	10		5	10	2	6	16			2	4	3	2	1	32	3	1
Moose		1					1					5		3	33		
Woolly rhino	9		1	1	2	5	4	1	3	3							7
Bighorn sheep												3		6?	4		
Musk ox												1				1?	
Red deer				1?													
Small Herbivores																	
Hare											66	61	20	11	38		827
Rodent, indet.											21	38	191	32	39		
Collared lemming												1	17				
Brown lemming											1	3	7				
Carnivores																	
Arctic fox						1					6	30	1	12	12		
Fox						1					1	3		5	9		
Wolf				1						1		4			4		
Cave lion												6		3			
Birds											23	24	4		2		92
Fishes						2					3	7			2		2
Unidentified	360	3	9	33**	32	75	105	8	4	3	1081	2958	2309	545	1429	7	10
Totals	415	10	30	127	64	202	254	23	14	49	1346	3771	2570	627	1631	27	1020

technique of northern Japan (see Morlan 1978a), although Mochanov interprets it as an oval biface knife. With these stone tools was an antler billet tentatively identified as from the red deer (*Cervus elaphus*) (Mochanov 1977:58).

Ikhine I and Ikhine II are located on opposite sides of a small creek on Aldan Terrace III (Mochanov 1977: 39-49). There are differing views on the age and number of cultural levels at Ikhine I. Tseitlin (1979: 231-234) mentions only two cultural levels, estimates the age of the lower at 12,000 to 13,000 years old, and argues that a woolly rhinoceros bone in this level must be intrusive since this taxon is not known elsewhere after 18,000 years B.P. (Tseitlin 1979: 237). Tseitlin's "lower" level is Mochanov's Layer II (1977) or Horizon B (1978a), and Mochanov assigns it to the Lipovsko-Novoselov warm phase of the Karginsk Interstadial (22,000 - 30,000 years ago). Only twenty stone tools were found in Horizon B at Ikhine I, and they include pebble tools, a scraper, a notched flake, a wedge-shaped core reminiscent of the Yubetsu technique, and several flakes.

Beneath Ikhine IB is a third cultural manifestation (Ikhine IC) not mentioned by Tseitlin (1979) but said by Mochanov (1977:44) to be "of great significance for elucidating the time of man's appearance in Northeast Asia" (translation mine). The humerus of a young mammoth was artificially fractured and flaked on one long side and one transverse end was associated with three bison bones, two horse bones, a moose bone and three unidentifiable bone fragments (Mochanov 1977: Fig. 12-3). According to Mochanov, the transverse end of the humerus exhibits traces of wear, including scratches and slight smoothing as if it were used as a scraper or polisher. Like the finds at Ezhantsy, the Ikhine IC (or Layer III) materials lay at the base of the floodplain alluvium, at the contact with the underlying gravel.

At Ikhine II, five stratigraphic levels within the floodplain alluvium on Terrace III were assigned to two cultural layers. The lowest level, IID, did not yield tools, but bones of mammoth, bison, horse and reindeer were found. The next level, IIG, comprises the oldest cultural layer (Ikhine IIC) and is represented by only two pebble choppers or scrapers and a worked mammoth rib that Mochanov (1977: 49) speculates might have been part of a bow drill.

The level has a radiocarbon date of 27,800 ±500 B.P. (IM-206).

Cultural layer B at Ikhine II consists of three stratigraphic levels (IIa, IIb, and IIv) in which eleven, six and two stone tools were found, respectively. The tools include pebble choppers, flakes and microblades. Five radiocarbon dates on wood from level IIb range from 24,300 to 30,200 B.P. Two dates on the middle part of level IIv are 26,600 ±900 (IM-201) and 31,200 ±500 (GIN-1020). Two dates on the lower part of Level IIv are 26,030 ±200 (IM-239) and 26,500 ±540 (IM-202). It is obvious that some of these dates are out of sequence, even if their two sigma limits are considered. Tseitlin (1979) prefers the older dates so that the predominantly "cold" fauna corresponds to the Konoshchel' cold period of the Karginsk Interstadial, and he considers that the younger dates might have been contaminated. A more parsimonious explanation, as noted by Schweger et al. (1982: 438; also Hopkins 1985), is that the older dates were obtained on wood that had been redeposited -- a problem often encountered in northern permafrost areas (see Hughes et al. 1981). This interpretation suggests an age range of between 24,000 and 27,500 years ago for Ikhine IIB and IIC, and the "cold" fauna could be attributed to the cooling trend that marks the transition from Boutellier to Duvanny Yar time. I do not find the evidence for cool conditions especially persuasive, because the fauna resembles all other proto-Dyuktai assemblages except for the single specimen of Arctic fox in Level IIb (Table 1).

Ust'-Mil' II has produced two proto-Dyuktai levels, B and C (Mochanov 1977: 34-39). Level C yielded 13 stone artifacts among which are pebble cores, bifacially worked pebbles resembling biface preforms, and a wedge-shaped core reminiscent of the Horoka technique of northern Japan (Mochanov 1977: Fig. 9-3; see Morlan 1978a). Semenov reported that the epiphysis of a mammoth bone from this level had been chopped off (in Mochanov 1977: Fig. 9-8). Radiocarbon dates range from 30,000 to 35,000 years ago. Level B, dated to 23,500 ±500 B.P. (LE-999), yielded only seven chert flakes. Other artifacts from Ust'-Mil' II illustrated by Mochanov (1977: Fig. 10) apparently were recovered from Level A (12,200 ±170 B.P.) or from the surface of the site (Ackerman 1985a: 401).

According to Mochanov (1977, 1978a, 1980), the proto-Dyuktai industry is older than the Dyuktai industry, and the former differs from the latter in having a larger number and variety of pebble tools and cores, relatively few bifaces, and apparently no finished bifaces that might be construed as projectile points. A developmental sequence within the Dyuktai tradition is implied by the definition of a proto-Dyuktai industry as an antecedent to Dyuktai *sensu strictu*, and it is interesting that none of the recent commentators (e.g., Michael 1984; Dolitsky 1985, with comments; Yi and Clark 1985, with comments) mentioned this. Most of the discussion has focused on small sample size, lack of standardized terminology, questionable radiocarbon dates, possible cryoturbation of deposits, and absence of detailed site reports. As Hopkins (1985) notes, only Ezhantsy appears to represent a living floor, and this site is dated only by means of "Tseitlin's (1979) Siberia-wide terrace correlations"; Ust'-Mil' and Ikhine "appear to consist of detrital accumulations of wood, fossil bones, and artifacts enclosed in alluvium" (Hopkins 1985: 372). In addition to redeposition, Ikhine and Ust'-Mil' "have been subjected to erosion, solifluction, and tectonic and, most important, cryogenic disturbances...," raising the possibility "that the small numbers of artifacts reported in the 'early' strata were introduced to these levels by geological agencies" (Yi and Clark 1985: 10). They conclude that "evidence for a Siberian Upper Paleolithic industry predating 20,000 B.P. is practically nonexistent" (Yi and Clark 1985: 13).

Further work at Ezhantsy, and more new sites, would seem to hold the key to an assessment of the proto-Dyuktai industry and its chronological position. Ezhantsy is a relatively rich site with 622 artifacts and 415 faunal specimens recovered from 90 square meters. All the identified faunal remains represent large mammals, but no data on anatomical parts or distributions within the site have been published. If these animals were hunted and killed by the Ezhantsy occupants, some kind of weaponry must have been required. The large number of burins, associated with microblades and possible microcores suggests that organic weapon tips armed with microblade insets may have been prepared at Ezhantsy. I shall return to this later.

WESTERN BERINGIA: THE DUVANNY YAR INTERVAL

Sites on Terrace II of the Aldan River represent Dyuktai *sensu strictu* and have been dated or assigned to various ages within the Duvanny Yar Interval (Mochanov 1977; Tseitlin 1979). Probably the oldest known Dyuktai sites are Verkhne Troitskaya and Nizhne Troitskaya in which cultural remains were recovered from the lower portions of the floodplain alluvium on Terrace II. Radiocarbon dates ranging from 14,530 to 17,680 are in correct stratigraphic order just above the Verkhne Troitskaya assemblage, and a date of 18,300 ± 180 B.P. (LE-905) is from one of the cultural levels. Fifty-two stone artifacts and one bone awl or needle were found with 49 animal bones (Table 1; Mochanov 1977: 58-69). Nearly the entire artifact inventory is illustrated by Mochanov (1977: Fig. 19-21), and it includes end and side scrapers, wedge-like flakes, wedge-shaped cores and microblades, at least one tabular core, a few irregular cores, a pebble chopper, several transverse burins reminiscent of the Araya burin of northern Japan and the Anangula burin of southwestern Alaska (Mochanov 1977: Fig. 21b, 2-3), and a variety of well-made more-or-less asymmetrical bifaces that may have functioned as knives, scrapers, or, in a few cases, points.

Mochanov (1977: 32-34; 1978a: 62) includes the open air site of Ust'-Dyuktai I among the Dyuktai sites, because some of the 397 stone artifacts resemble those from Dyuktai Cave VIIv. This assignment is questionable, however, because the artifacts, including one sherd, were found only 10 to 25 cm below the turf on Aldan Terrace IV, and the site was not directly dated. It is interesting that Tseitlin (1979) does not mention this site or two others, Tumulur and Leten Novyi, that Mochanov (1977: 69-76) includes in the Dyuktai tradition on typological grounds (see Michael 1984: 17-18).

The last site in the Aldan River sequence to be mentioned is Dyuktai Cave, the type site of the Dyuktai tradition. Stratigraphic nomenclature for this site is somewhat confusing with three cultural horizons (A-C) recognized in stratigraphic levels VIIa, VIIb, VIIv, VIII, and IX (Mochanov 1977, 1978a). In an earlier report (Mochanov 1970; see also Powers 1973), the sequence was described in terms of six stratigraphic levels containing 14 cultural levels.

Tseitlin (1979: 224) includes all these levels in his Layer 3 and states that neither artifact typology nor faunal assemblage composition requires subdivision of the deposit. Perhaps this view is influenced by the fact that radiocarbon dates ranging from 12,100 to 14,000 years ago are not entirely in stratigraphic order.

Judging from Mochanov's (1977: Figs. 1-7) illustrations, cultural layer C from stratigraphic level IX, the oldest, contains large retouched flakes, a side scraper, and a possible projectile point base. Cultural layer B, in levels VIIb and VIII, contains bifacial points and knives, a bifacial chopper, unifacially trimmed scrapers, a notched flake, wedge-shaped cores and microblades, and a retouched mammoth tusk flake interpreted as a lanceolate point. From cultural layer A, in level VIIa, there are flaked pebbles, tabular cores, a core rejuvenation flake, preforms for wedge-shaped cores, finished wedge-shaped cores, microblades, a complete lanceolate point, a grooved limb bone, and an antler billet.

It is clear, even from these selective and possibly biased glimpses of the Dyuktai tradition, that excellent bifacial workmanship developed along with a continuing emphasis on microblade technology and at least casual reliance on a variety of pebble tools. Technologically similar assemblages have been found outside the Aldan valley in northeastern Siberia in sites that Mochanov (1977) assigns to the Dyuktai tradition even when they do not always contain the association of bifaces and microblades. The Maiorych site on the Kolyma River, for example, yielded a small assemblage of retouched flakes, a retouched blade, and a wedge-shaped core thought to represent a Dyuktai occupation dated to approximately 16,000 years ago on typological grounds (Mochanov 1977: 90-93; Powers 1973: 77). On the coastal plain of the Okhotsk Sea, the Kukhtui III site yielded a series of flake tools, a retouched blade, ovate bifacial knives, an adz-like tool, and a delicately flaked straight-based projectile point but no microblades (Mochanov 1977: 87-90); it is interpreted as a Dyuktai occupation perhaps 13,000 years old.

Both microblades and bifacial points are found in Layers 6 and 7 of Ushki I on the Kamchatka Peninsula (Dikov 1979). The wedge-shaped cores in Layer 7 are said to be underdeveloped, and the most outstanding single category of tools in this 14,000 year old site is a series of stemmed points, some of which appear nearly side-notched (Dikov 1979: Figs. 3-6). Surface dwellings, consisting of adjacent huts sharing unlined hearths, and a variety of beads, pendants and other goods in a burial pit make this a unique site for this time level. I agree with Powers that to include Layer 7 in the Dyuktai tradition, as Mochanov (1977) does, "increases the variability of the components making up this culture to such a degree that the concept is not operationally useful" (Powers 1973: 92). Layer 6 at Ushki fits much more comfortably into the Dyuktai concept and demonstrates that Dyuktai reached a very wide distribution in northeast Asia (Dikov and Titov 1984).

Apparently Dyuktai reached its northern limit only 150 km above the mouth of the Indigirka River at 71 degrees north latitude. This is the Berelekh site (Mochanov 1977: 76-87), where a so-called mammoth cemetery is situated just downstream from a Paleolithic occupation site. At least 140 individual mammoths are represented by the more than 8000 bones collected from the cemetery (Vereshchagin 1977), but only 79 mammoth bones representing two individuals were found in the archaeological deposit where the blue hare (*Lepus tanaiticus*) was the most abundant species (850 bones/28 individuals). The people at Berelekh could have utilized a large amount of mammoth meat brought to the site after removal from the bones. The artifacts from Berelekh include microblades, wedge-shaped cores, bifacial points and knives, and drilled pendants, some of which are made on jade pebbles. In addition, there is a series of flaked mammoth bone and tusk artifacts, some with cut marks, and residents of the modern settlement of Berelekh had found at the site a fragment of tusk bearing an engraving of a mammoth (Bader and Flint 1977). A remarkable ivory thrusting spear was found washed out of the site on the riverbank (Vereshchagin 1974: 10; 1977: 40, Fig. 32).

Even assuming exclusion of Ushki I Layer 7, it is not yet clear whether all these manifestations should be included in the single tradition known as Dyuktai. Mochanov (1984: 717) suggests that the Dyuktai tradition might be represented by "two independent variants ...: one with bifaces and wedge-shaped cores, the other with bifaces devoid of wedge-shaped cores." Kukhtui III is seen as a possible representative of the latter, but the

Pleistocene archaeology of Chukotka, the region closest to Bering Strait, is poorly known, and not a single reliable archaeological site can contribute to the picture. Ackerman (1985b: 369) lists a number of "Paleolithic" finds in the region, but Dolitsky (1985: 375) remarks that most of them were not found *in situ*.

The stemmed points of Ushki I Layer 7 suggest that some other cultural tradition may have been present in Northeast Asia, but the find is surrounded by a heated controversy over the integrity and significance of the deposits (Mochanov 1977: 224, 1978a: 64; Dikov and Titov 1984: 69; Dolitsky 1985: 374; Guthrie 1985b). Regardless of this controversy, the Dyuktai tradition is seen as distinct from the Mal'ta-Afontova tradition that existed to the south and west in the Yenisei and Angara regions (Mochanov 1977, 1978b, 1980). Abramova (1979a, 1979b, 1980) describes important differences among the major sites of the latter grouping (Mal'ta, Afontova Gora, Kokorevo), but the absence of leaf-shaped bifacial points on the Yenisei (Abramova 1980: 137) is a major difference that distinguishes these sites from those of the Dyuktai tradition (Mochanov 1978b). In the place of bifacial stone points, weapon tips were of bone and antler in which long slots held rows of microblade segments, whose sharp edges would have caused massive hemorrhage in an animal, especially if it continued to run after impact. Such points are best known from Kokorevo I (Abramova 1979a: Fig. 50-53), but they were probably widespread. There is a growing consensus (e.g., Aigner 1985; Guthrie 1983; Larsen 1968) that insets in organic weapon tips represent a major, although not the only, use of microblades throughout Beringia. It may be that the abundance of microblades and microcores signifies the use of such weapon tips not only in the Mal'ta-Afontova but also in the Dyuktai tradition, even though the latter sites have failed to yield the bone and antler points.

According to Mochanov (1978b, 1980, 1984), the Mal'ta-Afontova tradition gave rise to a unifacial industry named Sumnagin that spread northeastward to replace Dyuktai toward the end of the Duvanny Yar Interval. As Anderson (1980) has noted, this change is parallel to technological changes in parts of Alaska, and I now turn our attention to that area.

EASTERN BERINGIA: THE BOUTELLIER INTERVAL

In addition to possible but unproven artificially altered bones on Disconformity A and in older deposits (see above), Old Crow Basin has produced a series of redeposited fossils on the modern river banks from which a variety of artifacts have been identified or proposed. Definite evidence of people can be seen in a caribou tibia fleshing tool, a caribou antler billet, a loon-bone awl, several cut-marked bones, a dog mandible, and a human mandible (Bonnichsen 1979; Fitzgerald 1980; Harington et al. 1975; Irving and Harington 1973; Irving et al. 1977; Morlan 1980). Of these, the fleshing tool and billet have been dated to late Holocene time (Nelson et al. 1986), and the others have been considered as possible Pleistocene only because of their dark staining and apparent but possibly spurious associations with demonstrably Pleistocene fossils. Also a variety of stone tools found among the fossils are of unknown age (see Morlan 1980: Chap. 8).

Among the redeposited fossils are hundreds of proboscidean limb bone fragments that were fractured when fresh, some of which exhibit conchoidal flake scars adjacent to prepared platforms while others are struck flakes with platform remnants. The close resemblance of these specimens to cores and flakes of stone encouraged Bonnichsen (1978, 1979; Bonnichsen and Young 1980; Bonnichsen and Will 1980) to propose that the bones represent a little known aspect of technology (see also Irving 1978b, 1978c; Morlan 1978b, 1980, 1983a). An experiment utilizing an African elephant as an analogue of the mammoth (Stanford et al. 1981) showed that proboscidean bones are very resistant to fracture but that controlled fractures of the leg bones offer access to large reservoirs of rich marrow, and the resulting pieces can be used to produce butchering knives by percussion flaking. Several of the mammoth bone cores and flakes from Old Crow Basin were well matched by the experimental replicas, and the usefulness of the flakes in butchering demonstrates the feasibility of this technology without, however, proving that the Old Crow specimens were man-made.

The possible artificial origin of the specimens is suggested by recoveries in archaeological sites elsewhere in the world where mammoth bone cores and flakes have been found in direct association with stone tools and the butchered remains of game animals. Such widespread sites include, for example, Vogelherd in Germany, Kulna Cave and Predmostí in Czechoslovakia, Bluefish Caves in the Yukon (see below), and Owl Cave and Lange-Ferguson in the United States (Hannus 1984; Miller 1983, 1984; Valoch 1980, 1982). I have already mentioned the fractured and flaked bones from Ikhine IC and Berelekh in Siberia, and Mochanov (1977: 93) reports other examples from undated exposures near Bochanut Lake. These occurrences do not prove that the Old Crow specimens were made by people, but they show that mammoth bone cores and flakes represent a widespread and often overlooked aspect of Paleolithic technology in both the Old and New Worlds.

Other studies have considered various natural processes that might fracture and flake fresh proboscidean bone. Such studies have been stimulated and guided by critics of the technological interpretation (e.g., Myers et al. 1980; Binford 1981; Guthrie 1984b; Thorson and Guthrie 1984), and attention has focused on sediment compaction, redeposition in ice-laden rivers, accidents during the lives of the mammoths, gnawing by large carnivores, trampling by mammoths and other large animals, and even carrying and throwing by mammoth trunks (Morlan 1983b, 1984a). Although none of these processes can be definitely ruled out in every case, none has been demonstrated to alter proboscidean bones like those from Old Crow Basin. Consideration of specific fracture patterns and available evidence concerning the actual results of these processes suggest that they do not adequately account for the fresh-fractured and flaked mammoth bones either collectively or singly.

Recently I have taken another approach to this interpretive dilemma. The "best" candidates for a technological interpretation were selected for radiocarbon dating by accelerator mass spectrometry (AMS). "Best" was defined as fresh-fractured bones with a well-defined scar representing a point of impact and those cores and flakes with prepared platforms and platform remnants. Twenty-four such specimens were sampled, and all yielded AMS dates of between 25,000 and 40,000 years ago (Morlan 1986b). This temporal interval corresponds to stratigraphic Unit 2b in Old Crow Basin overlying Disconformity A and underlying a thick deposit of glaciolacustrine clay from the last major glacial advance in Duvanny Yar time (Hughes et al. 1981; Morlan 1984b). I have excavated dozens of large mammoth bones from older deposits that underlie Disconformity A, some of them as much as 1.5 million years old, and none exhibiting the distinctive fresh bone fractures and flake scars seen on the dated specimens. Although the dated specimens were found out of context, the AMS dates suggest that they were eroded from Unit 2b. Therefore it appears that some new process began to alter mammoth bones in Old Crow Basin shortly after 40,000 years ago.

To suggest that such a new process is due to a natural cause would violate the principle of uniformitarianism if one considered only those processes that had long been operative in Old Crow Basin such as carnivore gnawing, animal trampling and the effects of river ice. There are two possible sources of new processes that could have begun in the basin shortly after 40,000 years ago: unstable lakes subject to repeated lowerings, and the arrival of man. Old Crow Basin is one of three northern Yukon basins that were flooded by glacial meltwater during the late Wisconsinan Hungry Creek Glaciation (Hughes et al. 1981). A thick layer of clay was deposited in both Old Crow and Bluefish basins during a deepwater phase that began around 25,000 years ago, but an earlier glacial meltwater event is recorded in Bell Basin and is older than 36,880 ±390 B.P. (TO-218; J.V. Matthews, Jr., pers. comm. in 1986). Glacial meltwater reached these basins through the Eagle River discharge channel, "a canyon-like feature more than 1 km wide incised into siltstone and mudstone of Mississippian age" (Hughes et al. 1981: 339). Mississippian spores appear abruptly in pollen assemblages from Old Crow River Locality 15 beginning just above Disconformity A and continuing throughout Unit 2b (Walde 1985), suggesting that the erosion that produced the disconformity resulted from transgression by glacial meltwater. Especially during its early history the lake level would have been inherently unstable due to the great distance from glacial meltwater sources (ca. 200 km) which themselves were among the most remote reaches of the Laurentide ice complex. I cannot specify a particular process associated with such a lake system that might fracture and flake fresh proboscidean bones, but the problem deserves study because the onset of lacustrine inundation appears to coincide with the earliest dated mammoth bones that exhibit these kinds of breakage.

The other cause of mammoth bone fracture could be the arrival of human hunters who fractured bones for marrow extraction and tool production. This kind of technology is present, although rare, in the known sites of the Dyuktai tradition in western Beringia. If all 24 of the dated proboscidean bones from Old Crow Basin are artifacts, then the oldest of them (39,500 ± 1600 B.P.) suggests that sites will be found in both western and eastern Beringia that are older than the disputed earliest date set forth by Mochanov for Ezhantsy and Ikhine IC (ca. 35,000 B.P.).

Elsewhere in eastern Beringia, altered bones and antlers that may have been artificially modified have been recovered from placer deposits in the Klondike and Fairbanks areas and at Jack Wade and Lost Chicken Creeks (Bonnichsen 1979; Harington 1975, 1980c). The Jack Wade assemblage, of the late Boutellier Interval, is dominated by sheep, although it also contains bison, caribou, a small Yukon horse, and a moose (Porter 1979). The Klondike, Fairbanks and Lost Chicken assemblages may span considerable time periods that include the cold Duvanny Yar Interval as well as earlier intervals. None of these bones can be said to be *in situ* in these muck deposits, and only one specimen possibly fractured by man has been dated: 10,370 ± 160 (I-8582) from Lost Chicken Creek (Harington 1980c: 170).

EASTERN BERINGIA: THE DUVANNY YAR INTERVAL

The oldest known archaeological site in eastern Beringia is the Bluefish Caves, situated in the northern edge of the Keele Range about 65 km southwest of Old Crow in an extensive tract of karst topography (Cinq-Mars 1979, 1982; Cinq-Mars and Lauriol 1985; Morlan and Cinq-Mars 1982). Caves 1 and 2 have similar stratigraphic records: limestone overlain by late Pleistocene loess in turn overlain by Holocene humus-rich rubble and humus. Faunal remains are found throughout the sequence but are much more abundant in the Pleistocene than in the Holocene sediment. The loess contains the remains of mammoth, horse, bison, caribou, moose, Dall sheep, wapiti, red fox, Arctic fox, bear, wolf, a large cat, mustelids, Arctic hare, snowshoe hare, marmot, ground squirrel, many species of microtine rodents, shrews, birds, and a few fish remains. The Holocene deposits contain fewer bones and fewer species of small mammals, birds and fish; the only large mammals are caribou and sheep. Pollen samples from the two caves show similar sequences of herb-tundra at the bottom, a birch zone within the loess, and spruce forest beginning at the lower boundary of the Holocene humus-rich rubble (Ritchie et al. 1982). This pattern of environmental change is precisely reflected by changes in the frequencies of microtine rodent species (Morlan 1983c, 1985).

Sixteen radiocarbon dates on bones from the loess range between 12,000 and 25,000 years ago (Cinq-Mars, pers. comm. in 1986), and span the Duvanny Yar Interval and the Birch zone of Hopkins (1982). Nearly all evidence of human occupation is found in the loess, and it is most abundant in the lower portion of loess where Pleistocene megafauna is also most abundant. Stone artifacts in Cave 1 -- a burin spall, a microblade, and a variety of chert flakes some of which may have been detached from bifaces (Cinq-Mars 1979; Morlan and Cinq-Mars 1982) -- occur in sediments representing the Birch zone. A cobble-sized hammerstone was associated with a large pile of Pleistocene megafaunal remains outside Cave 1. Dall sheep and caribou tibiae found in Cave 1 and outside Cave 2, respectively, have been split longitudinally and whittled and polished, suggesting use as fleshing tools. The lower part of the loess within Cave 2, likely representing the herb tundra of Duvanny Yar time, yielded a wedge-shaped core, core tablets, microblades, burins, burin spalls and flakes that appear not to have been detached from a biface (Cinq-Mars 1984). An angle burin on an edge-retouched chert blade was found in the loess outside Cave 2 (Morlan and Cinq-Mars 1982: Fig. 9). Found in the lower part of the loess outside Cave 2, and dated to the early part of the Duvanny Yar Interval, were a mammoth bone core and a flake that can be refitted on the core (Cinq-Mars, pers. comm. in 1986). In both caves many bones from the loess have butchering marks that are absent on bones from the Holocene deposits (Morlan and Cinq-Mars 1982: Fig. 10). Upon completion of the analysis, it is expected that patterns of bone distribution, such as the large pile outside Cave 1, and selective absence of anatomical parts, such as cranial and most axial elements of mammoths, will provide additional evidence of human activity.

It is not yet possible to identify discrete, datable occupations at the Bluefish Caves, but the microblades, microcore, burin and byproducts within Cave 2 were very near a mammoth scapula dated to approximately 20,000 years ago. Cinq-Mars (1984) has characterized the site as a palimpsest where people and carnivores preyed and scavenged on the diverse fauna. Further analysis, however, is needed to ascertain the time and conditions of human occupation during the Duvanny Yar interval.

Selected late Pleistocene sites in eastern Beringia are listed in Table 2. These, and others, have been extensively discussed by several authors (Anderson 1978, 1980; Clark 1981, 1984; Clark and Morlan 1982; Dumond 1977, 1980a, 1980b; Haynes 1980, 1982; Morlan and Cinq-Mars 1982; Workman 1980; and references in Table 2), and I will not describe them individually. Instead I will focus on broad technological categories and on questions raised by already published discussions.

One such question concerns the presence or absence of bifaces and microblades and their associations with one another. Bluefish Cave 2, apparently the oldest known assemblage, contains microblades and cores but not bifaces. In the next three sites, according to their radiocarbon dates (Table 2), Dry Creek I contains bifaces but lacks microblades while Chindadn at Healy Lake and Putu have yielded both bifaces and microblades. Dry Creek II has a series of artifact clusters, some with bifaces and others with microblades but none occurring together. A similar pattern is seen in the surficial clusters at Batza Tena. Mt. Hayes 111 and most other Tangle Lakes localities yield blades or microblades together with bifaces. The Akmak complex at Onion Portage contains both, but the slightly later Kobuk complex lacks bifaces. Layer III at Trail Creek Cave 2, apparently similar in age to the Kobuk complex, has produced the only eastern Beringian examples of antler points slotted for microblade insets. Larsen (1968: 72-75) traces the distribution of such points not only to the Mal'ta-Afontova sites but also across Eurasia all the way to the Danish Mesolithic.

There are many ways to integrate these sites into categories of varying

Table 2. Selected Late Pleistocene and Early Holocene sites in unglaciated Eastern Beringia.

Site Name	Date (B.P.)	Material	Reference(s)
Bifaces and microblades			
Akmak	9,570 ± 150 (K-1583)	Collagen	Minimum date; Anderson 1968; Tauber 1973: 107
Mt. Hayes 111	10,150 ± 280 (UGa-572)	Soil charcoal	West 1975, 1981a
Chindadn	11,090 ± 170 (GX-1341)	Apatite	McKennan and Cook 1968; Cook 1969
Putu	11,470 ± 500 (SI-2382)	Charcoal	Alexander 1974; Haynes 1980: 119
Bifaces or microblades in clusters			
Dry Creek II	10,690 ± 250 (SI-1561)	Charcoal	Minimum date; Thorson and Hamilton 1977; Powers and Hamilton 1978
Batza Tena	not dated		Fluted points presume late Pleistocene; Clark 1972; Clark and Clark 1975, 1980, 1983
Bifaces, no microblades			
Dry Creek I	11,120 ± 85 (SI-2880)	Charcoal	Thorson and Hamilton 1977; Powers and Hamilton 1978
Microblades or blades, no bifaces			
Trail Creek Cave 2	9,070 ± 150 (K-980)	Collagen	Larsen 1968; Tauber 1968: 318
Gallagher Loc. 1	10,540 ± 150 (SI-974)	Charcoal	Dixon 1975

complexity. Anderson (1968, 1970) includes Akmak, Kobuk and some manifestations of the Denali complex (West 1967), as represented at Mt. Hayes 111 and similar localities, in the American Paleoarctic tradition, but he explicitly excludes Chindadn and the Aleutian Island site of Anangula as representing "quite different cultural configurations" (Anderson 1980: 237). Anderson (1980) argues that in eastern Beringia bifaces and microblades occur together as an eastern facies or counterpart of the Dyuktai Tradition. Slightly later assemblages are unifacial and counterpart to Sumnagin. Dumond (1977, 1980a) considers Gallagher Loc. 1 and Anangula, on the one hand, and Putu and Chindadn, on the other hand, to be "independent examples of one limit of the normal range of variation within the Palaeo-arctic tradition" (Dumond 1977: 43). Regional descendants of this enlarged tradition in southern Alaska are seen around 9000 B.P. in newly deglaciated terrain at Ugashik Narrows, Hidden Falls 1, and Ground Hog Bay 2 (see Ackerman et al. 1979, 1985; Dumond et al. 1976; Henn 1978). West's (1981a) scheme is even more inclusive, embracing all the Alaskan and Siberian sites discussed here as well as many others in a single "Beringian tradition." In my opinion such an all-encompassing framework has little analytic or heuristic value.

An interesting belief in the published discussions of these sites is that a particular style of bifacial projectile point -- the fluted point -- does not belong with microblades. Perhaps this viewpoint is biased by the occurrence of fluted point complexes in midcontinental North America where microblades are lacking. For example, Haynes' (1980: 119-120; 1982: 396) perspective from the midcontinent is that there are two cultural traditions in late Pleistocene Alaska and that fluted points must have developed in "the one without microblades." Dumond (1980b: 990) states about the Putu site that "fluted points aside, this assemblage can be recognized on formal grounds as another manifestation of the Paleo-Arctic tradition." Clark and Clark (1983: 289) conclude that "prior to the spread of the Asian Dyuktai, or any other, microblade producing technological tradition into the eastern Beringian region of Alaska and the Yukon about 11,000 B.P., there was in this area a population whose lithic tools emphasized bifacially prepared projectile points. Application of the fluting technique to projectile points occurred within this context in

Alaska approximately 12,000 years ago." Only Alexander (1974) and West (1981a, 1982) have eschewed a dichotomy between fluted points and microblades, but their reasoning has sent them in opposite directions to account for northern fluted points.

We have seen that the microblade-producing Dyuktai tradition also produced a variety of bifaces including projectile points. Cinq-Mars' new data from Bluefish Caves suggest that at least the microblade component of this technology was present in eastern Beringia long before 11,000 years ago, perhaps closer to 20,000. We should not ignore the fluted points in defining the Paleoarctic tradition, because it is in the context of that tradition that the fluting technique may have been invented. The Dyuktai and Paleoarctic technology represents a well established tradition of removing elongate flakes from the edges, corners and faces of stone tools. The wide variety of core forms in this technology, including irregular, tabular and wedge-shaped, represents solutions to the technical problems that must be overcome in order to flute a point. The technical problems of thinning and shaping bifaces were solved during the development of bifacial wedge-shaped cores, knives and unfluted projectile points. Thus, I agree with West (1981a: 186 ff.) that fluted points can be construed as a natural development within the Beringian or Paleoarctic tradition.

From this point the published arguments take disparate and sometimes surprising turns. For example, West (1981a: 185-186) asserts that Beringian fluted points "have nothing whatever to do with the origins of the Clovis points," and he proposes that the ancestors of Paleoindian fluted point makers lived in *central* Beringia, in sites now under the Bering and Chukchi Seas. It is hard to imagine how this idea could gain support from empirical evidence, but it could be falsified by sites currently on land. West's (1983: 374) dubious assertion that blade technology is "virtually ubiquitous in Clovis collections" could suggest that northern and southern fluted point complexes might be closely related as components of a pan-continental technological tradition.

Bryan (1980: 81; also Schweger et al. 1982: 441-442) attempts to explain why microblade technology did not accompany fluted points as they were carried to the midcontinent from the north. He suggests that the introduction, or invention, of the atlatl favored the adoption of heavier stone points, although he does not demonstrate that a hafted fluted point is heavier than a microblade-armed antler projectile. In fact, there is no evidence that the atlatl was used to propel fluted points, while measurements of archaeological specimens (Gramly 1984) and of experimental replicas (Frison 1978: 333 ff.; Huckell 1982) suggest that thrusting spears may have been employed.

If we assume that armature in antler points represented one of the principal functions of microblades, then nonavailability of suitable antler might explain the absence of microblades on the High Plains (West 1983: 376; Morlan 1984c). This argument is based on Guthrie's (1983) experiment showing that caribou antler is superior to all other available materials for the construction of microblade-armed osseous points. As a projectile point, antler is superior to both bone and wood; it is less brittle than bone, harder than any wood in the north, susceptible to softening by soaking in water, easy to shape by cutting and bending, and restored to strength by drying. Among the antlers, caribou is superior because of its thicker cortex and long straight segments. These characteristics make caribou antler ideally suited to the manufacture of armed organic projectile tips, perhaps so superior to other materials that this technique of point production was abandoned in areas outside the range of caribou. One difficulty with this scenario is that eastern fluted point makers are widely believed to have preyed upon caribou (Spiess 1984) and yet their lithic tool kits lack the microblades that might have armed caribou antler points.

Nevertheless, Guthrie's (1983) observations are worth exploring. He suggests that late Pleistocene hunters in Beringia may have made at least three major kinds of projectile points: stone bifaces, microblades inset in antler, and simple points of ivory, antler or bone. These may have been alternatives in the production of weapons. It is not known why a given alternative might be selected -- whether, for example, it was a seasonal phenomenon or perhaps related to the kind of animal hunted -- but the widespread association of bifacial points and microblades raises the possibility that both lithic and osseous weapon tips were used by a single hunting band or by even a single hunter. Those sites or clusters with only one of the alternatives may represent season-specific or activity-specific occupations.

I have argued previously (Morlan 1977) that the dissolution of the steppe-tundra and its ungulate fauna might have prompted the abandonment of interior eastern Beringia and resulted in the sudden appearance of widespread fluted point complexes in the midcontinent of North America. If so, the technology would have spread beyond the range of the caribou, and the maintainence of armed projectile points would have depended on the trading of antler or on a resort to other, less satisfactory raw material. Given the high quality of the cherts and other rocks that are known from midcontinental Paleoindian sites, the microblade-armed antler point may have been too costly to maintain. The presence of unbarbed, unarmed points of ivory and bone in several Clovis and Folsom sites on the High Plains (Frison and Zeimens 1980; Haynes 1982) suggests that this is a plausible, if unproven, viewpoint.

DISCUSSION

Nearly all of the sites discussed in this paper have been found during the past two decades. They represent a significant advance in knowledge of the Pleistocene archaeology of Beringia, but there remain many major impediments to a full understanding of the subject. The actual time span of human occupation in Beringia is not known. The claims of extremely early occupation in both Siberia and the northern Yukon cannot yet be evaluated, and even the data and proposals mustered for the Boutellier Interval (25,000 - 40,000 B.P.) remain unconvincing. There are reasonable doubts concerning the dating of sites in western Beringia that contain stone and bone artifacts, and associated fauna in alluvium. In eastern Beringia all purported artifacts for this period are redeposited and lack context that would aid interpretation.

Data for the Duvanny Yar Interval (14,000 - 25,000 B.P.) are extremely scanty, with only the Verkhne Troitskaya site in the Aldan valley firmly dated to about 18,000 years ago, and Bluefish Cave 2 in the northern Yukon, with possible dates of 18,000 to 25,000 B.P. Given that the dating of Maiorych (16,000 B.P.) is based on typological comparison, there is no evidence of human occupation between the western and eastern extremes of Megaberingia, as here defined, during the Duvanny Yar Interval.

On present evidence, relatively widespread, presumably continuous, human occupation on both sides of Bering Strait begins during the Birch zone (10,000 - 14,000 B.P.). Sites have been securely dated to this period in the Aldan and Indigirka valleys, in Kamchatka, in central and northern Alaska, and in the northern Yukon. Many other sites are assigned to this period on the basis of typological comparisons and geological settings, but even the best sites are beset with problems or are poorly described. None of the Alaskan sites contains well-preserved animal remains, and many are in shallow or disturbed soils in which an admixture of diverse cultural materials cannot be ruled out.

This is a small and uneven data base, but because nearly all of it has been developed during the past two decades, it would appear that the future holds promise for major advances in knowledge. Therefore, I believe it premature to cite negative evidence in support of the idea that Beringia was uninhabitable during the height of the Duvanny Yar Interval (e.g., Müller-Beck 1966; Hamilton 1978; Tseitlin 1979: 261; Schweger et al. 1982: 439; Fladmark 1983: 22-23). As Hopkins (1983: 346) notes, any human populations then resident in Beringia "must have been almost entirely dependent upon land mammals for food as well as for other essential requirements," but there is an ongoing debate among paleoecologists as to the richness and diversity of both flora and fauna in Beringia during full glacial time (see Cwynar and Ritchie 1980; Guthrie 1982, 1985a; Matthews 1982; Ritchie 1984; Ritchie and Cwynar 1982). Understanding of the potential of Beringia for human occupation during the Duvanny Yar Interval must await clarification of this problem as well as new archaeological data.

But even now there are several ways to improve our interpretation of the data for Pleistocene Beringia. One approach is the ongoing taphonomic analysis of the effects of redeposition. Critics of the Old Crow interpretations have cited the lack of associated stone tools (West 1981b: 108) and human bones (Dumond 1982: 889) as arguments against the identification of fossil bones as artifacts. Both stone tools (Morlan 1980: 235-246) and a human bone (Irving et al. 1977) have been found along Old Crow River, but they cannot be assumed to belong with the Pleistocene fossils. Furthermore, given that the putative Boutellier Interval artifacts are redeposited proboscidean bone fragments, we should not expect, because of variations in their hydrodynamic properties, the stone tools and human bones to occur together (Pavlish et al. 1986). Fluvial transportation has differentially sorted them, as surely as it has destroyed the archaeological contexts on which we customarily depend.

It is not generally appreciated that occurrence of objects in an archaeological site confers no more than circumstantial evidence of their artificial origin. The Ikhine and Ust'-Mil' localities are interpreted in the literature as archaeological sites (Mochanov 1977, 1978a; Michael 1984; Yi and Clark 1985), but, as Hopkins (1985: 372) notes, they may be merely "detrital accumulations of wood, fossil bones, and artifacts enclosed in alluvium." If all these objects have undergone fluvial transportation, then it is quite likely that the stone tools bear no functional relationship to the animal remains (see Table 1), and no one should read into the frequencies of identified taxa a societal predilection toward big game hunting or a preference for a particular species. In fact, the Beringian data reviewed here are essentially devoid of any demonstrated functional relationship between artifacts and animal remains. Such a relationship has merely been assumed for the western Beringian sites while in eastern Beringia only Bluefish Caves contains a well-preserved fauna, which is still being analyzed.

Even if the bones and artifacts from Ikhine, Ust'-Mil', and Old Crow have been transported away from their original archaeological contexts, they remain important to our understanding of the past. We should not expect to learn very much about prehistory from transported material, but the mere presence of human societies is of profound importance from the standpoint of paleoecology. "Two qualitatively different sets of questions...often become confused in this discussion. The simple but important question whether people were present at a given time and place needs to be separated from the more complex questions of who they might have been, from where they came, how they might have lived, and with what they might have been equipped" (Schweger et al. 1982: 437). Individual artifacts, even if they have been redeposited, can provide answers to the "simple but important question" if criteria can be developed to warrant their identification and if a means can be found to date them. My plea is not to soften or to suspend standards (Dincauze 1984: 294; Owen 1984: 534) but to strengthen those pertaining to redeposited specimens.

Another interpretive approach is to consider the implications of data from one area or time period for those of another. For example, if the Old Crow mammoth bone cores and flakes are artifacts representing the arrival of people shortly after 40,000 years ago in eastern Beringia, then the archaeological record of western Beringia is woefully incomplete. Even Mochanov's most generous estimate for the age of the proto-Dyuktai industry cannot provide adequate time depth to account for the Old Crow Basin specimens. Likewise, if the Bluefish Caves can be shown to have been occupied during the height of the Duvanny Yar Interval, as they appear to have been, then, given that harsh continental conditions increased eastward from central Beringia (Hopkins 1983), the ancient Beringians demonstrably passed the most severe test that the region could offer. More generally, considerable cultural diversity is already evident among the Birch zone sites, even if some sites are season-specific or activity-specific manifestations of a single cultural tradition. Such diversity, scattered across an area of subcontinental proportions, "implies a complex history merely glimpsed in the data now on hand" (Morlan and Cinq-Mars 1982:376).

A third approach is to consider the Beringian archaeological data in relation to other data bases. In my opinion, Meadowcroft Rockshelter in Pennsylvania, suggests that people were south of the Laurentide ice sheet before the last glacial maximum (see Adovasio et al. 1985; Morlan 1986c), implying that human occupation of Beringia must have begun during or before the Duvanny Yar Interval. Rogers (1985) argues that a similar time span is required to account for the distribution and diversity of native languages in North America. Such a time span is compatible with calculations of divergence among various circum-Pacific populations based on dental (Turner 1985) and serological traits (Harper 1980; Williams et al. 1985). Some of these studies have attempted to calculate the time since bifurcation among populations, but the calculations require rather arbitrary decisions concerning the time of divergence between Aleut and Eskimo (Harper 1980) or the time since flooding of the Bering Land Bridge (Turner 1985).

Of more general importance in all studies of genetic traits is the recognition of three major categories of native North Americans -- Paleo-Indian, Na-dene, and Aleut-Eskimo in Turner's (1985) terms -- presumably representing three migrations into the continent. That all came from Northeast Asia is shown by their Sinodont dental pattern, as contrasted with Sundadonty of Southeast Asia and a Caucasoid pattern west of Lake Baikal (Turner 1985). There are many ways to model movements of people, distributions of technological traits, and possible origins and affinities of linguistic stocks, but some studies seem to overlook the potentially independent variables of race, language and culture.

I will not review all the models that have been suggested but will describe one that best fits the data as I understand it. Sinodont bearers of the Dyuktai tradition in the Lena Basin spread northeastward across Beringia and became the Paleoindian ancestors of most North and South American Indians at a time when Caucasoid bearers of the Mal'ta-Afontova tradition lived in the Baikal area and westward along the Yenisei and Angara Rivers. The Paleoindians spread across North America and established themselves south of the Laurentide ice before the last major advance. The proximity of the Laurentide and Cordilleran ice sheets isolated the Paleoindians from residents of Beringia for at least several millennia, during which a narrow and inhospitable corridor connected Beringia with the midcontinent (Fladmark 1983; Schweger and Mandryk 1986). Their Dyuktai tradition had equipped the Paleoindians intellectually and technologically to exploit a wide variety of resources, including late Pleistocene megafauna, and to invent such devices as fluted points either in Beringia or in areas to the south.

Szathmary (1985: 82) notes that genetic differentiation could occur between populations isolated for only a few millennia north and south of the ice sheets. This may account for the distinctiveness of the Na-dene group whose ancestors were residents of interior Beringia by the end of the Duvanny Yar Interval; I will call them the Paleoarctics. Differentiation may have been aided by continued gene flow across Beringia. New cultural influences from the expanding Sumnagin culture of northeast Asia arrived in Alaska. The Paleoarctics spread south and east into newly deglaciated terrain where they encountered northward expanding populations of Paleoindians. Like the latter, the Paleoarctics had the intellectual and technological background to invent the fluted point, which, if it had but a single origin, could have diffused either northward or southward. More importantly, the Paleoindians and Paleoarctics shared not only a common genetic origin but also a common cultural origin, and it may not have been necessary for large numbers of traits to be exchanged to create the impression of great similarity between the two regions.

The Aleut-Eskimo group apparently had a different origin, possibly moving along the coast from the vicinity of the Amur River to southwestern Alaska from whence they spread (Laughlin et al. 1979; Turner 1985). Whether they arrived before the Paleoarctics ("Na-dene," Turner 1985) may be difficult to establish if their earliest coastal sites are now under water.

This speculative reconstruction depends on extrapolation between distant data points rather than on interpretation of a coherent data base. Undoubtedly it oversimplifies a complex prehistoric record that should prove even more interesting in the future. The paucity of our data is brought into sharp relief when we reiterate some of the prerequisites for human life in Beringia discussed in the introduction to this paper. The evidence for substantial dwellings and tailored skin clothing has been found for the neighboring Mal'ta-Afontova tradition but not for Dyuktai or its eastern Beringian descendants except at the Ushki sites. The Dyuktai sites contain large mammal remains, but we have no idea how they were hunted, how exploitation of seasonal resources was scheduled, or how food was stored for the winter. Most of the eastern Beringian sites do not even contain any animal remains. When we contemplate the daily decisions that must have been required of these hunting people, the meagerness of our data seems even more dramatic. Consider, for example, the detailed knowledge of prey behavior including idiosyncratic habits (see Haynes 1982: 392; Marks 1982; Olivier 1982) that must have been required for a successful chase. As we study our fragments of stone and bone we must remember that colonization of Beringia and the New World during the late Pleistocene was a remarkable technological accomplishment.

ACKNOWLEDGEMENTS

I thank Jacques Cinq-Mars, Matthew Nitecki, and James V. Wright for helpful criticisms of an earlier draft of this paper; Raymond Le Blanc for assistance in transferring the text from Apple to IBM; Erle Nelson, John Southon and John Vogel for the AMS dates from Old Crow Basin; and Owen L. Hughes, John V. Matthews, Jr., and Charles Schweger for discussions of the geological implications of the AMS dates. This research was supported by the Archaeological Survey of Canada, Canadian Museum of Civilization. I am solely responsible for any errors or oversights in this presentation.

REFERENCES CITED

Abramova, Z.A. 1979a. *Paleolit Eniseia: Kokorevskaia kul'tura. (The Yenisei Paleolithic: Kokorevo Culture)*. 200 pp. Novosibirsk: Nauka.

Abramova, Z.A. 1979b. *Paleolit Eniseia: Afontovskaia kul'tura. (The Yenisei Paleolithic: Afontova Culture)*. 157 pp. Novosibirsk: Nauka.

Abramova, Z.A. 1980. Concerning the cultural contacts between Asia and America in the Late Paleolithic. In: *Early Native Americans*, ed. D.L. Browman, 133-139. The Hague: Mouton.

Ackerman, R.E. 1985a. Comment on Yi and Clark (1985). *Current Anthropology* 26(3):401.

Ackerman, R.E. 1985b. Comment on Dolitsky (1985). *Current Anthropology* 26(3):368.

Ackerman, R.E., T.D. Hamilton, and R. Stuckenrath. 1979. Early cultural complexes on the northern Northwest Coast. *Canadian Journal of Archaeology* 3:195-209.

Ackerman, R.E., K.C. Reid, J.D. Gallison, and M.E. Roe. 1985. *Archaeology of Heceta Island.* Center for Northwest Anthropology Project Report No. 3. Pullman: Washington State University.

Adovasio, J.M., R.C. Carlisle, K.A. Cushman, J. Donahue, J.E. Gilday, W.C. Johnson, K. Lord, P.W. Parmalee, R. Stuckenrath, and P.W. Wiegman. 1985. Paleoenvironmental reconstruction at Meadowcroft Rockshelter, Washington County, Pennsylvania. In: *Environments and Extinctions: Man in Late Glacial North America*, ed. J.I. Mead and D.J. Meltzer, 73-110. Center for the Study of Early Man. Orono: University of Maine.

Aigner, J.S. 1985. Comment on Dolitsky (1985). *Current Anthropology* 26(3):369.

Alexander, H.L. 1974. The association of Aurignacoid elements with fluted point complexes in North America. In: *International Conference on the Prehistory and Paleoecology of Western North American Arctic and Subarctic*, ed. S. Raymond and P. Schledermann, 21-31. Calgary: University of Calgary Archaeological Association.

Anderson, D.D. 1968. A stone age campsite on the gateway to America. *Scientific American* 218(6):24-33.

Anderson, D.D. 1970. Akmak: An early archaeological assemblage from Onion Portage, northwest Alaska. *Acta Arctica* 16, 80 pp.

Anderson, D.D. 1978. Western Arctic and Sub-Arctic. In: *Chronologies in New World Archaeology*, ed. R.E. Taylor and C.W. Meighan, 29-50. New York: Academic Press.

Anderson, D.D. 1980. Continuity and change in the prehistoric record from North Alaska. *Senri Ethnological Studies* 4:233-251.

Andrews, P., and J. Cook. 1985. Natural modifications to bones in a temperate setting. *Man* 20(4):675-691.

Bader, O.N., and V.E. Flint. 1977. Gravirovka na bivne mamonta s Berelekha. (An engraving on a mammoth tusk from Berelekh). *Trudy Zoologicheskogo Instituta* 72:68-72.

Binford, L.R. 1981. *Bones: Ancient Men and Modern Myths*. 320 pp. New York: Academic Press.

Bonnichsen, R. 1978. Critical arguments for Pleistocene artifacts from the Old Crow Basin, Yukon: A preliminary statement. *Occasional Paper, Dept. of Anthropology, University of Alberta* 1:102-118.

Bonnichsen, R. 1979. *Pleistocene Bone Technology in the Beringian Refugium*. 280 pp. Ottawa: National Museum of Man (Mercury Series: Archaeological Survey of Canada Paper No. 89).

Bonnichsen, R., and R. Will. 1980. Cultural modification of bone: The experimental approach in faunal analysis. In: *Mammalian Osteology*, ed. B.M. Gilbert, 7-30. Laramie: Modern Printing Co.

Bonnichsen, R., and D. Young. 1980. Early technological repertoires: Bone to stone. *Canadian Journal of Anthropology* 1(1):123-128.

Briggs, N.D., and J.A. Westgate. 1978. Fission-track age of tephra marker beds in Beringia. *Fifth American Quaternary Association Abstracts*, p. 190.

Bryan, A.L. 1978. An overview of Paleo-American prehistory from a circum-Pacific perspective. *Occasional Paper, Dept. of Anthropology, University of Alberta* 1:306-327.

Bryan, A.L. 1980. The stemmed point tradition: An early technological tradition in western North America. In: *Anthropological Papers in Memory of Earl H. Swanson, Jr.*, ed. L.B. Harten, C.N. Warren, and D.R. Touhy, 77-107. Pocatello: Idaho State Museum of Natural History.

Bryan, A.L. 1986. Paleoamerican prehistory as seen from South America. In: *New Evidence for the Pleistocene Peopling of the Americas*, ed. A.L. Bryan, 1-14. Center for the Study of Early Man. Orono: University of Maine.

Budinger, F.E., Jr. 1983. The Calico early man site. *California Geology*, April (1983):75-82.

Cinq-Mars, J. 1979. Bluefish Cave I: A late Pleistocene Eastern Beringian cave deposit in the northern Yukon. *Canadian Journal of Archaeology* 3:1-32.

Cinq-Mars, J. 1982. Les grottes du Poisson-Bleu. *GEOS* 11(1):19-21.

Cinq-Mars, J. 1984. A palimpsest in the making: Taphonomic note from the Bluefish Caves, Yukon Territory. Paper presented to the First International Conference on Bone Modification, Carson City.

Cinq-Mars, J., and B. Lauriol. 1985. Le karst de Tsi-it-toh-choh: Notes preliminaires sur quelques phénomènes karstiques du Yukon septentrional, Canada. *Annales de la Société Geologique de Belgique* 108:185-195.

Clark, D.W. 1972. Archaeology of the Batza Tena obsidian source, west-central Alaska. *Anthropological Papers of the University of Alaska* 15(2):1-21.

Clark, D.W. 1981. Prehistory of the western Subarctic. In: *Handbook of North American Indians*, Vol. 6, *Subarctic*, ed. J. Helm, 107-129. Washington, D.C.: Smithsonian Institution.

Clark, D.W. 1984. Northern fluted points: Paleo-Eskimo, Paleo-Arctic, or Paleo-Indian. *Canadian Journal of Anthropology* 4(1):65-81.

Clark, D.W., and A.McF. Clark. 1975. Fluted points from the Batza Tena obsidian source of the Koyukuk River region, Alaska. *Anthropological Papers of the University of Alaska* 17(2): 31-38.

Clark, D.W., and A.McF. Clark. 1980. Fluted points from the Batza Tena obsidian source, northwestern interior Alaska. In: *Early Native Americans*, ed. D.L. Browman, 141-159. The Hague: Mouton.

Clark, D.W., and A.McF. Clark. 1983. Paleo-Indians and fluted points: Subarctic alternatives. *Plains Anthropologist* 28-102:283-292.

Clark, D.W., and R.E. Morlan. 1982. Western Subarctic prehistory: Twenty years later. *Canadian Journal of Archaeology* 6:79-93.

Cook, J.P. 1969. The early prehistory of Healy Lake, Alaska. Ph.D. dissertation, University of Wisconsin, Madison, 352 pp.

Cwynar, L.C., and J.C. Ritchie. 1980. Arctic steppe-tundra: A Yukon perspective. *Science* 208: 1375-1377.

Dikov, N.N. 1979. *Arkheologicheskie pamiatniki Kamchatki, Chukotki i Verkhnei Kolymy (Archaeological sites in Kamchatka, Chukotka and the upper reaches of the Kolyma)*. 391 pp. Moscow: Nauka.

Dikov, N.N., and E.E. Titov. 1984. Problems of the stratification and periodization of the Ushki sites. *Arctic Anthropology* 21(2):69-80.

Dincauze, D.F. 1984. An archaeological evaluation of the case for Pre-Clovis occupations. In: *Advances in World Archaeology* 3, ed. F. Wendorf and A. Close, 275-323. New York: Academic Press.

Dixon, E.J., Jr. 1975. The Gallagher Flint Station: An early man site on the North Slope, Arctic Alaska, and its role in relation to the Bering Land Bridge. *Arctic Anthropology* 12(1):68-75.

Dolitsky, A.B. 1985. Siberian Paleolithic archaeology: Approaches and analytic methods. *Current Anthropology* 26(3):361-378.

Dumond, D.E. 1977. *The Eskimos and Aleuts*. 180 pp. London: Thames and Hudson.

Dumond, D.E. 1980a. A chronology of native Alaskan subsistence systems. *Senri Ethnological Series* 4:23-47.

Dumond, D.E. 1980b. The archeology of Alaska and the peopling of America. *Science* 209: 984-991.

Dumond, D.E. 1982. Colonization of the American Arctic and the New World. *American Antiquity* 47(4):885-895.

Dumond, D.E., W. Henn, and R. Stuckenrath. 1976. Archaeology and prehistory on the Alaska Peninsula. *Anthropological Papers of the University of Alaska* 18(1):17-29.

Fiorillo, A.R. 1984. An introduction to the identification of trample marks. *Current Research* 1:47-48.

Fitzgerald, G. 1980. Pleistocene loons of the Old Crow Basin, Yukon Territory, Canada. *Canadian Journal of Earth Sciences* 17(11):1593-1598.

Fladmark, K.R. 1983. Times and places: Environmental correlates of mid- to late-Wisconsinan human population expansion in North America. In: *Early Man in the New World*, ed. R. Shutler, Jr., 13-41. Beverly Hills: Sage Publications.

Frison, G.C. 1978. *Prehistoric Hunters of the High Plains*. 457 pp. New York: Academic Press.

Frison, G.C., and G.M. Zeimens. 1980. Bone projectile points: An addition to the Folsom cultural complex. *American Antiquity* 45(2):231-237.

Giddings, J.L., Jr. 1954. Early man in the Arctic. *Scientific American* 190(6):82-88.

Giterman, R.E., A.V. Sher, and J.V. Matthews, Jr. 1982. Comparison of the development of steppe-tundra environments in West and East Beringia: Pollen and macrofossil evidence from key sections. In: *Paleoecology of Beringia*, ed. D.M. Hopkins, J.V. Matthews, Jr., C.E. Schweger, and S.B. Young, 43-73. New York: Academic Press.

Gramly, R.M. 1984. Kill sites, killing ground and fluted points at the Vail site. *Archaeology of Eastern North America* 12:110-121.

Guthrie, R.D. 1982. Mammals of the mammoth steppe as paleoenvironmental indicators. In: *Paleoecology of Beringia*, ed. D.M. Hopkins, J.V. Matthews, Jr., C.E. Schweger, and S.B. Young, 307-326. New York: Academic Press.

Guthrie, R.D. 1983. Osseous projectile points: Biological considerations affecting raw material selection and design among Paleolithic and Paleoindian peoples. *BAR International Series* 163:273-294.

Guthrie, R.D. 1984a. Alaskan megabucks, megabulls, and megarams: The issue of Pleistocene gigantism. *Carnegie Museum of Natural History Special Publication* 8:482-510.

Guthrie, R.D. 1984b. The evidence for Middle-Wisconsin peopling of Beringia: An evaluation. *Quaternary Research* 22(2):231-241.

Guthrie, R.D. 1985a. Woolly arguments against the mammoth steppe: A new look at the palynological data. *Quarterly Review of Archaeology* 6(3):9-16.

Guthrie, R.D. 1985b. Comments on Dolitsky (1985). *Current Anthropology* 26(3):370-371.

Hamilton, T.D. 1978. Geologic perspectives on early man in Alaska. *Fifth American Quaternary Association Abstracts*, pp. 123-125.

Hamilton, T.D., and R.M. Thorson. 1983. The Cordilleran ice sheet in Alaska. In: *Late-Quaternary environments of the United States, Vol. 1: The Late Pleistocene*, ed. S.C. Porter, 38-52. Minneapolis: University of Minnesota Press.

Hannus, L.A. 1984. Flaked Mammoth Bone from the Lange/Ferguson Site, White River Badlands, South Dakota. Paper presented to the First International Conference on Bone Modification, Carson City.

Harington, C.R. 1975. A bone tool found with Ice Age mammal remains near Dawson City, Yukon Territory. *The Arctic Circular* 23(1):2-5.

Harington, C.R. 1977. Pleistocene Mammals of the Yukon Territory. 1060 pp. Ph.D. dissertation, University of Alberta, Edmonton.

Harington, C.R. 1978. *Quaternary Vertebrate Faunas of Canada and Alaska and their Suggested Chronological Sequence*. 105 pp. Ottawa: National Museum of Natural Sciences (Syllogeus No. 15).

Harington, C.R. 1980a. Faunal exchanges between Siberia and North America: Evidence from Quaternary land mammal remains in Siberia, Alaska and the Yukon Territory. *Canadian Journal of Anthropology* 1(1):45-49.

Harington, C.R. 1980b. Radiocarbon dates on some Quaternary mammals and artifacts from northern North America. *Arctic* 33(4):815-832.

Harington, C.R. 1980c. Pleistocene mammals from Lost Chicken Creek, Alaska. *Canadian Journal of Earth Sciences* 17(2):168-198.

Harington, C.R., R. Bonnichsen, and R.E. Morlan. 1975. Bones say man lived in Yukon 27,000 years ago. *Canadian Geographic Journal* 91(1-2):42-48.

Harper, A.B. 1980. Origins and divergence of Aleuts, Eskimos and American Indians. *Annals of Human Biology* 7(6):547-554.

Haynes, C.V. 1980. The Clovis culture. *Canadian Journal of Anthropology* 1(1):115-121.

Haynes, C.V. 1982. Were Clovis progenitors in Beringia? In: *Paleoecology of Beringia* ed. D.M. Hopkins, J.V. Matthews, Jr., C.E. Schweger, and S.B. Young, 383-398. New York: Academic Press.

Henn, W. 1978. Archaeology on the Alaska Peninsula: The Ugashik drainage, 1973-1975. *University of Oregon Anthropological Papers* 14. 183 pp.

Hopkins, D.M., ed. 1967. *The Bering Land Bridge*. Stanford: Stanford University Press.

Hopkins, D.M. 1972. The paleogeography and climatic history of Beringia during late Cenozoic time. *Inter-Nord* 12:121-150.

Hopkins, D.M. 1982. Aspects of the paleogeography of Beringia during the late Pleistocene. In: *Paleoecology of Beringia*, ed. D.M. Hopkins, J.V. Matthews, Jr., C.E. Schweger, and S.B. Young, 3-28. New York: Academic Press.

Hopkins, D.M. 1983. Hard times in Beringia: A short note. In: *Quaternary Coastlines and Marine Archaeology*, ed. P.M. Masters and N.C. Flemming, 345-346. New York: Academic Press.

Hopkins, D.M. 1985. Comments on Dolitsky (1985). *Current Anthropology* 26(3):371-372.

Hopkins, D.M., J.V. Matthews, Jr., C.E. Schweger, and S.B. Young, ed. 1982. *Paleoecology of Beringia*. 489 pp. New York: Academic Press.

Huckell, B.B. 1982. The Denver elephant project: A report on experimentation with thrusting spears. *Plains Anthropologist* 27-97:217-224.

Hughes, O.L., C.R. Harington, J.A. Janssens, J.V. Matthews, Jr., R.E. Morlan, N.W. Rutter, and C.E. Schweger. 1981. Upper Pleistocene stratigraphy, paleoecology, and archaeology of the northern Yukon interior, Eastern Beringia, 1. Bonnet Plume Basin. *Arctic* 34(4):329-365.

Irving, W.N. 1978a. Reflections. *Asian Perspectives* 19(1):91-95.

Irving, W.N. 1978b. Pleistocene archaeology in eastern Beringia. *Occasional Paper, Dept. of Anthropology, University of Alberta* 1:96-101.

Irving, W.N. 1978c. An approach to the prehistory of the Far East, from farther east. In: *Early Paleolithic in South and East Asia*, ed. F. Ikawa-Smith, 329-336. The Hague: Mouton.

Irving, W.N. 1982. Pleistocene cultures in Old Crow Basin: Interim report. In: *Peopling of the New World*, ed. J.E. Ericson, R.E. Taylor, and R. Berger, 69-79. Los Altos: Ballena Press.

Irving, W.N. 1985. Context and chronology of early man in the Americas. *Annual Review of Anthropology 1985* 14:529-555.

Irving, W.N., and C.R. Harington 1973. Upper Pleistocene radiocarbon-dated artifacts from the northern Yukon. *Science* 179:335-340.

Irving, W.N., J.T. Mayhall, F.J. Melbye, and B.F. Beebe. 1977. A human mandible in probable association with a Pleistocene faunal assemblage in eastern Beringia: A preliminary report. *Canadian Journal of Archaeology* 1:81-93.

Jopling, A.V., W.N. Irving and B.F. Beebe. 1981. Stratigraphic, sedimentological and faunal evidence for the occurrence of pre-Sangamonian artifacts in Northern Yukon. *Arctic* 34(1):3-33.

Julig, P., A. Jopling, B. Beebe, J. Alcock, C. D'Andrea, and W. Irving. 1983. Excavation report on an *in situ* bone assemblage from Locality 12, Old Crow River, northern Yukon. In: *Carnivores, Human Scavengers and Predators: A Question of Bone Technology*, ed. G.M. LeMoine and A.S. MacEachern, 15-37. Calgary: University of Calgary Archaeological Association.

Kontrimavichus, V.L., ed. 1984. *Beringia in the Cenozoic Era*. 724 pp. New Delhi: Amerind Publishing Co.

Larsen, H. 1968. Trail Creek: Final report on the excavation of two caves on Seward Peninsula, Alaska. *Acta Arctica* 15. 79 pp.

Laughlin, W.S., J.B. Jorgensen, and B. Frohlich. 1979. Aleuts and Eskimos: Survivors of the Bering Land Bridge coast. In: *The First Americans: Origins, Affinities, and Adaptations*, ed. W.S. Laughlin and A.B. Harper, 91-104. New York: Gustav Fischer.

MacNeish, R.S. 1976. Early man in the New World. *American Scientist* 63(3):316-327.

MacNeish, R.S. 1978. Late Pleistocene adaptations: A new look at early peopling of the New World as of 1976. *Journal of Anthropological Research* 34(4):475-496.

MacNeish, R.S. 1979. Earliest man in the New World and its implications for Soviet-American archaeology. *Arctic Anthropology* 16(1):2-15.

Marks, S.A. 1982. Arguing from the present to the past: A contemporary case study of human predation on African buffalo. In: *Paleoecology of Beringia*, ed. D.M. Hopkins, J.V. Matthews, Jr., C.E. Schweger, and S.B. Young, 409-423. New York: Academic Press.

Matthews, J.V., Jr. 1982. East Beringia during Late Wisconsin time: A review of the biotic evidence. In: *Paleoecology of Beringia*, ed. D.M. Hopkins, J.V. Matthews, Jr., C.E. Schweger, and S.B. Young, 127-150. New York: Academic Press.

McKennan, R.A., and J.P. Cook. 1968. Prehistory of Healy Lake, Alaska. *Proceedings of the VIIIth Congress of Anthropological and Ethnological Sciences*, Vol. 3, pp. 182-184.

Michael, H.N. 1984. Absolute chronologies of Late Pleistocene and Early Holocene cultures of northeastern Asia. *Arctic Anthropology* 21(2):1-68.

Miller, S.J. 1983. Osteo-Archaeology of the mammoth-bison assemblage at Owl Cave, the Wasden Site, Idaho. In: *Carnivores, Human Scavengers and Predators: A Question of Bone Technology*, ed. G.M. LeMoine and A.S. MacEachern, 39-53. Calgary: University of Calgary Archaeological Association.

Miller, S.J. 1984. Characteristics of Mammoth Bone Reduction at Owl Cave, the Wasden Site, Idaho. Paper presented to the First International Conference on Bone Modification, Carson City.

Mochanov, Y.A. 1970. Diuktaiskaia peshchera: Novyi paleoliticheskii pamiatnik Severo-Vostochnoi Azii (Diuktai Cave: A new Paleolithic site in Northeast Asia). *Po Sledam Drevnykh Kul'tur Yakutiia*, pp. 40-64. Yakutsk.

Mochanov, Y.A. 1977. *Drevneishie Etapy Zaseleniia Chelovekom Severo-Vostochnoi Azii (The Earliest Stages of Human Occupation of Northeastern Asia)*. 264 pp. Novosibirsk: Nauka.

Mochanov, Y.A. 1978a. Stratigraphy and absolute chronology of the Paleolithic of Northeast Asia. *Occasional Paper, Dept. of Anthropology, University of Alberta* 1:54-66.

Mochanov, Y.A. 1978b. The Paleolithic of Northeast Asia and the problem of the first peopling of America. *Occasional Paper, Dept. of Anthropology, University of Alberta* 1:67.

Mochanov, Y.A. 1980. Early migrations to America in the light of a study of the Dyuktai

Paleolithic culture in Northeast Asia. In: *Early Native Americans*, ed. D.L. Browman, 119-131. The Hague: Mouton.

Mochanov, Y.A. 1984. Paleolithic finds in Siberia (Resume of studies). In: *Beringia in the Cenozoic Era*, ed. V.L. Kontrimavichus, 694-724. New Delhi: Amerind Publishing Co.

Morlan, R.E. 1977. Fluted point makers and the extinction of the Arctic-Steppe biome in eastern Beringia. *Canadian Journal of Archaeology* 1:95-108.

Morlan, R.E. 1978a. Technological characteristics of some wedge-shaped cores in northwestern North America and northeast Asia. *Asian Perspectives* 19:96-106.

Morlan, R.E. 1978b. Early man in northern Yukon Territory: Perspectives as of 1977. *Occasional Paper, Dept. of Anthropology, University of Alberta* 1:78-95.

Morlan, R.E. 1979. A stratigraphic framework for Pleistocene artifacts from Old Crow River, northern Yukon Territory. In: *Pre-Llano Cultures of the Americas: Paradoxes and Possibilities*, ed. R.L. Humphrey and D. Stanford, 125-145. Washington, D.C.: The Anthropological Society of Washington.

Morlan, R.E. 1980. *Taphonomy and Archaeology in the Upper Pleistocene of the Northern Yukon Territory*. 380 pp. Ottawa: National Museum of Man (Mercury Series: Archaeological Survey of Canada Paper No. 94).

Morlan, R.E. 1983a. Pre-Clovis occupation north of the ice sheets. In: *Early Man in the New World*, ed. R. Shutler, Jr., 47-63. Beverly Hills: Sage Publications.

Morlan, R.E. 1983b. Spiral fractures on limb bones: Which ones are artificial? In: *Carnivores, Human Scavengers and Predator: A Question of Bone Technology*, ed. G.M. LeMoine and A.S. MacEachern, 241-269. Calgary: University of Calgary Archaeological Association.

Morlan, R.E. 1983c. Counts and estimates of taxonomic abundance in faunal remains: Microtine rodents from Bluefish Cave I. *Canadian Journal of Archaeology* 7(1):61-76.

Morlan, R.E. 1984a. Toward the definition of criteria for the recognition of artificial bone alterations. *Quaternary Research* 22:160-171.

Morlan, R.E. 1984b. Biostratigraphy and biogeography of Quaternary microtine rodents from northern Yukon Territory, Eastern Beringia. *Carnegie Museum of Natural History Special Publication* 8:184-199.

Morlan, R.E. 1984c. The Pleistocene archaeology of Beringia. Paper presented to the 17th Annual Meeting of the Canadian Archaeological Association, Victoria.

Morlan, R.E. 1985. Some paleoecological implications of microtine rodents from the Bluefish Caves, northern Yukon Territory. Paper presented to the 18th Annual Meeting of the Canadian Archaeological Association, Winnipeg.

Morlan, R.E. 1986a. Pleistocene archaeology in Old Crow Basin: A critical reappraisal. In: *New Evidence for the Pleistocene Peopling of the Americas*, ed. A.L. Bryan, 27-48. Center for the Study of Early Man. Orono: University of Maine.

Morlan, R.E. 1986b. New dates on the Old Crow Pleistocene. Paper presented to the 19th Annual Meeting of the Canadian Archaeological Association, Toronto.

Morlan, R.E. 1986c. Review of "Environments and Extinctions: Man in Late Glacial North America," ed. J.I. Mead and D.J. Meltzer. *Zooarchaeological Research News* 5(2):10-16.

Morlan, R.E., and J. Cinq-Mars. 1982. Ancient Beringians: Human occupation in the late Pleistocene of Alaska and the Yukon Territory. In: *Paleoecology of Beringia*, ed. D.M. Hopkins, J.V. Matthews, Jr., C.E. Schweger, and S.B. Young, 353-381. New York: Academic Press.

Morlan, R.E., and J.V. Matthews, Jr. 1978. New dates for early man. *GEOS*, Winter (1978):2-5.

Müller-Beck, H. 1966. Paleohunters in America: Origins and diffusion. *Science* 152:1191-1210.

Myers, T.P., M.R. Voorhies, and R.G. Corner. 1980. Spiral fractures and bone pseudotools at paleontological sites. *American Antiquity* 45(3):483-490.

Nelson, D.E., R.E. Morlan, J.S. Vogel, J.R. Southon, and C.R. Harington 1986. New dates on northern Yukon Artifacts: Holocene not Upper Pleistocene. *Science* 232:749-751.

Olivier, R.C.D. 1982. Ecology and behavior of living elephants: Bases for assumptions concerning the extinct woolly mammoths. In: *Paleoecology of Beringia*, ed. D.M. Hopkins, J.V. Matthews, Jr., C.E. Schweger, and S.B. Young, 291-305. New York: Academic Press.

Owen, R.C. 1984. The Americas: The case against an Ice-Age human population. In: *The Origins of Modern Humans: A World Survey of the Fossil Evidence*, ed. F.H. Smith and F. Spencer, 517-563. New York: Alan R. Liss, Inc.

Pavlish, L.A., A.V. Jopling, and Z. Zhenchun. 1986. Selective transport of stone and bone: An experimental approach from flume to field. Paper presented to the 19th Annual Meeting of the Canadian Archaeological Association, Toronto.

Porter, L. 1979. Ecology of a late Pleistocene (Wisconsin) ungulate community near Jack Wade, east-central Alaska. M.S. thesis, University of Washington.

Powers, W.R. 1973. Palaeolithic man in Northeast Asia. *Arctic Anthropology* 10(2):1-106.

Powers, W.R., and T.D. Hamilton. 1978. Dry Creek: A late Pleistocene human occupation in central Alaska. *Occasional Paper, Dept. of Anthropology, University of Alberta* 1:72-77.

Prest, V.K. 1969. Retreat of Wisconsin and Recent ice in North America. Geological Survey of Canada Map 1257A.

Repenning, C.A. 1980. Faunal exchanges between Siberia and North America. *Canadian Journal of Anthropology* 1(1):37-44.

Ritchie, J.C. 1984. *Past and Present Vegetation of the Far Northwest of Canada*. 251 pp. Toronto: University of Toronto Press.

Ritchie, J.C., J. Cinq-Mars et L.C. Cwynar. 1982. L'environnement tardiglaciaire du Yukon septentrional, Canada. *Geographie physique et Quaternaire* 36(1-2):241-250.

Ritchie, J.C., and L.C. Cwynar. 1982. The late Quaternary vegetation of the north Yukon. In: *Paleoecology of Beringia*, ed. D.M. Hopkins, J.V. Matthews, Jr., C.E. Schweger, and S.B. Young, 113-126. New York: Academic Press.

Rogers, R.A. 1985. Glacial geography and native North American languages. *Quaternary Research* 23(1):130-137.

Schweger, C.E., and C. Mandryk. 1986. The Goldeye Lake pollen record: Could man survive the ice-free corridor? *Ninth American Quaternary Association Abstracts*, p. 162.

Schweger, C.E., and J.V. Matthews, Jr. 1984. Mid-Wisconsinan climatic history for Alaska-Yukon as revealed by Old Crow tephra. *Eighth American Quaternary Association Abstracts*, p. 113.

Schweger, C.E., J.V. Matthews, Jr., D.M. Hopkins, and S.B. Young. 1982. Paleoecology of Beringia -- A synthesis. In: *Paleoecology of Beringia*, ed. D.M. Hopkins, J.V. Matthews, Jr., C.E. Schweger, and S.B. Young, 425-444. New York: Academic Press.

Shackleton, N.J., and N.D. Opdyke. 1973. Oxygen isotope and paleomagnetic stratigraphy of equatorial Pacific core V28-238: Oxygen isotope temperatures and ice volumes on a 10^5 and 10^6 year scale. *Quaternary Research* 3:39-55.

Spiess, A.E. 1984. Arctic garbage and New England Paleo-Indians: The single occupation option. *Archaeology of Eastern North America* 12:280-285.

Stanford, D., R. Bonnichsen, and R.E. Morlan. 1981. The Ginsberg experiment. *Science* 212: 438-440.

Steen-McIntyre, V., R. Fryxell, and H.E. Malde. 1981. Geologic evidence for age of deposits at Hueyatlaco archaeological site, Valsequillo, Mexico. *Quaternary Research* 16:1-17.

Szathmary, E. 1985. Peopling of North America: Clues from genetic studies. In: *Out of Asia: Peopling the Americas and the Pacific*, ed. R. Kirk and E. Szathmary, 79-104. Canberra: Australian National University Printery.

Tauber, H. 1968. Copenhagen radiocarbon dates IX. *Radiocarbon* 10(2):295-327.

Tauber, H. 1973. Copenhagen radiocarbon dates X. *Radiocarbon* 15(1):86-122.

Thorson, R.M., and R.D. Guthrie. 1984. River ice as a taphonomic agent: An alternative hypothesis for bone "artifacts." *Quaternary Research* 22:172-188.

Thorson, R.M., and T.D. Hamilton. 1977. Geology of the Dry Creek site: A stratified early man site in interior Alaska. *Quaternary Research* 7:149-176.

Tseitlin, S.M. 1979. *Geologiia Paleolita Severnoi Azii (Geology of the Paleolithic of Northern Asia)*. 286 pp. Moscow: Nauka.

Turner, C.G. II. 1985. The dental search for native American origins. In: *Out of Asia: Peopling the Americas and the Pacific*, ed. R. Kirk and E. Szathmary, 31-78. Canberra: Australian National University Printery.

Valoch, K. 1980. Knochenartefakte aus dem Micoquien (Schicht 7C) in der Kulna-Höhle im Märischen Karst. *Acta Musei Moraviae* 65:7-18.

Valoch, K. 1982. Die Beingeräte von Predmostí in Mähren (Tschechoslowakei). *Anthropologie* 20(1):57-69.

Velichko, A.A., L.L. Isayeva, V.M. Makeyev, G.G. Matishov, and M.A. Faustova. 1984. Late Pleistocene glaciation of the Arctic shelf, and the reconstruction of Eurasian ice sheets. In: *Late Quaternary Environments of the Soviet Union*, ed. A.A. Velichko, 35-41. Minneapolis: University of Minnesota Press.

Vereshchagin, N.K. 1974. The mammoth "cemeteries" of north-east Siberia. *Polar Record* 17(106):3-12.

Vereshchagin, N.K. 1977. Berelekhskoe "kladbishe" mamontov (The Berelekh "cemetery" of mammoths). *Trudy Zoologicheskogo Instituta* 72:5-50.

Walde, K. 1985. Pollen analysis and taphonomy of Locality 15 alluvial sediments, Old Crow Basin, Yukon. M.A. thesis, University of Alberta.

Waters, M.R. 1985. Early man in the New World: An evaluation of the radiocarbon dated Pre-Clovis sites in the Americas. In: *Environments and Extinctions: Man in Late Glacial North America*, ed. J.I. Mead and D.J. Meltzer, 125-143. Center for the Study of Early Man. Orono: University of Maine.

West, F.H. 1967. The Donnelly Ridge site and the definition of an early core and blade complex in central Alaska. *American Antiquity* 32:360-382.

West, F.H. 1975. Dating the Denali Complex. *Arctic Anthropology* 12(1):76-81.

West, F.H. 1981a. *The Archaeology of Beringia*. 268 pp. New York: Columbia University Press.

West, F.H. 1981b. Reflections on Beringian prehistory and early man in America. *Geoscience and Man* 22:103-114.

West, F.H. 1982. Making points with points. *Quarterly Review of Archaeology* 3(3): 6-7.

West, F.H. 1983. The antiquity of man in America. In: *Late-Quaternary Environments of the United States*, Vol. 1: *The Late Pleistocene*, ed. S.C. Porter, 364-382. Minneapolis: University of Minnesota Press.

Westgate, J.A., R.C. Walter, G.W. Pearce, and M.P. Gorton. 1985. Distribution, stratigraphy, petrochemistry, and palaeomagnetism of the late Pleistocene Old Crow tephra in Alaska and the Yukon. *Canadian Journal of Earth Sciences* 22(6):893-906.

Williams, R.C., A.G. Steinberg, H. Gershowitz, P.H. Bennett, W.C. Knowler, D.J. Pettitt, W. Butler, R.Baird, L. Dowda-Rea, T.A. Burch, H.G. Morse, and C.G. Smith. 1985. GM allotypes in native Americans: Evidence for three distinct migrations across the Bering Land Bridge. *American Journal of Physical Anthropology* 66(1):1-19.

Wilmsen, E.N. 1965. An outline of early man studies in the United States. *American Antiquity* 31(2):172-192.

Workman, W.B. 1980. Holocene peopling of the New World: Implications of the Arctic and Subarctic data. *Canadian Journal of Anthropology* 1(1):129-139.

Yi, S., and G. Clark. 1985. The "Dyuktai Culture" and New World origins. *Current Anthropology* 26(1):1-20.

Yurtsev, B.A. 1984a. Problems of the late Cenozoic paleogeography of Beringia in light of phytogeographic evidence. In: *Beringia in the Cenozoic Era*, ed. V.L. Kontrimavichus, 129-153. New Delhi: Amerind Publishing Co.

Yurtsev, B.A. 1984b. Beringia and its biota in the late Cenozoic: A synthesis. In: *Beringia in the Cenozoic Era*, ed. V.L. Kontrimavichus, 261-275. New Delhi: Amerind Publishing Co.

Mastodont Procurement by Paleoindians of the Great Lakes Region: Hunting or Scavenging?

Daniel C. Fisher

Museum of Paleontology
University of Michigan
Ann Arbor, Michigan

Investigations of the evolution of human hunting behavior are complex in many respects, but one of the most challenging issues they must address is the distinction between hunting and scavenging. The difficulty of deciding whether instances of carcass utilization represent hunting or scavenging exists in part because the means of carcass procurement frequently leaves no record on preserved materials. Even when it does, such a record may be obscured by, or confused with, the modification sustained during butchery or other postmortem events (papers by Klein and Behrensmeyer, this volume). The context and nature of an assemblage (as in many Great Plains bison kills; Frison 1978; Todd, this volume) or the preservation of structures showing a unique functional association with hunting (Frison, this volume) sometimes make it possible to reject scavenging as an explanation for a given site. However, judged over the whole history of human utilization of animal carcasses, such a clear choice is more the exception than the rule.

The hunting-scavenging dichotomy may be important not only for descriptions of human subsistence activity, but also for understanding issues such as the timing and causes of some of the changes in early human social organization (e.g., Isaac 1983, 1984; Potts 1983). Predictably, such wide-ranging inferences have attracted close scrutiny, focusing attention on practical and theoretical aspects of distinguishing hunting from scavenging (e.g., Brain 1981; Binford 1981; Shipman 1983, 1986; Potts 1984; papers by Klein, Behrensmeyer, and Binford, this volume). However, the paleoecological significance of the hunting-scavenging dichotomy is not restricted to human behavior. Hunting (by humans) implicates human activity as a factor controlling prey population dynamics and potentially influencing evolution and/or extinction of prey. In contrast, scavenging involves little or no direct impact on the biology of the species utilized.

The aim of this paper is to evaluate the relative incidence of hunting and scavenging associated with Paleoindian utilization of mastodonts in the Great Lakes region, during the late Pleistocene. This implies at least provisional acceptance of previously reported evidence that certain of the sites from this region and time period represent Paleoindian butchery of mastodonts (Fisher 1984a, b; Shipman et al. 1984). On a "local" level, this study may contribute to our understanding of Paleoindians and of the late Pleistocene megafaunal extinctions (Martin and Wright 1967; Martin and Klein 1984). Although this focus is admittedly distant both geographically and temporally from the evolutionary origins of human hunting and scavenging behavior, its relevance to such larger scale problems is two-fold. First, this study serves as an example of the difficulty of empirical discrimination between hunting and scavenging, even in a setting where the record is in many ways more complete than for earlier stages in the evolution of human behavior. More importantly, this work demonstrates arguments and sources of data that may be useful for evaluating the hunting vs. scavenging issue in other contexts, including cases involving proboscideans other than mastodonts. The importance of human/proboscidean associations in the development of ideas concerning human hunting (Klein, this volume), makes this of more than passing interest. In addition, similar approaches may be applied to taxa other than proboscideans.

NATURE AND BACKGROUND OF STUDY

This analysis compares the distribution of sex, age, and season of death within and between samples of butchered and non-butchered individuals. The relevance of these comparisons to the hunting vs. scavenging issue will be discussed below. Determinations of sex, age, and season of death are based on a combination of old and new data on proboscidean dental development, offering more detail on proboscidean growth and life history than has been available previously. Current results pertain primarily to mastodonts, but preliminary examination of mammoths and extant elephants demonstrates that the same approaches should be feasible with them.

My current focus on mastodonts may appear unconventional, given that the proboscideans most often discussed in conjunction with Paleoindians in North America are mammoths. Numerous sites now provide documentation of mammoth remains associated with fluted projectile points and other elements of Clovis technology (see Table 1 in Saunders 1980). Furthermore, considerations such as the suitability of fluted projectile points for use on spears, against large prey, have led many workers to conclude that these associations represent animals that were hunted (e.g., Haynes 1969, 1982; Frison 1978; Saunders 1980), although the possibility of scavenging is also recognized (e.g., Sellards 1938; Haynes 1985).

In contrast to mammoths, mastodonts have a somewhat checkered past in discussions of human/faunal association. They are sometimes cited as plausible prey of Paleoindians (Quimby 1960; Müller-Beck 1966). However, despite a number of mastodont sites described as offering evidence of human involvement (e.g., Sanford 1935; Mehl et al. 1966; Palmer and Stoltman 1976), few have shown unproblematic associations with lithic artifacts (Williams 1957). The exceptions to this pattern (e.g., Irwin-Williams 1967; Bryan et al. 1978; Gustafson et al. 1979; Graham et al. 1981) are of considerable interest but are still small in number. Against this background, treatment of the hunting vs. scavenging question for mastodonts may appear premature. However, comparative taphonomic analysis of several late Pleistocene mastodont sites from southern Michigan and neighboring areas suggests that there is indeed a problem to be addressed. In my assessment, the evidence for human association at some of these sites is strong enough to warrant investigation of the means of procurement of those carcasses that do appear to have been butchered. While extension of the approaches outlined here to mammoths is an obvious next goal, treatment of mastodonts is no less important and can be accomplished with data already in hand.

EVIDENCE OF PALEOINDIAN BUTCHERY OF MASTODONTS

The sites discussed here (Table 1) are late-glacial occurrences of *Mammut americanum*. Radiocarbon dates fall between 10,000 and 12,000 B.P. (except for an erroneously young date on Russell Farm I; Crane and Griffin 1959; cf. Meltzer and Mead 1983), and nondated sites occur in stratigraphic contexts suggesting a similar age. This is a period for which there is clear evidence of Paleoindian presence in the Great Lakes region (Farrand 1977). Depositional environments include a range of stream, pond, bog, fen, and lake settings, and all occurrences (except for Russell Farm, where two individuals were found) are of isolated individuals.

Taphonomic analysis suggests that these sites fall into two categories: those that do, and those that do not, show evidence of butchery (Fisher 1981, 1984a, b). Nine of the sites in Table 1 were treated previously (Fisher 1984b), and the others listed there have been analyzed subsequently by the same methods. The sites in Table 1 include all those excavated within the study area from 1979 to 1985, plus previously excavated specimens from this region, taken in order of decreasing completeness and quality of preservation. This protocol probably introduces no bias with regard to butchery, or a slight bias in favor of including non-butchered individuals (Fisher 1984b). Each assignment to the "butchered" or "non-butchered" categories is supported by at least two independent lines of evidence. Conflicting lines of evidence are treated as cancelling one another, but within this sample, conflicts have been few and have not resulted in indeterminacy. Rather than discuss each site in detail, I will restrict myself here to a summary of the evidence for butchery.

The clearest evidence consists of types of bone modification that are distinct from known styles of nonhuman modification but that bear a recognized functional relationship to the demands of processing a large mammalian carcass. These types of bone modification appear unlikely to be associated with a carcass that has *not* been butchered, but at least moderately

Table 1. Specimen identification and sex and age data for butchered and non-butchered individuals in this study. Age is designated in terms of Laws' (1966) age classes and his interpretation of the associated age of African elephants (A. E. Y., African elephant years). Repositories: BGSU, Bowling Green State University; CIS, Cranbrook Institute of Science; MSU, Michigan State University; UM, University of Michigan; WMU, Western Michigan University.

SITE	REPOSITORY	SEX	AGE	(A.E.Y.)
Butchered:				
Wattles	UM14404	M	IX	(10)
Heisler	MSU-UM	M	XII	(18)
Elkhart	UM 34302	M	XIV	(22)
Van Sickle	UM 58028	M	XVI	(26)
Russell Farm I	UM 37811	M	XVII	(28)
Russell Farm II	UM 57022	F	?	
Winameg	BGSU	M	XX	(34)
Cole	UM	M?	XXI	(36)
Pleasant Lake	UM 57705	M	XXI	(36)
New Hudson	UM 57856	F?	?	
Not Butchered:				
Shelton	CIS	?	XIII	(20)
Sheathelm	MSU-UM	F	XVI	(26)
Quagaman	UM 24240	M	XXI	(36)
Powers	WMU	F	XXII	(39)
Owosso	UM 23498	F	XXII	(39)
Johnson	UM 57648	M	XXII	(39)
Sakstrup	UM 54910	F	XXIV	(45)
Kuhl	UM 59936	M	XXIV	(45)
Taylor	UM 61246	M	?	

likely on one that *has* been butchered. These include superposable pairs of marks on the conarticular surfaces of disarticulated bones, apparently produced during disarticulation of the carcass into manageable units. Scanning electron microscopic (SEM) analysis of such marks demonstrates the similarity of the superposable elements of a pair, relating each to the same "event." It also shows that the stresses associated with the formation of these marks correspond to those produced by driving a bone wedge between the conarticular surfaces of an intact joint, a practice that greatly facilitates disarticulation (Fisher 1984b; Shipman et al. 1984). In addition, some bone fragments show microstructural modifications produced only by exposure to temperatures hotter than those usually encountered in forest fires, but routinely maintained in controlled cooking fires (Shipman et al. 1984). Bones from several sites also preserve marks that appear on SEM examination to be cutmarks produced by stone or bone tools. Recognition of such marks on microscopic grounds alone is difficult due to their similarity to trampling marks (Behrensmeyer et al. 1986). However, contextual evidence (e.g., consistent patterns of anatomical distribution of putative cutmarks, within and between sites; lack of association with medium- to course-grained clastics; rarity of comparable marks on individuals thought on other grounds to be non-butchered) suggests that rejection of all putative cutmarks on this basis would be unwarranted. Finally, some fragments of bone show patterns of fracture that suggest the production of expedient bone tools. A number of these fragments show localized wear that seems most likely to have resulted from use in skinning or in cutting or scraping soft tissues.

Another class of evidence distinguishing butchered individuals consists of patterns of bone distribution and condition that show clear association with each other and with the evidence discussed above, but that seem less strictly determined by structural and functional exigencies of proboscidean butchery. This includes patterns of bone disarticulation and distribution, presence and absence of bones from particular anatomical regions, and the distribution and intensity of carnivore gnawing (Fisher 1984b). These patterns are susceptible to interpretation a posteriori (Fisher 1984b), but the relative likelihood of observing them in natural and human associations cannot at present be determined through a priori arguments. This implies that these patterns may be more open to question *as individual diagnostic criteria.* Nevertheless, their co-occurrence at least corroborates the hypothesis that we are dealing with two types of site with different histories of formation.

In addition to evidence such as that summarized above, the recently discovered Heisler site, in Calhoun County Michigan, shows a distinctive pattern of transverse gouges (unlike toothmarks or clawmarks) on a young mastodont's skull roof. These gouges were apparently all made by the same object and could have been produced during removal of the skin or a layer of subcutaneous fat such as is known from (frozen) mammoths (Augusta and Burian 1962). In addition, the Heisler site shows a pattern of bone distribution that warrants further analysis. The bones occur within a sequence of pond sediments, in multiple, discrete clusters. Each cluster is composed of the presently disarticulated remains of several portions of the carcass that are not necessarily derived from anatomically contiguous regions. In general, butchering damage is limited to points at which the carcass was disarticulated into major segments (i.e., little evidence of meat removal from most segments). These patterns may represent, in part, preliminary butchery followed by caching of portions of the carcass by submersion in the cold waters of the pond (season of death, discussed later, was late autumn or early winter). Similar styles of caching are known from ethnographic reports (Taylor 1969). The patterns on which this interpretation is based require replication at other sites before their significance can be fully evaluated. However, a first step in this direction is the observation that the suites of bones recovered from several of the other putative butchery sites correspond approximately, in anatomical composition, to the contents of one or more of the Heisler clusters. This suggests that the complementary patterns of bone presence and absence noted earlier (Fisher 1984b) might correspond not to "primary" and "secondary" butchery sites, but rather to caches of relatively standard anatomical composition, some of which were recovered by Paleoindians (presumably with meat attached) and some of which are recovered by paleontologists.

The arguments summarized here build on a large body of literature, including studies of modified proboscidean bone from other localities (e.g.,

Bonnichsen 1979; Stanford 1979; Morlan 1980), bone tool production and use (e.g., Frison 1978; Stanford et al. 1981; Johnson 1982), carnivore processing of bone (e.g., Haynes 1980, 1983; Binford 1981; Brain 1981), butchery vs. natural disarticulation (e.g., Coe 1978; Hill 1979; Hill and Behrensmeyer 1985), and diagnostic microstructural features produced by various agents of bone modification (Potts and Shipman 1981; Shipman 1981). Moreover, the patterns that suggest mastodont butchery in the Great Lakes region show similarities to other cases of proboscidean butchery (e.g., Frison 1978; Miller and Dort 1978; Hannus 1983).

Despite the evidence of butchery discussed here, the apparent absence of lithic artifacts (and/or debitage) may be regarded by some workers as anomalous. However, it may not be appropriate to take this negative finding at face value. Of the sites in Table 1, only Johnson, Powers (Garland and Cogswell 1985), and Shelton have been excavated under conditions that allowed extensive sieving or screen washing, and none of these is interpreted here as showing evidence of butchery. Most of the other excavations have been essentially salvage efforts, with serious limitations of time and personnel. Some sediment has been wet-sieved from Pleasant Lake, New Hudson, Van Sickle, Sheathelm, and Taylor, and a larger sampling program is underway at Heisler, but neither lithic artifacts nor unambiguous debitage have been recovered. Although a positive outcome from any of these attempts might have been significant, we have sampled such a small amount of sediment that a negative outcome is not necessarily significant. Two additional considerations also bear on this negative finding. First, as pointed out previously (Fisher 1984b), there is independent archeological evidence of a high degree of curation and conservation of lithic materials by Paleoindians of the Great Lakes region (Fitting 1966). In conjunction with this behavior, discard of artifacts at butchery sites may have been less common in this region than in some others. Second, and perhaps more importantly, *if* caching of portions of butchered mastodont carcasses in ponds or bogs is responsible for the preservation and spatial patterning of skeletal elements, the in situ location of bones may consistently differ from the point at which carcass processing took place -- the most probable location of lithic artifacts or debitage. At the

Heisler site, for example, processing probably occurred along or beyond the pond margin, rather than within the pond, at currently recognized bone clusters. Sampling of the pond margins and peripheral areas has begun, but its outcome is still unclear. In any event, if this interpretation is applicable to other butchery sites, much previous effort may have been expended looking in the "wrong" place for lithic material.

DEFINITIONS OF HUNTING AND SCAVENGING

Hunting and scavenging are activities in which many species participate, but the focus of this paper is on hunting and scavenging by humans. Where hunting and scavenging are mentioned without explicit qualification, "by humans" is usually implied. In certain instances, a statement may hold for both human and nonhuman activity, but little inaccuracy is anticipated from focusing on the human application. If it is important to restrict a statement to nonhuman predators or scavengers, I will do so explicitly.

In order to discriminate consistently between hunting and scavenging, these terms must be defined so as to be mutually exclusive. In conventional usage, this is accomplished in part by distinguishing hunting and scavenging according to the proximate cause of death. An animal killed by direct assault of a predator is said to have been hunted, whereas an animal utilized after death, by individuals not associated with its cause of death, is said to have been scavenged. Thus, a given animal may be both hunted and scavenged (in turn), but a given interaction is classified as either hunting or scavenging -- not both. This usage suffices for most purposes, but a potentially intermediate condition arises when a predator (human or otherwise) kills and utilizes an animal that is moribund (i.e., already has a negligible likelihood of survival) as a result of injury or some other agency independent of that predator. This does not represent scavenging in the sense of gaining access to the carcass of an animal that is already dead, but neither does it represent the logistic (and for humans, technological) challenge of hunting active and otherwise viable (i.e., nonmoribund) prey. In addition, utilizing moribund prey is incidental to prey population dynamics, since the die, so to speak, has already been cast for moribund individuals. In both these respects, utilizing moribund prey resembles scavenging more than hunting. This complication is familiar to ecologists who recognize that the impact of predation on prey populations depends on the distribution of reproductive potential among the individual prey selected by a given species of predator. Indeed, morbidity represents only one (extreme) interval along a continuous spectrum of prey condition and reproductive potential.

To ignore this complication implies either that the primary phenomena of interest are "Who kills whom?" and "Who eats what?" or that the "intermediate condition" can be assumed to be rare relative to more clearcut instances of hunting and scavenging. Neither of these generalizations seems appropriate for the present investigation. We are clearly more interested in the overall predatory capability of a potential predator (humans), or in its impact on potential prey (proboscideans), than in the specific cause of death of an individual. Furthermore, although the incidence of humans killing and utilizing moribund proboscideans is unknown, discussions of human/proboscidean association often refer to this type of interaction. Thus, a decision regarding its classification cannot be avoided. Ultimately, the most important consideration is pragmatic: with respect to the evidence of hunting and scavenging discussed below, utilization of moribund prey produces a record distinct from "other" instances of hunting, but similar to instances of scavenging. I do not regard this as unfortunate, since it prescribes a categorization of interactions that corresponds closely, as noted above, to our interests as anthropologists and paleontologists. I will therefore treat utilization of moribund prey as a subset of scavenging rather than a subset of hunting. Distinguishing hunting from scavenging on the basis of prey viability, rather than the proximate cause of death, implies a "fuzzier" boundary between these categories (due to uncertainty in judging likelihood of prey survival). However, it also results in categories that are more homogeneous in their implications for both human behavior and prey populations.

Conventional definitions of hunting and scavenging may still be preferable for other investigations, but for this analysis I will use the following definitions. Expressed in terms of humans, scavenging includes all cases of organisms whose proximate cause of death did not involve human activity, but were subsequently encountered and utilized by humans, or that were dying by a proximate cause (i.e., had a proximate cause of morbidity) that did not involve human activity, but were encountered, killed, and utilized by humans who thus became a jointly proximate cause of death. Hunting by humans includes all cases of organisms whose proximate cause of death involved only human activity. This definition does not mean that only assaults on healthy, vigorous prey will be classified as hunting. Prey may indeed be weakened by some other cause and still "qualify" for hunting; the critical point is that they have a non-negligible chance of survival just prior to their encounter with humans. Hunting by humans will be contrasted with "natural death," which includes all cases of organisms whose proximate cause of death did not involve human activity.

PLAN OF ARGUMENT

Before evaluating the evidence for hunting or scavenging, it is important to clarify the type of argument that will be used. The major problem faced in this analysis, aside from the general absence of a direct record of the means of carcass procurement, is that there are no cases of "known" (i.e., observed or strongly inferred) hunting or scavenging *from a sufficiently similar context* to guarantee that their attributes might be generalized and taken as diagnostic of hunting or scavenging in the context at issue here. The contextual variables that I suspect may be significant include (without being limited to): the identity and behavior of the predator/scavenger and prey/carcass species; available technology (in the case of human predator/scavengers); climate; and the identity, behavior, and abundance of competing predator/scavengers. A "known" case that might be comparable to Paleoindians and mastodonts of the late Pleistocene Great Lakes region would be the more or less contemporary Paleoindian/mammoth associations of the Great Plains and southwestern United States. As noted above, these are generally considered to represent hunting.

However, even this relatively close comparison is complicated by differences in climate and "target" species, and by possible cultural differences between Paleoindians of these two regions. Other human/proboscidean associations are less well constrained with regard to procurement, offer even dimmer prospects of contextual comparability, or both.

The problem of potential confounding variables threatens any attempt to infer means of procurement by comparisons between samples derived from different contexts. However, even if these risks are judged acceptable, such an approach leaves the inference for the case in question dependent on the "known" status of the case being compared with it. While I do not have particular reason to doubt the conventional interpretation of the classic Paleoindian/mammoth sites (but see Haynes 1985), I would prefer in principle to evaluate each case using evidence as nearly intrinsic to it as possible. Among other desirable outcomes, this minimizes the threat of logical circularity in subsequent comparative studies. The ideal realization of this strategy would treat each butchered individual as a separate case and would rely on recognition of features that are unambiguously diagnostic of either hunting or scavenging. A relatively close approach to this goal is the use of cutmarks crosscutting toothmarks to refute hunting and thereby support scavenging (e.g., Potts and Shipman 1981; discussed further in the Appendix). The butchered individuals in Table 1 have been examined for such marks, but none have been found. This outcome is compatible, however, both with hunting and with some scavenging interpretations (e.g., scavenging an animal that died of an injury that was not predator-related). We furthermore have no constraint on the relative frequency of the types of scavenging for which cutmarks crosscutting toothmarks would be expected and those for which they would not. Therefore, the failure to find such marks does not constitute a basis for choosing between hunting and scavenging.

Searching for attributes with diagnostic potential comparable to that of cutmarks crosscutting toothmarks, yet applicable to the butchered individuals in Table 1, is complicated by the very fact that makes this investigation so intriguing -- that we currently know little about how Paleoindians might have hunted or scavenged mastodonts, let alone the relative or absolute frequency with which they might have engaged in either activity. It is therefore difficult to predict on a priori grounds the attributes that would be expected to characterize cases of hunting or scavenging. Although this may appear to thwart any attempt at rigorous evaluation of competing hypotheses regarding mastodont procurement, our prospects are not in fact so bleak.

The approach used here departs in several respects from the ideal strategy described above. One compromise is to relax the insistence on treating individuals separately and focus instead on the means of procurement of the whole sample of butchered carcasses. This allows the study to treat attributes whose relationship to procurement is more complex than that of attributes that imply a decision between support and refutation, support and indeterminacy, or refutation and indeterminacy, based on observation of one of two identifiable states of the attribute. In particular, I will deal with attributes for which only the relative frequency of states of the attribute within a *sample* of butchered carcasses is supposed to reflect means of procurement. While this allows a greater number of attributes to be used, it also complicates the problem by requiring consideration of a larger set of hypotheses, ranging from "all hunted," through various "mixed" interpretations (i.e., some individuals having been hunted and others scavenged), to "all scavenged."

Another shift in strategy is that rather than compare the butchered sample (Table 1) of unknown procurement history to a butchered sample of "known" procurement history, I will compare the "unknown" sample to the non-butchered sample in Table 1. I do not claim to be able to predict the attributes of either sample a priori, but aspects of the similarity or difference *between* butchered and non-butchered samples can be interpreted in light of general ecological principles that are testable in a wide range of contexts. Most importantly, the outcomes of comparisons between butchered and non-butchered samples can be shown to have different probabilities under competing procurement hypotheses. The tractability of this approach is enhanced by the derivation of both butchered and non-butchered samples from the "same" context -- at least to the extent that relevant biological, cultural, and climatic factors can be resolved. Although this expands the analysis beyond evidence intrinsic to single butchered individuals, it at least retains the focus on evidence intrinsic to a single, broadly-defined context.

The importance of comparisons between butchered and non-butchered samples may not be immediately obvious. The butchered sample represents those individuals for which the means of procurement is in question, but what do non-butchered individuals have to do with this issue? In short, non-butchered individuals are in all probability natural deaths. As such, they are to some degree representative of the original sample of natural deaths and/or moribund individuals from which butchered individuals would have been selected, *if* the latter were procured by scavenging. On the other hand, *if* butchered individuals were procured by hunting, their attributes would reflect a selection process (i.e., the suite of factors controlling prey selection and hunting success by Paleoindians) acting independently of the processes associated with natural death. The significance of this difference will be developed further below.

The argument that non-butchered individuals represent natural deaths proceeds by ruling out, at least provisionally, alternative explanations. For instance, one possibility is that the "non-butchered" interpretation is in error and that these individuals were in fact hunted and butchered, but without any (perceived) trace of such activity. However, this seems unlikely given that evidence for butchery shows up clearly in other cases from the same general population. If the "non-butchered" interpretation is accepted, these individuals might still have been hunted: if they represent "ones that got away" but died later from wounds received during their encounter with humans; if they were killed directly but usurped by nonhuman scavengers prior to butchery; or if they were killed directly but not butchered because of time constraints or poor physical condition (e.g., low fat content; Speth 1983). The first of these possibilities has been proposed, on very reasonable grounds, for the Naco mammoth (Haynes 1966), but no comparable evidence supports interpreting the non-butchered individuals in Table 1 in this manner. The second possibility might be more of an issue if we were dealing with early hominids. However, given the technological sophistication of Paleoindians, if they were indeed capable of killing a mastodont, I suspect that they could have kept potential scavengers at bay long enough to proceed with butchery. Finally, the third possibility may be important in explaining incomplete butchery or non-butchery of certain carcasses in mass kill settings (Speth 1983; Todd, this volume). However, as noted above, the carcasses in Table 1 appear to represent isolated individuals. In such situations, physical condition of the prey and the amount of time available for processing could probably have been assessed before killing the animal. Assuming that the risk of hunting would not be undertaken without a high likelihood that the hunt, if successful, would provide important resources, this explanation would not be likely to apply to many of the non-butchered carcasses in this study. For these reasons, it seems safe to identify non-butchered individuals as natural deaths.

The potential relationships between butchered and non-butchered samples can be elucidated further by reference to the flowchart in Fig. 1. This diagram is divided into "fields" bounded by dashed lines. Horizontal boundaries distinguish a series of conditions through which organisms (or their remains) pass on their way toward representation in a fossil sample, and vertical boundaries distinguish some of the agents responsible for transitions along this series. The transitions relevant to this analysis are indicated by arrows. Some conceivable transitions are omitted because they are judged to occur infrequently (e.g., a carcass of an individual hunted by humans but stolen by nonhuman scavengers prior to butchery) or are irrelevant to evaluation of the competing hypotheses of mastodont procurement (e.g., nonhuman scavenging of butchered remains). All individuals eventually undergo the transitions involving death and decomposition, but preservation and collection exhibit no such comprehensiveness. In Fig. 1, hypothetical samples located within each "field" are characterized by numbered histograms describing the frequency distribution for a hypothetical multi-state attribute, assessed separately for each sample. The attributes of interest here include any "premortem" feature of individual organisms that is recorded in fossil material. By "premortem," I mean any feature manifested within members of the living population and "fixed" at or before the time of death. With the exception of age at death, the attributes examined in this study (sex, age, and season of death) have little conventional association with evaluating alternative procurement hypotheses. They are all

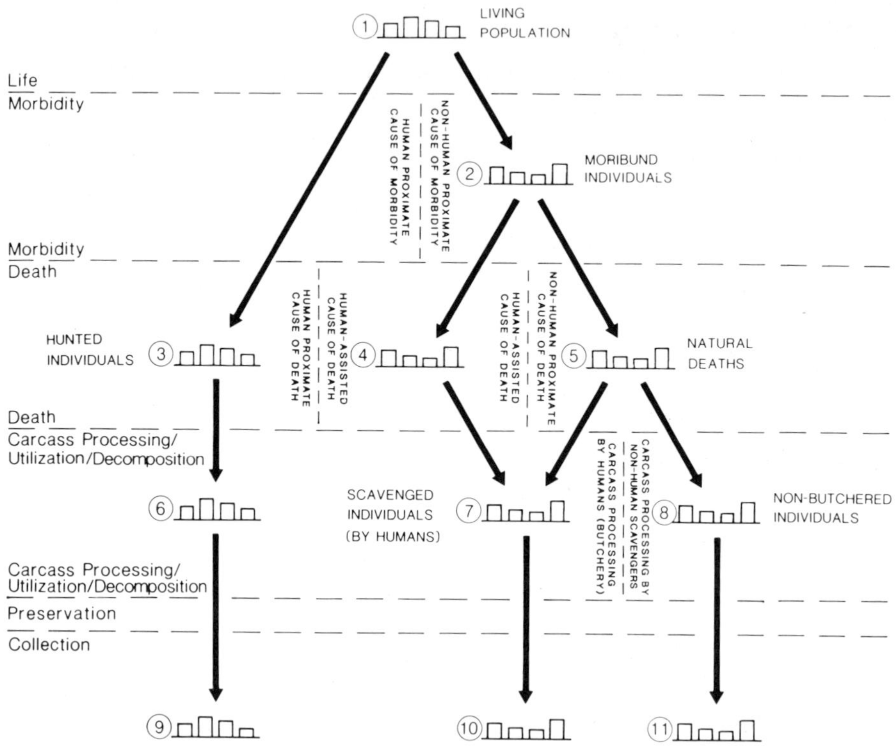

Figure 1. Flowchart illustrating possible pathways of progressive derivation of samples from a living population and consequent opportunities for alteration of the distribution of an attribute within samples. The diagram is partitioned into regions (bounded by dashed lines and identified by text along boundaries) representing conditions through which samples typically pass. Arrows represent the major transitions between samples shown here. Histograms represent the distribution of a given attribute within successive samples and have been constructed (arbitrarily) to resemble an age profile (the relative frequency of individuals in four age classes). Numbers at left of histograms are used in the text to refer to particular samples or transitions between samples. In addition, commonly cited samples are given titles at left or right.

nonetheless appropriate for the type of argument proposed here.

Each transition in Fig. 1 represents a set of events or processes that have the potential to alter the distribution of a given attribute in the succeeding sample. The nature and degree of such alteration -- or bias -- depends on the selectivity with which that transition operates. However, if a transition is comprehensive (i.e., entrains essentially all individuals present in the preceding sample) or if it operates in an effectively random manner with respect to a given attribute, the distribution of that attribute will be conserved across that transition. The samples available for analysis emerge from such a network of interactions. The non-butchered individuals in Table 1 comprise, as argued above, a sample that is probably equivalent to sample 11. They thus reflect the cumulative effects of the transitions (and their associated events and processes) leading from the original living population (sample 1) to sample 11. On the other hand, the butchered individuals in Table 1 comprise a sample that is either hunted, scavenged, or some combination of both. If hunted, the butchered sample would be equivalent to sample 9; if scavenged, it would be equivalent to sample 10; and if some individuals were hunted and others scavenged, it would represent a combination (in some particular proportion) of samples 9 and 10. Choosing among these possibilities is precisely the problem at hand. In any event, the attribute distributions of the butchered sample reflect the cumulative effects of the transitions intervening between it and the original living population. Of more immediate relevance, however, is that the outcome of comparing the distributions of a given attribute in butchered and non-butchered samples depends on the *cumulative differential* effects of the transitions linking these two samples through their most recent common source. If butchered individuals were scavenged, this most recent common source is the original "pool" of natural deaths and/or naturally moribund individuals (samples 2 and/or 5). However, if butchered individuals were hunted, their most recent source in common with non-butchered individuals is the original living population (sample 1).

If butchered individuals were procured by scavenging, a first-order expectation might be that they would represent a random sample (with respect to a given attribute) of natural deaths and/or naturally moribund individuals. If, in addition, the processes of preservation and collection do not *differentially* alter the distributions of that attribute in butchered (whether hunted or scavenged) and non-butchered samples -- an assumption that I will refer to below as the "taphonomic auxiliary hypothesis" -- it follows that the butchered sample (here interpreted as scavenged) should show an attribute distribution similar to that seen in the non-butchered sample. In contrast, if butchered individuals were procured by hunting -- i.e., by a process operating independently of the processes of natural morbidity and death -- then there is no compelling reason to expect an attribute's distribution in the butchered sample (here interpreted as hunted) to resemble that in the non-butchered sample.

The first-order assumption of random scavenging may, however, be too simplistic. It is indeed possible that either hunting or biased scavenging could lead to disparate distributions of a given attribute in butchered and non-butchered samples. In addition, hunting could coincidentally produce an attribute distribution similar to that seen in the non-butchered sample. Nevertheless, arguments drawn from theoretical ecology, taken in conjunction with the taphonomic auxiliary hypothesis, suggest that even without assuming that scavenging by humans is random, the probability of disparate attribute distributions in butchered and non-butchered samples is greater if butchered individuals were hunted than if they were scavenged. Likewise, the probability of identical attribute distributions (within sampling error) in butchered and non-butchered samples is greater if butchered individuals were scavenged than if they were hunted. The arguments leading to these conclusions are exemplified in the sections discussing data and are developed at greater length in the Appendix.

Regardless of whether scavenging by humans is random with respect to a given attribute, it is unlikely to be comprehensive for any one (or more) states of that attribute. In other words, even when there is a bias in favor of scavenging individuals that show certain attribute states, some individuals showing those states may be expected to go unscavenged (by humans), if only because humans do not always locate them soon enough to make them worth salvaging. This point is also developed further in the data sections and Appendix, and it leads to a relatively straightforward expectation (which again requires the taphonomic auxiliary hypothesis). If butchered individuals were scavenged, a large enough non-butchered sample should include representatives of each attribute state found in the butchered sample (and possibly, but not necessarily, other attribute states as well). On the other hand, if butchered individuals were hunted, there is no intrinsic reason to expect this type of overlap in attribute distributions, since butchered and non-butchered samples would then have been derived from the living population by separate processes.

The usefulness of overlap in attribute distributions for evaluating alternative procurement hypotheses depends on the observed representation of attribute states in the non-butchered sample. If the non-butchered sample includes representatives of all recognizable attribute states, the type of overlap described above is inevitable under both hunting and scavenging interpretations. If the non-butchered sample includes representatives of most attribute states, extensive overlap may still be expected under a hunting interpretation. It may develop, for instance, either because hunting and natural death operate similarly with respect to a given attribute, or because there remain few options involving failure of overlap. However, the more uneven the attribute distribution seen in the non-butchered sample and the greater the number of resolvable attribute states, the more likely it is that the separate processes of hunting and natural death will lead to some failure of overlap of butchered by non-butchered samples. In particular, a case involving complete failure of overlap (i.e., for a given attribute, no states occupied by both samples) is unlikely under any hypothesis involving scavenging, but entirely plausible if all butchered individuals were hunted.

In summary, an outcome in which an attribute's distribution in the butchered sample is identical (within sampling error) to its distribution in the non-butchered sample is more likely under the scavenging hypothesis than under any hypothesis involving hunting. It thus tends to corroborate scavenging and refute both hunting and a combination of hunting and scavenging. By similar reasoning, an outcome exhibiting pronounced disparity of attribute distributions, including complete failure of overlap in occupation of attribute states, tends to corroborate hunting and refute both scavenging and a combination of hunting and scavenging. Intermediate outcomes have less extreme implications, though they may still favor some interpretations over others. All possible outcomes are ultimately subject to multiple conceivable interpretations, but by examining several attributes, the range of strongly supported interpretations may be substantially narrowed.

The pattern of reasoning outlined above can, with strategically chosen assumptions and qualifications, be phrased so as to *resemble* a standard, "falsificationist" style of argument. However, its actual structure corresponds more closely to the reasoning involved in "Bayes' Theorem," or the field of Bayesian statistical inference in general (Burks 1977; Rosenkrantz 1977; Iversen 1984). An underlying thesis of this analysis is that Bayes' Theorem provides a more accurate and explicit portrayal of the interaction between assumptions, data, and conclusions than is evident in the superficially simple syllogisms of falsificationist logic. The differences between these two styles of argument are relatively minor in some instances but become progressively more important with increasing complexity of the issues. Fortunately, Bayesian insights are already implicit in many of our "common sense" patterns of reasoning. I will therefore discuss the data and their implications without further belaboring epistomological concerns. Nevertheless, a Bayesian formulation helps to "discipline" what might otherwise appear to be a rather freewheeling interplay of evidence and interpretation. In order to clarify this perspective and demonstrate the potential generality of the arguments used here, I will discuss Bayes' Theorem and other aspects of methodology in the Appendix.

SEX RATIO AS EVIDENCE OF HUNTING OR SCAVENGING

The first attribute to be discussed here is the sex of individuals. This is a "premortem" attribute in the sense described above (determined well before death), and its state can be inferred from analysis of sexual dimorphism. Since we recognize only two states, the "distribution" of sexual identity in a sample is reducible to a single value, the sex ratio. This fundamental simplicity limits the possible configurations of data, and hence the likely degree of resolution among interpretations (based on these data). Nevertheless, comparison of sex ratios in butchered and non-butchered samples does bear on the relative probability of competing procurement hypotheses.

Source of Data

Identification of the sex of individuals in Table 1 is based mostly on widely recognized criteria, discussed in some detail by Osborn (1936) and thus not treated at length here. The most important of these are a striking dimorphism in tusk size (especially tusk basal diameter, considered relative to tusk length) and a relatively pronounced dimorphism in body size among sexually mature individuals. Both of these relationships may be variable from one time and region to another. However, in samples aggregated on the scale used here, these criteria suffice to sort most individuals into two groups that can be identified to sex by analogy with extant elephants (larger tusk and body size in males). In addition, these identifications have been corroborated using the shape of the tusk pulp cavity, a character recognized as sexually dimorphic in elephants by Sikes (1971), and the thickness of annual increments of tusk dentine, discussed below. Identifications of sex are not yet possible for animals that died while sexually immature (except for males, which *may* have mandibular tusks), but most individuals in Table 1 were apparently sexually mature at death (see life history discussion).

Documentation of sexual dimorphism and of the identifications in Table 1 will be presented elsewhere. Although different degrees of corroboration are available for different individuals, most of the identifications do not appear to be problematic. My only reservations are as follows: Wattles has been sexed only by the presence of two, small alveoli in the position of mandibular tusks; Shelton has not yet been analyzed for sex; and Cole and New Hudson have been sexed using skeletal measurements for which good comparative data are as yet lacking.

Attribute-Specific Statement of Argument

Before turning to comparison of sex ratios of butchered and non-butchered samples, the argument to be used should be framed explicitly in terms of this attribute. By the approach proposed in the *Plan of Argument*, identical sex ratios (within sampling error) in butchered and non-butchered samples would be interpreted as support for scavenging, and a significant difference between these sex ratios would be at least suggestive of hunting. But what particular considerations having to do with the sex of individuals allow such inferences? This question may be addressed by first supposing that all butchered individuals were procured by scavenging, then by supposing them all to have been procured by hunting, and finally by comparing these two interpretations.

As indicated in Fig. 1 and as explained in the *Plan of Argument*, non-butchered individuals are interpreted as natural deaths that were not scavenged by humans. They are thus representative of the "complement" of those natural deaths and naturally moribund individuals (if any) that humans do scavenge (since all natural deaths and naturally moribund individuals either are, or are not, scavenged by humans). If the butchered sample was procured by scavenging, any bias presumed to have originated between samples 7 and 8 in Fig. 1 implies that the incidence of scavenging by humans was disproportionate relative to the abundance of the two sexes in the "pool" of natural deaths and moribund individuals from which scavenged individuals were drawn. Biased scavenging may result from a positive bias on the part of humans (i.e., *for* one sex or the other), from a negative bias on the part of humans (*against* one sex or the other), from disproportionate availability of one sex or the other due to negative or positive bias on the part of nonhuman predator/scavengers, or from some unbalanced combination of these factors. Scavenging biases are indeed conceivable (e.g., *for* males, because of their larger body size; *against* females, if they are in poor physical condition after calving). However, with incomplete knowledge of the behavior of Paleoindians and competing nonhuman predator/scavengers, it is difficult to determine on a priori grounds whether bias between a scavenged and a non-butchered natural death sample is probable or improbable.

On the other hand, if all butchered individuals were procured by hunting, then in the case of an attribute that is fixed throughout life (as sexual identity in mastodonts would have been) the non-butchered sample represents a complement of human hunting activity. Under such circumstances, a bias between butchered and non-butchered samples may result from any of the types of interaction noted above: a bias (positive or negative) intrinsic to human hunting behavior, disproportionate availability of the sexes due to a bias intrinsic to some agent of natural death, or some combination of these factors. However, incomplete knowledge of Paleoindian hunting behavior and of the relevant dynamics of natural death in mastodonts makes it difficult to predict whether bias between a hunted and a non-butchered natural death sample is probable or improbable.

Despite the apparent intractability of this problem when hunting and scavenging are considered separately, *comparison* of these two activities (and assessment of the relative conditional probabilities of possible outcomes) yields a more nearly determinate solution. Any of the anatomical correlates of sex that might be construed as plausible grounds for biased scavenging would also be potential biasing factors in hunting. However, dealing with a live and at least relatively healthy animal, rather than a moribund or dead one, implies that behavioral correlates of sex (e.g., differences in temperament, degree of spatial isolation from conspecifics) would be added to the list of potential biasing factors in hunting (whereas they would generally be irrelevant to scavenging). Since sex-specific behavioral attributes could significantly alter the cost/benefit ratio of hunting members of a particular sex, they increase the number of opportunities for causal interactions that could result in bias. From this perspective, although it is still not possible to specify whether sex bias in hunting is probable or improbable, it is at least *more* probable than a sex bias in scavenging.

Another argument leading to this same conclusion concerns the spatiotemporal distribution of live mastodonts of either sex, as compared to moribund or dead ones in scavengeable condition. It seems reasonable to assume that live mastodonts would almost always have been both more abundant and more spatiotemporally predictable than scavengeable mastodonts. Hunting would thus have allowed and required greater selectivity than is usually possible in scavenging. In other words, as a relatively rare and unpredictable resource, a scavengeable carcass would be more likely to be utilized regardless of sex, while live prey would be more likely to induce a selective response toward one sex relative to the other. Two other arguments supporting this view are presented in the Appendix.

Based on these considerations, if we could census sex ratios in "fresh," roughly contemporaneous samples of butchered and non-butchered carcasses (equivalent to samples 6/7 and 8 in Fig. 1) from the same general region, the probability that the sex ratio in the butchered sample would be significantly biased relative to that of the non-butchered sample appears to be greater if the butchered individuals were procured by hunting than if they were procured by scavenging. I do not assert this as a self-evident truth, nor do I wish to rely on it as a blind assumption. Rather, it is a generalization based on observations of natural history and principles of theoretical ecology. As a general proposition (i.e., not bound to Paleoindians and mastodonts in particular), it is testable in a wide range of ecological contexts. In any event, I will refer to this statement of relative conditional probability as a "behavioral auxiliary hypothesis." It is "behavioral" in that it summarizes the effects of human and nonhuman behavior, as they bear on the development of similarities and differences in attribute distributions between butchered and non-butchered samples. By "auxiliary hypothesis" I mean to imply that I take this proposition to be relatively unproblematic and therefore accept it provisionally for use in evaluating competing "focal hypotheses" (the alternative interpretations of the means of procurement of butchered carcasses).

As mentioned in the *Plan of Argument*, I will also use a "taphonomic auxiliary hypothesis" to summarize the effects of taphonomic processes that act on, and sometimes distort, attribute distributions. The taphonomic processes

7-10, and 8-11 transitions in Fig. 1. Framed in terms of sex ratios, the taphonomic auxiliary states that a sex bias in butchered samples relative to non-butchered samples is equiprobable, whether the butchered individuals are hunted or scavenged. Given the *overall* similarity of the sexes in anatomy and gross body size (considered over the whole lifespan), I would actually expect little or no sex-related taphonomic bias. However, even if there is a sex-related taphonomic bias, I would not expect it to operate in a different fashion depending on whether or not the animals were butchered, and certainly not depending on whether the butchered animals had been hunted or scavenged. Only the last of these three points is essential to the taphonomic auxiliary, and this seems quite secure.

The conjunction of the above behavioral and taphonomic auxiliary hypotheses allows us to conclude that the probability of finding a statistically significant difference between the sex ratios of fossil samples composed of butchered individuals (on one hand) and non-butchered individuals (on the other) is greater if the butchered sample was procured by hunting than if it was procured by scavenging. Given this, finding a significant difference tends to support hunting, whereas finding no significant difference tends to support scavenging.

In the *Plan of Argument*, another approach to evaluating competing procurement hypotheses was proposed, involving the pattern of overlap in occupation of attribute states between butchered and non-butchered samples. However, this is unlikely to be applicable to premortem attributes that are constant throughout the lifespan (such as sex in mammals). In dealing with such attributes, it is overwhelmingly probable that *some* individuals representing *each* attribute state will die of natural causes (i.e., other than by human predation) and go unscavenged by humans, thus entering the non-butchered sample. Indeed, as will be seen below, the non-butchered sample in Table 1 includes both sexes. In this situation, overlap in occupation of attribute states of butchered by non-butchered samples is assured, whether butchered individuals were hunted or scavenged. Overlap pattern therefore cannot in this instance be used to evaluate competing procurement hypotheses.

Data Analysis

The sex ratio of non-butchered individuals in Table 1 is as close as it could be to 50% in a sample of this size (4 males, 4 females, and 1 undetermined). In contrast, the butchered sample shows a relatively strong predominance of males (probably 8 males and 2 females). The two samples thus clearly differ in terms of actual numbers. Interpreted at face value, this could be judged as support for hunting. However, the sex ratio of the butchered sample is really only an estimate of that parameter in the total population of butchered individuals. In analogy to the question typically raised in coin-tossing experiments, we might ask, "What is the probability of getting a distribution as uneven as this (or worse) if the actual sex ratio of the butchered population was 50%?" Using the binomial probability distribution, $P = 0.109$ for 8 out of 10 individuals belonging to the same sex. In other words, this result is moderately improbable under a null hypothesis that the sex ratio of the total butchered population was 50% (i.e., identical to that of the non-butchered sample), but not improbable enough to reject this null hypothesis at the .05, or even the .10 significance level. It is of course possible that the apparent sex ratio of the butchered sample is an accurate reflection of that parameter, that it differs from the sex ratio of the non-butchered sample, and that it only failed significance testing because of small sample size. Nonetheless, present results are inconclusive. The outcome suggests some slight preference for hunting (of at least some individuals), but the cautious approach would be to consider the question unresolved.

Discussion

It is part of the nature of a Bayesian approach to inference (see Appendix for further discussion), and thus of the arguments proposed here, that preference for one hypothesis relative to a second is not equivalent to definitive rejection of the second in favor of the first. What emerges instead is a statement regarding the differential support provided to various hypotheses by available data. Furthermore, this support is conditional on background information and assumptions. Most analyses therefore allow a number of interpretations. Far from being a sign of inferential incompetence, these options point out real interdependencies in the data, assumptions, and conclusions.

In the case of sex ratios, if accumulation of additional data shows that the sex ratios of butchered and non-butchered mastodonts are indistinguishable, the choice presented would be between (1a) considering scavenging to be corroborated in preference to hunting or (1b) considering one or both auxiliary hypotheses, however valid in general, to have been misleading in this instance (for any of a particular range of possible reasons). If, on the other hand, additional data show that the sex ratios are significantly different, the choice would be between: (2a) considering hunting preferable to scavenging or (2b) again considering one or both auxiliaries to have been misleading (though for different possible reasons than those addressed in 1b). These options may be clarified further by fleshing them out with a (nonexhaustive) variety of illustrative interpretations. 1a means that procurement of all butchered carcasses by scavenging is a more likely explanation of identical sex ratios in butchered and non-butchered samples than are alternative explanations allowing some or all butchered individuals to have been hunted. 1b, in contrast, continues to entertain hunting hypotheses despite identical sex ratios. Possible interpretations rationalizing this include: that whatever the relative probability of sex bias in human hunting and scavenging, sex bias in either activity was so improbable within the context at issue that no significant difference in sex ratios could be expected; that in the case of sex ratios, the combination of anatomical and behavioral attributes relevant to human hunting makes a sex bias less likely (rather than more likely) in hunting than in scavenging; or that an unanticipated taphonomic bias distinguishing hunted from scavenged animals has coincidentally obscured a real sex bias associated with human hunting (either intrinsic to hunting itself, or to natural death). 2a means that procurement of all butchered carcasses by hunting is a more likely explanation of disparate sex ratios in butchered and non-butchered samples than are alternative explanations allowing all butchered individuals to have been scavenged; mixed interpretations in this instance show a wide range of probabilities, depending on their assignment of particular individuals, but most are of intermediate value or even lower than all-scavenging. 2b, in contrast, continues to entertain the all-scavenging hypothesis despite disparate sex ratios. Possible interpretations rationalizing this include: that there was a sex bias in scavenging, but that this bias was not expressed in hunting because Paleoindians could not or did not hunt mastodonts; that there was a sex bias in scavenging and hunting, but that, by chance, no hunted individuals are included in this butchered sample; or that sex bias in scavenging was minimal, but an unanticipated taphonomic bias is responsible for the disparate sex ratios. In the absence of new background information or conflicting data drawn from a different attribute of the case under study, options 1b and 2b would be considered unacceptably ad hoc. However, the questions they raise do remain and can direct the course of further research.

It should also be noted that this approach to evaluating hunting and scavenging hypotheses gains much of its generality, but undoubtedly loses some resolution, through the broad character of its statements regarding conditional probability of the data. It is, for instance, agnostic about the precise cause of a disparity in sex ratios, should one emerge. Nevertheless, questions of this latter sort are tempting to address in retrospect, once the data for a given case are available and their pattern is evident. This must, however, be considered a completely separate operation from the primary analysis. This kind of follow-up discussion may be thought of as generating "retrospective" (or more properly, a posteriori) auxiliary hypotheses. Retrospective auxiliaries suggest a finer scale interpretation by integrating the observed data more directly with the focal hypothesis they already support and by explaining this integration in relation to available background information. This can be exemplified by supposing that additional sex ratio data sustain the patterns shown by present samples. Accepting the behavioral and taphonomic auxiliaries would then imply support for hunting. The non-butchered sex ratio (50%), by corresponding to a common value for populations of living mammals, would tend to suggest that there had been little overall sex bias in natural death, scavenging by humans, and the preservation and collection of natural deaths --

though other interpretations are possible. A retrospective auxiliary might then suggest that hunting was in fact sex-biased and focused on males, and that this was because postadolescent male mastodonts, like male elephants, were more frequently (than females) solitary and thus easier to approach and kill without risk of interference from conspecifics. The apparent lack of female bias in natural deaths is compatible with a male bias in human hunting if hunting accounted for substantially fewer deaths (integrated over this sampling interval) than did natural causes. This in turn is compatible with the relative numbers of butchered and non-butchered carcasses if there was a preservational bias--due to the type of meat caching proposed for the Heisler site--favoring butchered individuals (but not of one sex or the other).

The effect of this complex, retrospective auxiliary is to enhance the conditional probability attributed to the observed data and thus apparently strengthen the hunting interpretation. It clearly does so, however, at the expense of a more severe set of assumptions. At their worst, retrospective auxiliaries involve little more than opportunistic foraging through the miscellany of background information, searching for relationships that would bear on the validity of the preexisting auxiliaries. At their best, they point toward testable relationships that do indeed support, in secondary fashion, the primary analysis. This particular example is probably at an intermediate position on this spectrum, but more fully developed retrospective auxiliaries will be treated below.

Retrospective auxiliaries are clearly elicited by the data themselves and in this sense resemble ad hoc hypotheses. Both have highly selective support within background information. The difference is that ad hoc hypotheses (as the term is most commonly used) serve to protect focal hypotheses that would otherwise be *rejected* by the data. Additional analysis may place either type of hypothesis on a less selective footing (essentially removing retrospective or ad hoc status), or it may reject them altogether. To the extent that retrospective auxiliaries encourage exploration of new relationships, they are entirely beneficial. The only danger is that they may overrepresent the conclusiveness of a result and appear to offer greater resolution on unknown aspects of a case than is warranted by the evidence.

AGE PROFILE AS EVIDENCE OF HUNTING OR SCAVENGING

Individual age at death and the distribution of age within a sample (i.e., the age profile) represent the next attribute to be considered. Since age allows a greater number of states to be recognized than does sex, it has the potential to offer greater resolution on issues of similarity or dissimilarity of distributions.

Source of Data

Analysis of incremental lamination of tusk and molar dentine (see *Season of Death...*) has the potential to provide relatively precise absolute ages for individuals preserved with dental remains. In addition, comparative studies should allow extension of absolute age estimates (based on associated schedules of bone growth and epiphysis fusion) to specimens with postcranials only (see also Roth 1984). However, much work remains before absolute ages are available for all individuals in Table 1. Thus for consistency, the age assignments in Table 1 are (with one exception, noted below) relative rather than absolute. They follow the procedure suggested by Saunders (1977), who based age classes for mastodonts on the stages of molar eruption and wear defined by Laws (1966) for African elephants. Because of differences between mastodonts and African elephants in the number of, and extent of wear on, simultaneously functioning molars within a single jaw quadrant, the application of Laws' age classes to mastodonts requires some breadth of interpretation. Assignments emphasize the central portion of the occlusal field in each quadrant, thus optimizing the comparison with respect to the principal molar(s) functioning at the time of death. Pending further data, I have refrained from any age statement on individuals for which neither a whole tusk nor molars are available ("?" in Table 1 for Russell Farm II, New Hudson, and Taylor).

Laws' (1966) estimates of the absolute equivalents of his age classes for African elephants are shown in parentheses in Table 1 and used as the basis for the histograms in Fig. 3, to allow more direct comprehension of the age profiles. These estimates have been criticized by other workers (e.g., Sikes 1971; Fatti et al. 1980) as being inaccurate, but they are used here because preliminary counts of annual dentine increments in whole tusks of individuals for which molar eruption and wear can also be assessed show somewhat better concordance (at least early in life) with Laws' age scale than with alternative African elephant age scales. Since I do not dispute the criticism of Laws, I assume this result is coincidental. The only substantive application of Laws' age scale was to the Van Sickle mastodont, which offered an absolute age (based on a whole tusk) but no Laws' age class (only a fragment of one molar was found). Its associated Laws' age class has therefore been assigned "in reverse."

Attribute-Specific Statement of Argument

The analysis of age profiles follows a pattern parallel to that of sex ratios. The behavioral auxiliary hypothesis covering similarity or dissimilarity of attribute distributions may be framed in terms of age data as follows: the probability that the age profile of a butchered sample will be biased relative to that of a non-butchered sample from the same context is greater if the butchered sample was procured by hunting than if it was procured by scavenging. This proposition is based once again (translating from the sex ratio argument) on the probable relevance, during hunting, of both behavioral and anatomical correlates of age, whereas only anatomical correlates of age are likely to be relevant during scavenging. It is also supported by the probable contrasts in spatiotemporal distribution of live prey of any age vs. scavengeable carcasses of that age and by the other arguments presented in the Appendix. The taphonomic auxiliary for age profiles is that the probability of differential expression of an age-related taphonomic bias between butchered and non-butchered samples is more or less equal, whether the butchered individuals were hunted or scavenged. In other words, we may grant that taphonomic processes are biased against preservation of young individuals, and even that this effect may differ for butchered and non-butchered individuals, but it still seems unlikely that this or any other such effect would be expressed differentially on hunted and scavenged samples. The conjunction of these behavioral and taphonomic auxiliaries allows us to conclude that the probability of finding a statistically significant difference between the age profiles of butchered and non-butchered fossil samples is greater if the butchered sample was procured by hunting than if it was procured by scavenging. Thus, finding a significant difference tends to support hunting, whereas finding no significant difference tends to support scavenging.

The argument summarized above is in fact only a more general -- and "relativized" -- version of hypotheses that have been well recognized as underlying previous applications of age profiles to archeological problems (e.g., Frison 1978; Klein 1982a, 1982b, this volume). However, some comparison of terms is needed to demonstrate this. As illustrated by Klein, in this volume (see also Kurtén 1953; Van Valen 1964; Voorhies 1969), age profiles in assemblages of mammals (and many other organisms) typically resemble one of two broad models, termed "catastrophic" and "attritional." Catastrophic assemblages result from sampling different-aged individuals within a living population approximately in proportion to their abundance (i.e., sampling individuals in age-independent fashion, as would occur in a mass mortality). In contrast, attritional assemblages result from the combined incidence of numerous sources of mortality, as they would impinge on a population showing a stable age structure throughout the interval of study. Some of these agents of mortality sample individuals in proportion to their abundance (certain types of accidental death), but most have a more age-dependent impact on individuals (usually most pronounced among the youngest and oldest age classes). From the standpoint of definitions then, "attrition" is not equivalent to "natural death," nor is "catastrophe" equivalent to "hunting." Natural death can occur in either an attritional or a catastrophic mode, and hunting can occur in either age-independent or age-dependent fashion (Klein and Cruz-Uribe 1984).

The conventional use of catastrophic and attritional age profiles in addressing the hunting vs. scavenging issue (Klein 1982b; Klein and Cruz-Uribe 1984) is that these two patterns, with liberal allowances for variation, are treated as diagnostic of hunted and scavenged assemblages, respectively. Differences in life history strategies can sometimes complicate analysis (Klein and Cruz-Uribe 1984), and similar patterns can arise in more than one way (Conybeare and Haynes 1984; Haynes 1985). Nevertheless, this approach has been both tractable and informative in many studies.

The conventional application of catastrophic and attritional age profiles to this problem can also be seen as a "special case" of the arguments proposed here. It is based on the assumption that, integrated over time, most of the natural deaths (and moribund individuals) available for scavenging will reflect attritional mortality. Catastrophic mass mortality, whatever proportion of total mortality it comprises, is implicitly treated as unlikely to provide a large portion of the carcasses scavenged by humans. This decision can frequently be defended by arguing either that populations are more often than not "in between" catastrophes, or that in the midst of a catastrophe carcasses occur in superfluity and many go unscavenged by humans. It is then further assumed that scavenging will be relatively nonselective (with respect to age) and that it will therefore reflect the attritional pattern of its source. This pattern, in turn, is usually recognized by comparison with attritional patterns for closely related groups and by assuming a stable age structure for the population from which the "unknown" sample is derived (Haynes 1985). This line of reasoning converges on the arguments proposed here for systems in which the above assumptions are met, and in which catastrophic natural death comprises a relatively small portion of all natural death. Under these circumstances, scavenged and non-butchered samples will indeed show similar age profiles. In contrast to scavenging, human hunting of all but the most challenging prey (Klein and Cruz-Uribe 1984), whether through simultaneous cropping of many individuals or isolated assaults on single individuals, is usually treated as sampling individuals more nearly as they are encountered, according to their frequency in the living population. Because mammals show relatively consistent patterns of population structure, the resulting age profile may (again, assuming stable age structure) be recognized as distinct from the attritional pattern typical of scavenged assemblages and (in this scenario) most natural deaths. If these assumptions are met, fresh assemblages of hunted individuals and non-butchered natural deaths will clearly show disparate age profiles. Finally, conventional age profile analysis involves a counterpart to the taphonomic auxiliary proposed here. It is the high likelihood that age-related taphonomic biases have similar effects on hunted and scavenged samples that justifies the expectation that age profiles of "diagnostic" form may be recognized in fossil material.

The arguments proposed here differ from the conventional approach to age data by making less precise predictions. They also, however, make fewer and less severe assumptions, claiming less detailed knowledge of the context in which a given assemblage was produced. For instance, they allow a wider variety of possible combinations of attritional and catastrophic sources of scavengeable carcasses. They are compatible with an age-related bias in scavenging, either the bias against juveniles proposed by Vrba (1980), or some alternative. They also do not assume that hunting will occur in a "catastrophic" manner (i.e., proportional to live abundance), although "catastrophic" hunting is by no means unreasonable, given the Paleoindian technological repertoire. For age data, I am inclined to agree that the more severe assumptions associated with the conventional use of attritional and catastrophic profiles can be defended. These assumptions are more likely to yield determinate results, because they suggest not only *that* hunted and scavenged assemblages might be different, but also *how* they should differ. Nonetheless, if the more severe assumptions are ever judged to be risky, it may be preferable to retreat to the less severe assumptions associated with the arguments proposed here.

Another approach to evaluating competing procurement hypotheses, proposed in the *Plan of Argument*, involved the pattern of overlap in occupation of attribute states of butchered by non-butchered samples. Although this could not be used in the case of sex, age differs from sex in reflecting a transient attribute of individuals. In addition, there may be strongly age-dependent probabilities of death by either human hunting or natural causes. Although a large enough natural death sample might be expected to have some individuals in each recognizable age state, some states may typically be unoccupied in small samples of natural deaths, due to low probability of death at that age. This pattern is seen in the natural death sample provided by non-butchered mastodonts, although it undoubtedly has a combination of causes relating to both life history and taphonomy. In any event, it allows the overlap pattern of state occupation to be used in interpreting age data. This in turn requires a different behavioral auxiliary: the probability that the range of age states occupied by a butchered sample will extend beyond (i.e., fail to be a subset of) the range of states occupied by a non-butchered sample from the same context is greater if the butchered sample was hunted than if it was scavenged. This relationship obtains, in brief, because scavenged samples are directly derived from assemblages produced by natural death processes, while hunted samples represent a different agent of accumulation, with the potential for different age-dependent probabilities of death. For application to fossil samples, it is used in conjunction with the now-familiar taphonomic auxiliary.

Data Analysis

The age profiles for butchered and non-butchered individuals are indeed different in overall shape (Fig. 2). Interpreted at face value, and with respect to the discussion of similarity and difference of attribute distributions, this result favors hunting. In addition, these patterns may be interpreted according to the more conventional use of age profiles. Although a probable taphonomic bias against juveniles is evident in both profiles, the butchered profile looks at least moderately catastrophic (i.e., fewer old individuals and more young and prime-age adults) -- as would be expected for hunting. In comparison, the non-butchered sample looks more attritional. Rigorous interpretation of these results, however, requires a statistical test of significance to demonstrate that the disparity between butchered and non-butchered profiles is not likely to have resulted from sampling error alone. Following Klein and Cruz-Uribe (1984), a Kolmogorov-Smirnov test statistic was calculated, and the difference between profiles was not significant ($P > 0.1$). This is not surprising given the small sample size. As with sex ratios, we are left with results that show some

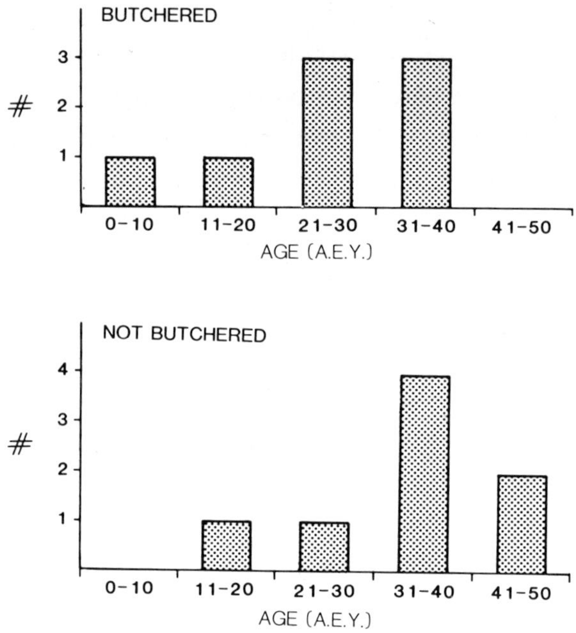

Figure 2. Age profiles for butchered and non-butchered individuals from Table 1. Age is indicated in African elephant years (A. E. Y.), according to Laws (1966).

qualitative support for hunting relative to scavenging but fall short of standard thresholds of significance.

Turning to the pattern of overlap of attribute distributions, the range of states occupied by the butchered sample does extend beyond (i.e., fails to be a subset of) the states occupied by the non-butchered sample. Since this outcome is more probable under hunting than under scavenging, it tends to support hunting as the means of procurement of carcasses for butchery. However, this is hardly an impressive instance of support for hunting. For the coarsely aggregated age data in Fig. 2 the failure of subset status is attributable to only one individual in one age state (the only one unoccupied by the non-butchered sample). Even using the finer grouping of data offered by Laws' (1966) age classes, only Wattles and Heisler are younger than the youngest non-butchered individual (Shelton), and only the younger of these (Wattles) differs by more than one age class. Statistical testing of this pattern, as explained in the Appendix, is not straightforward; but judged subjectively, the failure of subset status probably cannot in this instance be distinguished from sampling error.

Discussion

Although no strong preference for either hunting or scavenging can be claimed from this analysis of age profiles, the results are not entirely neutral either. There is at least tentative support for hunting, due principally to interaction between the conventional approach to analyzing age profiles and the more explicitly comparative approach proposed here. In other words, while the differences between the butchered and non-butchered age profiles are not statistically significant, and while the butchered profile might not stand out on its own as markedly catastrophic, the *direction* of differences between the two profiles supports the *relatively* catastrophic nature of the butchered profile. In this sense, the association between catastrophic age profiles and hunting, despite its historical priority, is functioning much like a retrospective auxiliary, enhancing the apparent probability of the observed data, conditional on hunting. In this instance, the separate logical status of the retrospective auxiliary allows its contribution to be substantial.

It is also important to retain alternative interpretations within focus, but since these were presented in some detail for sex ratios and would follow an identical pattern if rephrased in terms of age profiles, they need not be elaborated here. The tentative support for hunting noted above, even if it were sustained by sample sizes large enough to make the present patterns statistically significant, is mainly a statement about the relative probability of the all-hunting and all-scavenging procurement hypotheses. Hypotheses involving both hunting and scavenging would for the most part be intermediate in probability or lower than all-scavenging.

It is also notable that the data for sex ratios and age profiles are consistent in their support for procurement hypotheses. In addition to the corroborative value of this finding, it is useful in the further development of retrospective auxiliaries. This will be postponed, however, until seasonal mortality data can be incorporated.

SEASON OF DEATH AS EVIDENCE OF HUNTING OR SCAVENGING

The final attribute to be considered is the season of death of individuals and the seasonal distribution of mortality in a sample. Unlike age at death, season of death has little conventional association with the problem of distinguishing hunting from scavenging, although it has been used widely and successfully in characterizing seasonal patterns of utilization of resources by various human groups (e.g., Kay 1974; Spiess 1976). In addition, however, it is an appropriate, premortem attribute for use in the comparative arguments proposed here. I emphasize this because readers familiar with the use of season of death data in characterizing seasonal patterns in human subsistence behavior may understandably acquire the impression that its extension to the problem of distinguishing hunting from scavenging requires some assumption about the season or seasons in which hunting and/or scavenging typically occur. If available, even on a region-specific or taxon-specific basis, such information might be very useful. However, it is not available at present from contexts similar enough to that of the case under study to allow generalization of seasonal changes in the frequency of hunting or scavenging. The main object of interest here is thus the *comparison* of patterns of seasonal mortality between butchered and non-butchered samples from the *same* context.

Source of Data

A large body of research, particularly in wildlife management, has been devoted to development of techniques of absolute age determination using mammalian teeth (e.g., Klevezal and Kleinenberg 1969; Grue and Jensen 1979).

These methods involve counting periodically formed structural features that define incremental laminae within the cementum and/or dentine (in a few cases, the enamel). Provided the period of such a structural cycle is known, and provided some point within the cycle can be related to season of formation, the amount of time represented by the interval between the most recently formed reference feature and the last-formed cementum or dentine can be estimated, suggesting a season of death. The approach used here is essentially identical, but offers exceptional resolution as a result of several favorable characteristics of proboscidean dental histology.

Mastodont teeth are in many respects similar to those of any mammal. Enamel is laid down from the enamel-dentine interface outward, and primary dentine is deposited in successive layers, moving inward from the enamel-dentine interface, progressively defining the crown and root. In the molars, deposition of primary dentine fills in much of the pulp cavity, but ceases when the openings at the root apices are almost occluded. Thereafter, the pulp cavity tends to be filled in by secondary dentine, with a more convoluted and irregular structure. Cementum is deposited on the root, from the cementum-dentine interface outward but may also encroach on the enamel. Mammoth molars are topologically identical to this, despite their highly specialized morphology. Proboscidean tusks (highly modified second incisors; Sikes 1971) also show the same basic organization of tissues. At birth, extant elephants have both deciduous and permanent tusks, but the former are shed early in life. Associated with the distinctive morphology and size of permanent tusks is the restriction of enamel to the tip and the retention of an open, continuously growing root. Successive increments of primary dentine are added throughout life, along a conical surface facing onto the pulp cavity (Fig. 3). This extends the tusk proximally (like a stack of ice cream cones) while it erupts in a distal direction.

Tooth tissues of all mammals show some laminar structure, but proboscidean dentine is unusually complex (Fisher 1984b; Fisher and Koch 1983). Tusk dentine has a hierarchy of laminar features that are conformable to the surface of dentine accretion and are therefore almost undoubtedly incremental in nature (Fig. 4). The highest order of lamination (1°) is defined

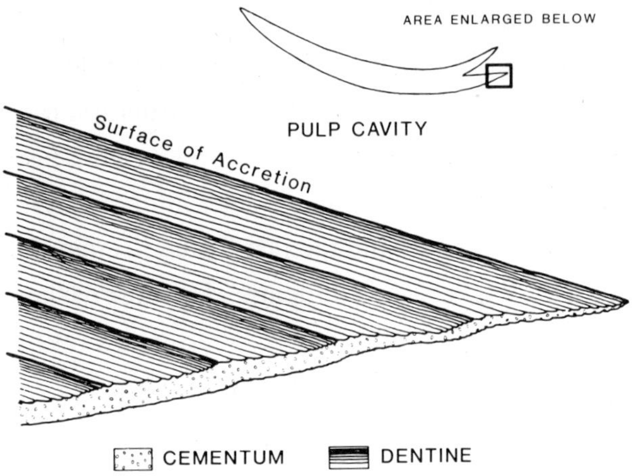

Figure 3. Diagrammatic longitudinal section of a proboscidean tusk near its proximal (growing) end. For ease of illustration, the angle between the dentinal lamination and the cementum-dentine interface is exaggerated, and the number of 2° laminae is reduced. 1° lamination is defined by variation in spacing of 2° laminae, variation in color (dark bands correspond to closely spaced 2° laminae; see also Fig. 6), and the largest-scale undulations of the cementum-dentine interface. The latter features are apparently produced by slight, regular variation in the timing and rate of dentine deposition, root elongation, and tusk eruption. They are seen in intact specimens as circumferential annulations, particularly near the base of the tusk. 2° laminae are indicated in this drawing by the lines bounding single 2° increments and by their step-like expression at the cementum-dentine interface.

most clearly in fossil material by a regular alternation between relatively thick, light-colored bands and thinner, dark bands of dentine (Fig. 4A). The same cycle is also manifested as a difference in density of calcification, as variation in the spacing of laminae of the next lower order (2°), and as undulation of the cementum-dentine interface (visible externally, especially near the proximal end of a tusk). 2° lamination is shown as the darkest approximately horizontal lines in Fig. 4C and D and as the single most continuous, horizontal dark line in Fig. 4E. These same micrographs illustrate in addition a third, still finer, order of lamination (3°). The enlargement in Fig. 4E shows that both the 2° and 3° "lines" are zones of tiny (< 1 m) vacuities, most of which branch off from dentinal tubules (vertical black lines in Fig. 4C-E). 2° and 3° lines are distinguished, at the smallest scale, by the concentration of vacuities. In some cases 2° features appear to be accentuations of a single 3° line, but in other cases they seem to involve accentuation of several adjacent 3° lines. The two orders of lamination are clearly different and yet may be formed by some modulation of the same process. Molar dentine is similar to that of tusks, although molars have thinner increments of comparable order and frequently fail to show well defined 3° lamination (Fig. 4B). These patterns have been studied most thoroughly in mastodonts, but examination of mammoths and extant elephants indicates similar organization.

To my knowledge, only two aspects of the hierarchy of lamination characteristic of proboscidean dentine have been noted by previous workers. Osborn (1936) described one specimen showing a series of circumferential annulations (i.e., ring-like topographic features) on the proximal portion of each tusk. He compared these to similar annulations on a tusk from an Asian elephant and speculated that they might be annual (i.e., yearly) in nature. These features correspond to the undulations of the cementum-dentine interface, described above as one aspect of 1° lamination (Fig. 3). In addition, Schour and Hoffman (1939), in a survey of dental lamination in various mammals, examined a mastodont molar and observed what I have referred to as 2° lamination. They suggested (incorrectly, I believe) that this was a daily incremental feature but were unaware of Osborn's finding or any other evidence of hierarchical organization in proboscidean dentinal lamination.

Osborn's suggestion that tusk annulations (and now, by extension, all of the 1° features described above) might be annual is supported by the pronounced regularity of 1° features (Fig. 4A) and by the fact that when summed along an entire, longitudinally sectioned permanent tusk (Fig. 4F), they

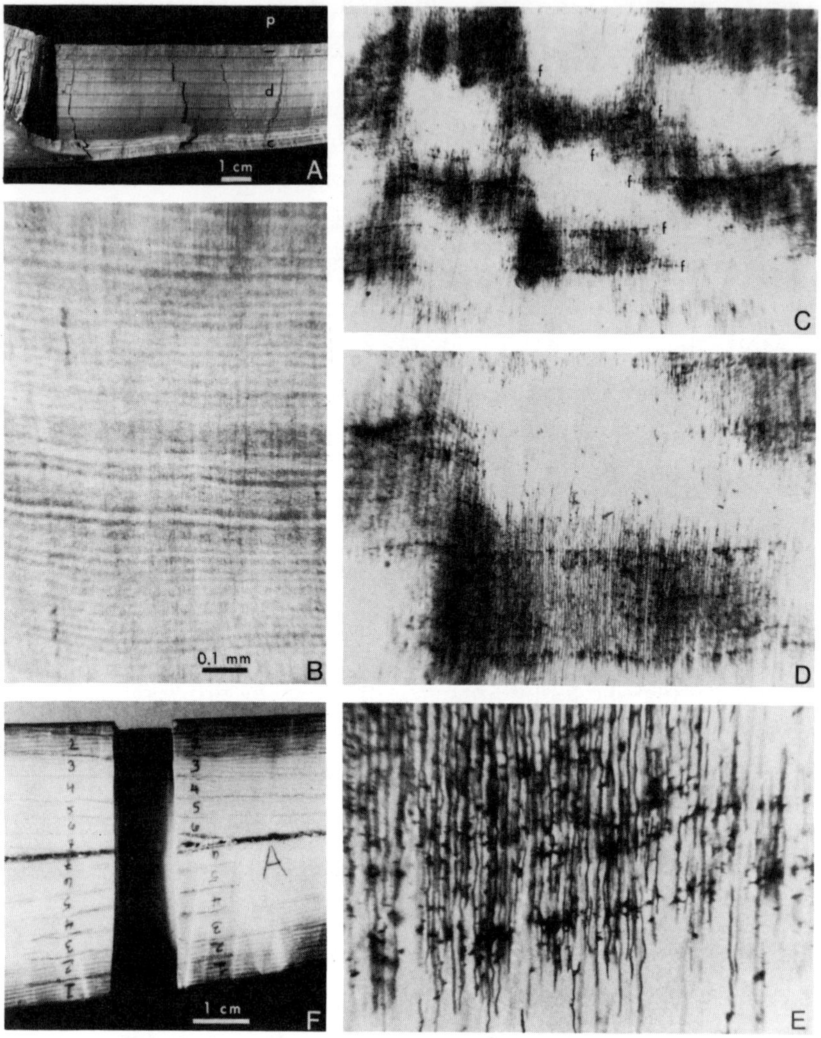

Figure 4. A, Longitudinal fracture section, near the proximal end of a mastodont tusk (Van Sickle); *p*, pulp cavity; *d*, dentine, showing prominent 1° color banding; *c*, cementum. B, Photomicrograph of thin section of mastodont molar dentine (Cole), showing two 1° cycles, each including about twenty-six 2° cycles. C, Photomicrograph of transverse thin section of mastodont tusk dentine (Van Sickle); crossbars of *f*'s located opposite 2° lines; 3° lines visible on finer scale; magnification ca. 35x. The diagonal series of irregular, dark and light regions

suggest an individual age that is reasonable by analogy with extant elephants of known age. Nevertheless, a more compelling argument for their annual nature involves numerical aspects of the hierarchy of lamination. The lines making up the 2° and 3° lamination tend to be somewhat discontinuous (Fig. 4E), but by scanning laterally and focusing up and down through a thin section (unfortunately not possible in a photograph) they can be counted in relatively routine fashion. Conventional SEM does not help to resolve these incremental features, apparently because the vacuities that define them are too far apart within a zone, relative to the distance between adjacent zones, to show up as distinct topographic lineations (even with etching). However, preliminary results of back-scattered electron (BSE) imaging with a scanning electron microscope appear to offer great improvement in resolving power. This

---

Figure 4 (*continued*) are produced by the pattern of undulation of dentinal tubules that is typical of proboscidean dentine (Saunders 1979). This is also the structural basis for the 'spiral checkerboard' pattern of light and dark regions that is commonly used to identify objects made of tusk dentine (i.e., ivory). The light, apparently structureless areas in the thin section represent domains of similarly oriented tubules that have been truncated by the plane of section. They, and any adjoining vacuities, have been flooded by the epoxy used to affix the dentine to the microscope slide. This causes the difference in refractive index between them and surrounding calcified areas to be lost, whereupon the incremental lamination seems to disappear. D, Enlargement of lower central region of C; pronounced horizontal lines are 2° lamination (indicated by *f* in C); fainter horizontal lines are 3° lamination; magnification ca. 90x (identical to B). E, Enlargement of lower central region of D, showing one 2° line and several 3° lines; individual lines defined by zones of vacuities branching off dentinal tubules (vertical); magnification ca. 320x. F, Longitudinal polished section of mastodont tusk (Powers), showing site of removal of a dentine sample for analysis; approximate boundaries of the first seven 1° increments recorded in this tusk are marked in pencil, based on their clearer appearance in the adjacent transverse section; conspicuous light and dark bands are not incremental (i.e., neither 1° nor 2°) but rather are the longitudinal manifestation of the pattern of undulation of dentinal tubules (also responsible for the 'spiral checkerboard' pattern of transverse sections; cf. C).

technique is apparently picking up fine-scale variations in the density of dentine calcification. Although this may turn out to provide a much better basis for quantitative analysis of the hierarchy of lamination, the results reported here are all based on light microscopic analysis of thin sections. Fig. 5 is a histogram showing the number of 2° increments found within each of thirty-three 1° increments, selected from six different individuals. Within a given 1° increment, individuals tend to show either about thirteen or about twenty-six 2° increments, yielding two clear modes in Fig. 5. Counting 3° increments is more dependent on finding patches of excellent preservation, but where these are large enough, cases comprising the mode near twenty-six consistently show fourteen 3° increments per 2° increment (i.e., between consecutive 2° lines; Fig. 4C, D). Even the best preserved cases comprising the mode near thirteen are difficult to assess since they require more continuous stretches of excellent preservation. However, their number of 3° increments can be estimated to be very close to twenty-eight (i.e., twice that of the cases in the other mode). The number of 3° increments per 1° increment thus appears to be about 26 x 14 (or 13 x 28), or 364. This is so nearly equal to the number of days in a year that the 1° laminations must be annual, and the 3° must be daily. This implies that the 2° increments of which there are about twenty-six represent fortnights, and those of which there are about thirteen represent lunar months.

I do not presume that the documentation given here for these patterns (Fig. 4) is sufficient to establish their general validity. This task will require significantly more discussion and analysis elsewhere. Nevertheless, these patterns have now been observed in many individuals. Although there is always some danger of subjectivity in visually-based counting and measurement, many cases are clear enough to offer little latitude for interpretation. Densitometry and Fourier analysis of both light and BSE micrographs are also being pursued to augment objectivity.

Given the numerical relationships described above, a "perverse" rendering of the argument suggested here can make the temporal interpretation of the hierarchy of lamination sound completely circular: if 1° increments are assumed to be years, then the 3° increments, of which there are about 365,

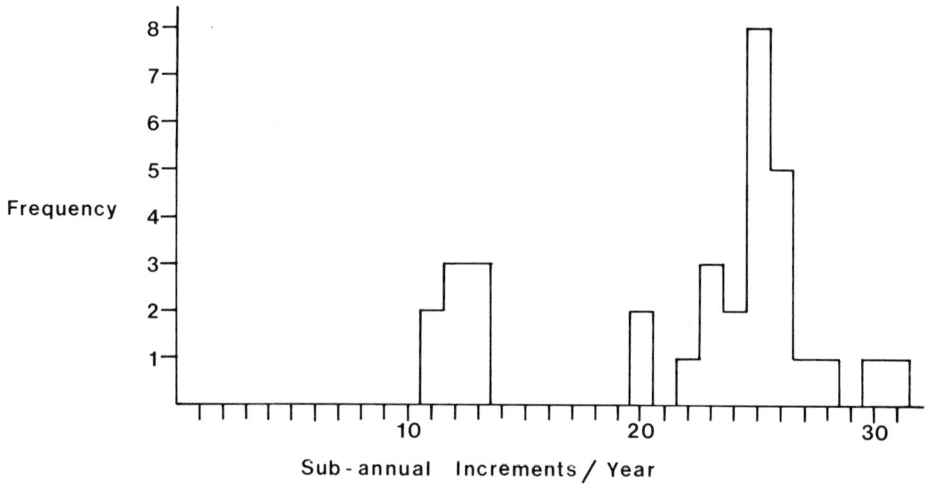

Figure 5. Histogram showing number of sub-annual (2°) increments per year (1° increment) in mastodont dentine. The mode near thirteen is formed by the last four years of a very old male (Kuhl) and four young-adult years of a younger male (Russell Farm I). These individuals, during these years, were showing 2° banding with lunar monthly periodicity. The mode near twenty-six is formed by two females (Owosso and Powers) and two males (UM 3493, not discussed here, and Pleasant Lake), all of which were in their reproductive prime. They show 2° banding with fortnightly periodicity.

must be days; but if 3° increments are days, then the 1° increments must be years. However, I do not consider the temporal interpretation to be circular at all -- only parsimonious. It looks at the hierarchical pattern as a whole "package" and asks, "What natural units occur in this approximate number within the lives of large mammals and have about 365 subunits?" If the answer is not "years," we shall have to discover some other unit that behaves similarly.

Tooth formation in proboscideans represents a large metabolic investment and might be expected to show variation reflecting daily and annual cycles, but the fortnightly and lunar periodicities are surprising. A biological interpretation of 2° banding is not a prerequisite for using the banding pattern to determine season of death--as long as the period of 2° banding is known. Nevertheless, the cause of 2° banding is an obvious problem whose solution would provide an independent check on the numerical argument above. My tentative interpretation of this pattern is based partly on evidence that both African and Asian elephants have an estrous cycle on the order of two-three weeks (Short 1966; Eisenberg et al. 1971). This suggests a plausible basis for a fortnightly cycle in females but does not explain its expression in dentine deposition. It also raises the question of how the same cycle occurs in males. No endogenous male cycle of this length has been reported, but adult male elephants show abundant behavioral response to the estrous status of females (Rasmussen et al. 1982). Mating in elephants involves several hours of courtship and is therefore a significant interruption in a normal daily activity pattern that usually includes foraging for 17-18 hours, moving between drinking and feeding areas for several hours, and only 2-3 hours of sleep (Sikes 1971; Douglas-Hamilton and Douglas-Hamilton 1975). Since the 2° lines defining fortnightly increments are essentially accentuated daily lines, they may be a facultative expression of sexual behavior, mediated by its attendant alteration of normal daily activity patterns. The lunar cycle could be an indication of dependence of the daily activity pattern (perhaps foraging) on level of nocturnal illumination and hence the phases of the moon. As such, it might be evident in individuals that were not sexually active. This would suggest that a lunar cycle might be superimposed on the fortnightly cycle in individuals showing the latter.

Although highly speculative, these interpretations of 2° banding are compatible with details of the age/sex composition of the two modes in Fig. 5. Mastodont males and females in their prime show the fortnightly pattern, but the single extremely old individual I have examined (Kuhl, a male) shows a shift from a fortnightly to a lunar monthly pattern at what might be interpreted as the onset of a postreproductive phase. Another individual (Russell Farm I) shows a lunar monthly pattern as a young adult male, interpretable as sexually mature but young enough that reproductive access to females was limited. The main apparent anomaly I have encountered is that juveniles (e.g., Wattles and the juvenile years of Powers) show a fortnightly pattern. Although unexpected, this may be explained if mastodont calves, like elephant calves, associated closely with their mother (Douglas-Hamilton and Douglas-Hamilton 1975). Their dental diaries may thus have recorded her reproductive activity (presumably through disruption of normal daily activity patterns) in a facultative manner comparable to that postulated for reproductively active males. If so, the shift, in males, from a juvenile fortnightly pattern to a lunar monthly pattern implies dissociation from their mothers, which occurs in male elephants at sexual maturity (Sikes 1971). The onset of sexual maturity and pregnancy in females might be expected to show up as a shift from fortnightly to lunar monthly banding during the gestation period (for elephants, twenty-two months), though a period of lactational anestrous might also be involved. This has not yet been observed, but the entire life has been checked for only one female. Furthermore, expectation of such a pattern is complicated by reports that elephants continue to show estrous cycles during pregnancy (Short 1966). This approach to analysis of female life histories, and other approaches as well (elemental analysis and interpretation of variations in tusk growth rate), require further testing.

This interpretation accommodates some of the irregularities of banding patterns (e.g., more complex and variable 2° banding in adult males than in adult females) as well as their overall regularity. It also retains greater plausibility for proboscideans (e.g., reproductive activity in elephants is year-round rather than seasonal; Sikes 1971) than it would for most other mammals. Nevertheless, a multitude of questions remain. Could storms have affected activity patterns and thereby 2° banding? Was season of calving regulated in any way? Is there a typical phase relationship between the fortnightly and lunar monthly cycles, and what, if anything, is its functional basis? Some of these and other issues should be possible to address using evidence intrinsic to mastodonts (and mammoths) themselves. However, it will also be important to carry out studies of dentinal structure on extant elephants with known life histories. Much more remains to be done, and these results can only be considered provisional, but they at least suggest directions for further work.

Whatever the uncertainties regarding the mechanism of formation of 2°

lines, their relationship to other incremental features suggests that they demarcate sub-annual intervals of at least approximately equal length. This allows rate of dentine deposition to be expressed, albeit indirectly, as 2° increment width (i.e., thickness). Fig. 6 shows a plot of 2° increment width through portions of three consecutive years, and there is a clear cycle on what was previously recognized as an annual period. Assuming rate of dentine deposition reflects overall growth rate, it seems most likely that the portion of the year represented by the thinnest increments (slow growth due to food shortage and thermal stress?) is winter, and that the portion represented by thicker increments (more rapid growth) is spring through autumn. The location of the dark color bands within series of thin 2° increments corroborates this, because where dark annual dentine bands are known in extant mammals, they are usually associated with winter (Klevezal and Kleinenberg 1969). In addition, both this seasonal interpretation and the annual nature of 1° lamination have recently been confirmed by oxygen isotope analyses of the carbonate fraction of dentinal apatite (Koch and Fisher 1986).

The detailed record described above allows season of death to be determined from any tooth depositing primary dentine at the time of death. Samples must abut the pulp cavity and should ideally cover several years. Season of death can be measured by counting 2° increments backward from the last-formed dentine to an arbitrary reference point within the year. The most consistently recognizable reference is the thinnest 2° increment within a sequence of thin increments, associated with a dark, annual color band. The end of this increment is treated as the winter-spring boundary. Although the calendar date of this reference is both unknowable and variable from year to year, as spring comes either early or late, it is adequate for studies of *relative* season of death. Cross-checks within individuals have always produced consistent results down to the finest observable details. By this means, season of death may be determined to within one fortnight, relative to the last winter-spring boundary. For individuals on a lunar monthly cycle, time of death is estimated to half-cycle precision so that it is commensurate with the more common, fortnightly data.

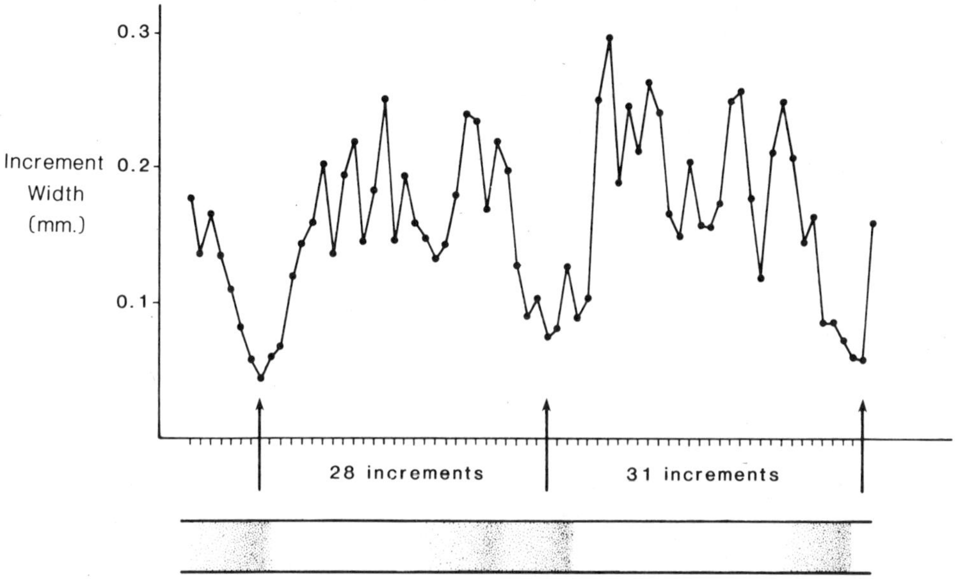

Figure 6. Time-series of sub-annual (2°) increment width, measured over a period slightly exceeding two years, from a mature male mastodont (Pleasant Lake). Dark color bands, indicated below the abscissa, are associated with thin 2° increments within each annual cycle. These are typically associated with an abrupt dark-light change in color and are used as an arbitrary reference point with respect to which season can be calculated.

Attribute-Specific Statement of Argument

The behavioral auxiliary hypothesis for analysis of the similarity or difference of season of death distributions is that the probability of a significant disparity between the patterns of seasonal mortality for butchered and non-butchered samples is greater if butchered individuals were procured by hunting than if they were procured by scavenging. Even if scavenging were selective with respect to season of death, the generally sparse and unpredictable spatiotemporal distribution of scavengeable carcasses (natural deaths), along with competition for carcasses from nonhuman scavengers, would tend to keep the seasonal distribution of carcasses scavenged by humans from diverging appreciably from the seasonal distribution of carcasses of natural deaths not scavenged by humans (i.e., non-butchered). In contrast, hunting -- an activity that may occur independently of the seasonal distribution of natural death -- is not constrained in this way.

Application of the taphonomic auxiliary hypothesis to data on season of death also seems unproblematic. Taphonomic processes may vary in their effects from one season to another, with different mean times till burial or different exposure to physical extremes of the environment. However, even if these variations differ in their impact on butchered and non-butchered carcasses, they would not be expected to act differentially according to whether butchered carcasses had been hunted or scavenged by humans.

Finally, the behavioral auxiliary for analysis of attribute state occupation can be applied to data on season of death. Since scavenging draws upon the "pool" of natural deaths and/or naturally moribund individuals, the seasons of death represented among carcasses scavenged by humans should be a subset of the seasons of death found among natural deaths not scavenged by humans. Only seasonally comprehensive scavenging by humans -- i.e., scavenging all natural deaths within particular seasons -- could disrupt this relationship. In contrast, the seasonal distribution of human hunting is not limited by the incidence of natural death in mastodonts.

Data Analysis

Season of death data for the individuals in Table 1 are plotted on a circular, seasonal histogram in Fig. 7. This histogram is divided into the twenty-six fortnightly periods that comprise an average annual interval of dentine deposition, and season of death is counted and plotted relative to the winter-spring boundary. As noted above, this is the most nearly comparable reference point from year to year and individual to individual, and it works well for deaths at most times of the year. However, deaths occurring in the latest portion of winter take place during the formation of a dark band and a series of decreasing 2° increment widths. Since the increment within which they occur follows the thinnest complete increment nearby, they are placed, by convention, in the first increment of spring. While I suspect this convention has influenced the placement of one or two individuals, its effect is not "felt" more than one or two fortnights from the winter-spring boundary.

Non-butchered individuals, interpreted as natural deaths, show a coherent pattern of seasonal mortality. On the grounds discussed above, I interpret this pattern as indicating highest incidence of death in the latest part of winter through the early portion of spring, tailing off into the summer. In contrast, all butchered individuals died in the middle to late autumn. These two patterns are strikingly different. A relatively conservative evaluation of the significance of this disparity uses (R. A.) Fisher's Exact Test on a 2x2 contingency table for butchered vs. non-butchered and autumn death vs. non-autumn death, and the null hypothesis of random association among these alternatives is definitively rejected ($P << 0.01$). The resolution available in the data is more fully utilized in tests based on circular distributions (Watson and Williams 1956; Wheeler and Watson 1964; Mardia 1972). With these, both butchered and non-butchered samples show decidedly nonrandom (i.e., highly seasonal) patterns of mortality, and the difference between their means is pronounced ($P << 0.01$). We need not doubt that the two samples differ in distributional properties. The most direct interpretation of this result is as clear support for hunting. Season of death has also been analyzed for fifteen isolated teeth in the collections of the University of Michigan Museum of Paleontology. Although these are not referable to butchered or non-butchered categories on taphonomic grounds, they collectively show the same total distribution seen in Fig. 7, with similar membership in each mode.

Assessed with respect to overlap of attribute states, the seasonal mortality patterns of butchered and non-butchered samples are completely disjunct. That is, the seasonal incidence of mortality associated with butchery fails in its entirety to be a subset of the seasonal distribution of "known" natural deaths. This constitutes strong support for procurement of all butchered carcasses by hunting. Although statistical tests of this relationship

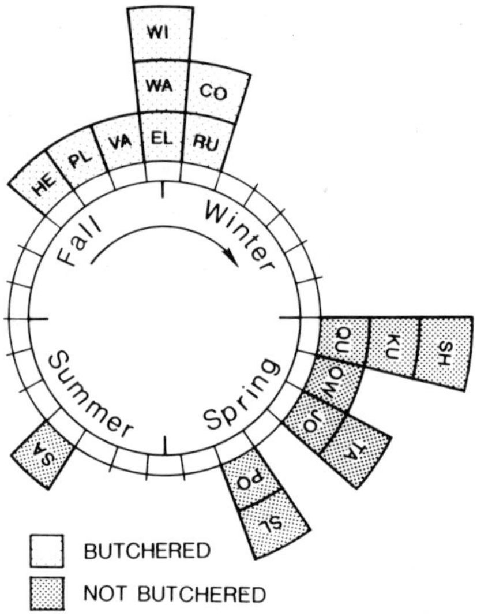

Figure 7. Seasonal mortality data for butchered and non-butchered mastodonts from Table 1. Season of death is plotted on a circular histogram representing an idealized year with twenty-six fortnightly intervals, divided among four seasons of (arbitrarily) equal duration. Season of death for each individual is measured relative to the last winter-spring boundary. HE, Heisler; PL, Pleasant Lake; VA, Van Sickle; EL, Elkhart, WA, Wattles; WI, Winameg; CO, Cole; RU, Russell Farm I; KU, Kuhl; QU, Quagaman; SH, Sheathelm; OW, Owosso; JO, Johnson; TA, Taylor; PO, Powers; SL, Shelton; SA, Sakstrup.

are not, to my knowledge, available, the high degree of "cohesion" within each sample (tested above) indicates that random sampling error cannot easily account for the absence of non-butchered individuals that died in the autumn. I also consider it unlikely, though this judgment is admittedly subjective, that the butchered individuals represent scavenged natural deaths that were appropriated so comprehensively (despite random lapses and competition for carcasses from nonhuman scavengers) from the seasonal "pool" of natural deaths that no non-butchered natural deaths remain from that season, in a sample of this size.

Discussion

Seasonal mortality data provide the clearest evidence yet that the butchered individuals in Table 1 were procured by hunting. With respect to these data, even the various "mixed" procurement hypotheses have intermediate probabilities (i.e., lower than all-hunting), and the all-scavenging hypothesis is least attractive. Still, as always, questions remain regarding both the causes of observed patterns and the viability of alternative explanations of observed patterns. To address the first of these issues, a multi-part, retrospective auxiliary will now be proposed. Its first component is that the distribution seen in the non-butchered sample is the expected pattern of natural mortality, typical for the majority of extant large mammals in highly seasonal, mid- to high-latitude climates. It reflects death in a season of high physiological stress (when food is scarce and fat reserves are depleted) and perhaps, increased probability of accidental death (e.g., falling through a lake or pond ice-cover in an early thaw, or drowning in spring floods). The autumn, in contrast, should have been a season when mastodonts were in the best physical condition, after a full summer's feeding and growth, with fat reserves intact for the winter. With a reproductive pattern not constrained to highly seasonal encounters, autumn would not be a time of especially high reproductively associated mortality, if indeed this is ever an issue for proboscideans. The autumn peak in mortality that is seen among butchered individuals is then interpreted as representing highly seasonal hunting, intentionally focused on the time of year when the fat content of meat is highest (Speth 1983; Speth and Spielmann 1983) and when supplies are being stocked for the winter. The caching hypothesis elicited by the Heisler site may be brought into this picture as a method of storing at least part of the winter's food supply. In addition, caching carcasses in ponds would tend to promote a relatively abundant record of butchery and associated behavior, since occasional surpluses, planned or otherwise, could be expected, and any surplus left uneaten after the spring thaw would face a rapidly rising probability of permanent abandonment.

This retrospective auxiliary clearly enhances the perceived likelihood of observing exactly the results we do and simultaneously makes scavenging appear to be an even less plausible explanation. Although complexly "contrived," this suggestion should not attract undue criticism for "ad hoc-ness" as long as its status is recognized as separate from the main arguments proposed here. In addition to its merits with respect to seasonal mortality data, it is compatible with the retrospective auxiliaries proposed for other attributes -- postulating a hunting bias toward males (possibly because of their relatively solitary habits) and mature individuals in their prime. It thus represents an attractive synthesis of presently available data and background information. Some parts of it, such as the winter-spring pattern of natural mortality, or the seasonal trends in, and nutritional importance of, fat content, have been documented in a controlled fashion and shown to have a sound biological basis (Speth and Spielmann 1983). Other aspects are at least amenable to empirical analysis.

The issue of alternative explanations can also be addressed through the retrospective auxiliary described above. In particular, the possibility of caching presents a serious, even if somewhat ad hoc competitor to the hunting interpretation. This defense of scavenging postulates that the winter-spring mortality of non-butchered individuals is seen not because this is when natural deaths predominantly occurred, but because this is the season at which they occurred in close conjunction with depositional environments (through unanticipated spring thaws, drownings, etc.). It proposes that natural deaths occurred throughout the rest of the year and yet remained "invisible" to the fossil record (by their lack of association with depositional environments) *except* during the autumn, when Paleoindians scavenged fat-rich carcasses from the surrounding terrain and cached them in depositional settings. This proposal could also explain the displacement of the butchered age profile toward younger ages, since caching might increase the preservation potential of young individuals. However, a scavenging bias against the oldest individuals (possibly because they would tend to be lean) must be added if their absence from the butchered age profile is to be explained.

Again, for a proposal that is admittedly contrived to fit the data, this has some attractive points. Nevertheless, there are also shortcomings. For instance, injured or sick elephants commonly remain near sources of water

(Sikes 1971), so it seems unlikely that non-drowning, non-scavenged (by humans), natural deaths occurring at times of year other than winter-spring would have been excluded from depositional settings. Data from other attributes also have a bearing on this issue, since *if* present patterns persist in larger samples, scavenging would be required to involve simultaneous biases with respect to sex, age, and season of death. This seriously compromises its effectiveness as a subsistence strategy and as an explanation of observed data.

SUMMARY

The data discussed here range from mildly to strongly suggestive of hunting as the mode of procurement of mastodonts from the late Pleistocene of the Great Lakes region. While these data have certain weaknesses, among which small sample size is most notable, the three sets of attributes are at least independent of one another and consistent in their implications. Additional data will be needed to pursue further the questions raised here. However, my principal interest has been to develop a general approach for dealing with an inherently complex issue while relying on fewer and/or less severe assumptions than are typically required by direct appeals to broadly generalizable properties of hunted and scavenged assemblages. The conclusions of the argument do not offer absolute falsification of any of the alternative interpretations; the argument is not designed to operate in this fashion even with "compelling" data. Rather, it presents a more qualified and conditional (Duhemian or Lakatosian, rather than Popperian) view of the relative standing of competing hypotheses.

An especially important further test of the interpretations presented here will be to conduct similar comparisons across an expanded temporal scale (Fig. 8). If, for instance, the autumn mode in the season of death distribution is the result of hunting, then it should extend into the past only as far as the time of initiation of a significant human hunting presence in North America. In contrast, the winter-spring mode of natural death should persist through much of the Pleistocene. If the autumn mode persists as well, then either we

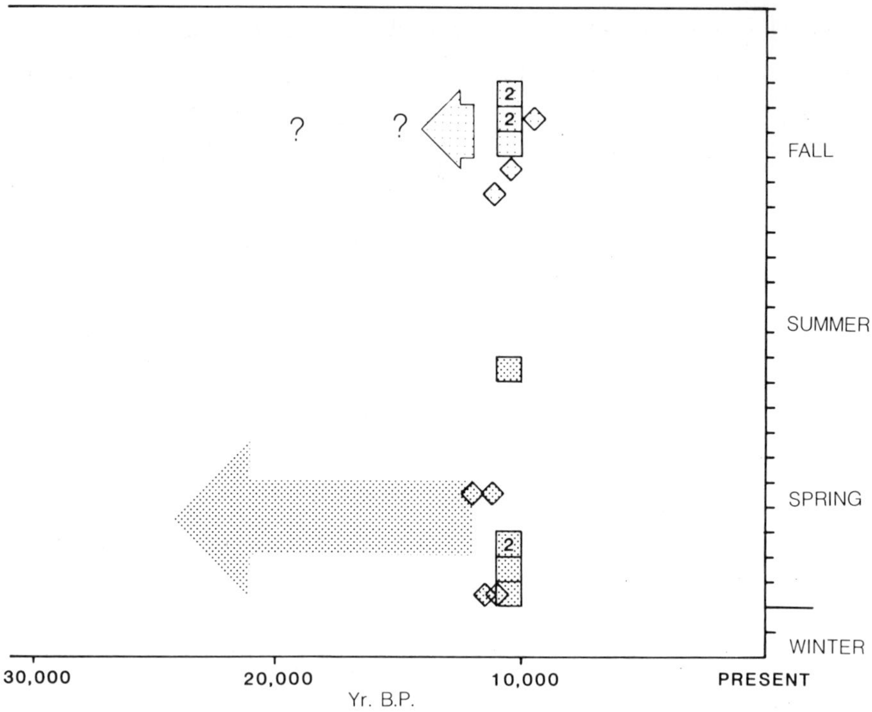

Figure 8. Predicted historical extension of patterns of seasonal mortality in mastodonts. Diamonds indicate specimens with radiocarbon dates and are placed accordingly. Squares represent specimens that are dated on geological and botanical evidence only. Numbers indicate overlapping plots, and specimen identification may be obtained from Fig. 7.

will have new evidence for bracketing the time of human entry into North America, or the hunting interpretation will have been refuted. The choice between these two alternatives would depend upon how far back the autumn mode persists and in conjunction with what other events it arises. Although a short exploration into the latest Pleistocene may be inconclusive, more extensive coverage should provide a relatively definitive answer -- at least to the question of the origin of the autumn mode in mortality. Furthermore, a non-butchered sample from a time interval just prior to the development of an autumn mode would offer a "demonstrably" precontact sample of known natural deaths. The context of such a sample would probably be closely comparable to that of butchered samples such as the one studied here, providing an alternative standard for the butchered/non-butchered comparison. Sex ratios and age profiles should also be examined historically, along with any other premortem attributes (including life history status and incidence of debilitating skeletal abnormalities, among others) that can be discerned in fossil material. The patterns that emerge through this type of multidimensional analysis should be much more informative than any attempt to focus on a single attribute.

Recent ecosystems and preservational environments also comprise an important domain for testing the arguments proposed here. The auxiliary hypotheses on which this analysis is founded have roots in both theory and observation, but they need to be checked in a more controlled and comprehensive fashion against the actual behavior of predators and scavengers, both human and nonhuman.

Part of the significance of the results reported here is their contribution to the ongoing debate regarding the causes of the late Pleistocene megafaunal extinction. If the cases of butchery discussed here had all turned out to represent scavenging, it would scarcely have helped, and might have been interpreted as quite damaging to, the overkill hypothesis. Demonstration of hunting, when and if this becomes secure, of another victim of this extinction is at least permissive of overkill, especially relative to the prior situation in which evidence of hunting (except of mammoths) has been most conspicuous by its absence. More importantly for this question, the life history data that derive from the analysis of incremental lamination in tusk and molar dentine provide access to information on the population dynamics of a species affected by the extinction. By documenting the "response" represented by changing patterns of life history traits, prior to and during the time of extinction, it should be possible to infer to some degree the nature of the "external stimulus" bringing about the decline in population size. Details of the design of such studies must be presented elsewhere, but they are a direct outgrowth of the investigations discussed here.

The data discussed here also provide insight into the behavior and lifestyle of Paleoindians, but this must be interpreted cautiously. Taking butchery as given, but assuming as little else as possible, all that can be said at the moment is that most of the preserved mastodont carcasses processed by Paleoindians from this time and region were those of young and middle-aged males in their prime, procured, probably by hunting, in the late autumn. The various retrospective auxiliaries discussed above suggest more richly detailed reconstructions, but these must be treated as highly provisional and based on additional assumptions.

Finally, application of the approach developed above may permit an improved understanding of a variety of other associations between humans and proboscideans, and even other taxa, in the archeological record. Especially intriguing in the context of this symposium are the localities of Torralba and Ambrona, from the middle Pleistocene of Spain (references in Klein, this volume). Problems of adequate preservation and sample size would of course complicate any such study, as they have this one. In addition, the results would have to be interpreted within a regional, historical framework. Even greater problems would accompany any extension of this approach to early hominids. Some of the arguments would have different degrees of support, and other taxa would be involved, possibly not providing as rich a reservoir of data as do proboscideans. Nevertheless, the principles underlying the analysis would be the same, and the potential insight into the early history of human subsistence behavior would be well worth the effort.

ACKNOWLEDGEMENTS

This work has benefitted from the assistance of a great many people. I am grateful to the symposium organizer, Matthew H. Nitecki, and to the other symposium participants, for their comments and encouragement. Consideration of the possible role of caching was prompted largely by discussion with George Frison and Arthur Spiess. The manuscript has been greatly improved by comments and suggestions from Catherine Badgley, Paul Koch, Arthur Spiess, and John Speth; the responsibility for its remaining unruliness is my own. Paul Koch has also provided steadfast assistance in excavation and data collection. For additional technical assistance, I am indebted to Bret Beall, Cynthia Darling-Fisher, and Karen Klitz. Elizabeth Garland, Alan Holman, Ron Kapp, Hezy Shoshani, and Daniel Howe have all provided generous access to specimens and information under their care. This work has been supported in part by the National Science Foundation (BNS-8511097).

APPENDIX

The preceding text is focused on an immediate problem: determining the means of procurement of the butchered mastodonts in Table 1. However, by pursuing this goal in the most direct fashion available, I have sacrificed several other significant objectives. The intent of this Appendix is to fill in some of these gaps.

One objective not dealt with in the text is a description of Bayesian inference, on which many of my arguments are based. I relied instead on a "common sense" notion that there is an association between the probability of observing certain data, given a particular hypothesis, and the probability of that hypothesis, given that those data are observed. "Common sense" can occasionally be misleading, however, and it is thus preferable to state clearly the conditions under which a particular inference is warranted. These qualifications are formalized, and the result compared to more familiar patterns of inference, in the sections on Bayes' Theorem and its application to this problem.

Another objective is to relate the arguments and general approach proposed here to previous methods of distinguishing hunted from scavenged assemblages. This is essential if the strengths and weaknesses of each are to be compared, or any conflicts resolved. This issue was avoided in the preceding text, because clarification of the differences in the logical structure of alternative approaches would have required an extended digression. I have still refrained from detailed methodological comparisons, but have at least attempted to explain why attributes such as a butchered individual's sex, age, or season of death might be expected to have anything whatsoever to do with whether it was hunted or scavenged.

A third objective of this Appendix is to convey the potential generality of the approach proposed here. Although the "front-line" arguments must always reduce to particular statements about real phenomena, the essential design of the arguments is independent of the taxa, environmental context, and attributes discussed above. This is emphasized in part by discussing attribute distributions in terms of an abstract causal model (based on Fig. 1) of the processes associated with hunting and scavenging. This model simultaneously provides a means and an occasion for more precise description of the factors controlling attribute distributions.

Finally, I will use some of the conceptual distinctions developed in this discussion to comment on the use of postmortem attributes (e.g., bone modification, patterns of disarticulation, or patterns of presence/absence) in distinguishing hunted and scavenged assemblages. Although postmortem attributes have not figured heavily in the present analysis (except to distinguish butchered from non-butchered carcasses) they have been the primary focus of several previous workers.

Bayes' Theorem

As noted above, Bayes' Theorem constitutes the model for many of the arguments involved in this analysis. Proponents of Bayesian statistical inference are sometimes criticized for their apparent dependence on highly quantitative representations of relationships about which there may be little certain knowledge. However, the following application of Bayes' Theorem is qualitative in nature. My intent is not to calculate probabilities, but rather to use Bayes' Theorem as a framework for clarifying the roles and relationships of different classes of evidence and assumptions. Bayes' Theorem may be presented as follows:

$$P(H_i|DX) = P(H_i|X)\frac{P(D|H_iX)}{P(D|X)} \quad (1)$$

$P(H_i|DX)$ is the posterior probability of hypothesis H_i, conditional on observation of data D and additional assumptions or relevant information X. The goal of this analysis is the assessment of the posterior probabilities of hunting (H_h) and scavenging (H_s) as alternative interpretations of the means of procurement of butchered mastodont carcasses. For the purposes of this analysis, I am assuming that butchery has been demonstrated satisfactorily, but this can always be reevaluated later.

$P(H_i|X)$ is the prior probability of hypothesis H_i, conditional only on assumptions and relevant information X. The prior probabilities of hunting and scavenging can be thought of as reflecting the relative incidence of these two modes of carcass acquisition in some larger universe of similar cases, considered apart from the particulars of the present case. Prior probabilities are sometimes interpreted as representing the general understanding, or estimate, of the probabilities of competing hypotheses "prior" to the current analysis. X then includes the information on which "general understanding" is based but excludes any new data brought by the present case. Prior probabilities may also be based on arguments from first principles and relevant background information about the present case and its contextual circumstances, all of which is then included in X. In either case, X represents the same thing in all expressions, within a given problem. Prior probabilities of H_h and H_s are influenced by many factors. Some of these, such as the nature of the Paleoindian technological repertoire, are susceptible to empirical investigation and are even moderately well understood. Other factors, such as the energetic and logistic costs and benefits of hunting and scavenging, are difficult to treat quantitatively and yet must have been influential. Information about the factors included in X has a bearing on the choice between H_h and H_s and thus has a role in analyzing the present problem.

Other things being equal, hypotheses with greater prior probabilities tend to be favored.

Although the information used to determine prior probabilities may be considered "evidence," in a broad sense, the evidential role of D is more direct. D consists of the data, or particular attributes, of the case under study, which comprise the overt basis for evaluating competing hypotheses as explanations of that case (i.e., not just "in general"). $P(D|H_iX)$ is the probability of this data, conditional on acceptance of a given hypothesis (H_i) and background information and assumptions (X); this expression reflects the ability of the hypothesis being evaluated to predict or explain the observed data. In a standard, quantitative Bayesian analysis, the probability conferred on the observed data by each hypothesis would be calculated or determined empirically from "known" cases. However, the procedure that will be followed here involves theoretical probabilities, determined on a relative basis only, by examining the causal relationships between hypotheses and data. Other things being equal, hypotheses that confer a higher degree of probability upon the observed data tend to be favored.

$P(D|X)$ is the probability of the observed data, conditional only on X, or in other words, considered across all competing hypotheses. It is evaluated as follows:

$$P(D|X) = \sum_{i=1}^{m} P(H_i|X)P(D|H_iX), \qquad (2)$$

where m is the total number of competing hypotheses. The impact of this expression is more difficult to describe, since its terms include both the prior probability of each hypothesis and the probability of the data under each hypothesis. However, data with a higher overall probability of being observed tend to offer a lower power of discrimination among hypotheses beyond that already provided by prior probabilities.

Hypothesis Set

It is usually desirable (and necessary for evaluating $P(D|X)$) to specify exhaustively the set of all H_i. If H_h and H_s refer to a single site at a time, their mutually exclusive and jointly exhaustive definitions guarantee that this goal has been met (again, assuming we are dealing with butchered animals). It follows that $P(H_h|DX) + P(H_s|DX) = 1$. However, if H_h and H_s refer to the complete set of butchery sites, as is necessary for most of the arguments proposed here, then a total of 2^n alternative hypotheses, where n is the number of butchery sites ($n = 10$ in Table 1), is needed to cover all conceivable interpretations (i.e., "all H_i"). H_h and H_s are then only the two homogeneous end-members, wherein all butchery sites are given the same interpretation (H_h = all hunted, and H_s = all scavenged). The other 2^n-2 hypotheses are various heterogeneous alternatives, according to which certain sites represent hunting and others, scavenging. In the multi-site case, the sum of the posterior probabilities of all 2^n hypotheses = 1. For either the single-site or the multi-site case, a relatively conservative decision criterion is to favor the hypothesis having the greatest posterior probability (Salmon 1975). In the single-site case, this hypothesis must have a posterior probability > 0.5, but this is not necessarily true of the multi-site case. Although only the single-site case really represents a choice between as few as two competing hypotheses, the multi-site case can in principle be handled as a series of comparisons between pairs of competing hypotheses. Since it is easier to evaluate terms in Bayes' Theorem using homogeneous, rather than heterogeneous hypotheses, I will treat the multi-site case initially as a determination of the posterior probabilities of the two end-member hypotheses H_h and H_s. Whether the homogeneous alternatives will prove to be the most and least attractive hypotheses overall (i.e., whether they will show maximal and minimal posterior probabilities relative to others in the 2^n hypothesis set) cannot be anticipated at this stage in the analysis but will be dealt with when the treatment of H_h and H_s is complete. To preserve generality, the following discussion is framed in terms of the multi-site case, but many points apply equally to the single-site case.

Simplifying the Problem

If all expressions in Bayes' Theorem were evaluated quantitatively, we would have not only a basis for choosing among hypotheses, but also an indication of the confidence that might be placed in a given hypothesis. However, when completely quantitative treatment is unwarranted, we may settle for the goal of hypothesis choice alone. This may be handled as an evaluation of the ratio of posterior probabilities. Because the value of $P(D|X)$ is the same for competing hypotheses from the same hypothesis set, it cancels, leaving the following equality:

$$\frac{P(H_h|DX)}{P(H_s|DX)} = \frac{P(H_h|X)P(D|H_hX)}{P(H_s|X)P(D|H_sX)}. \tag{3}$$

The problem thus reduces to an evaluation of prior probabilities and the probability conferred on the data by each competing hypothesis. For Paleoindians and proboscideans in late Pleistocene North America, my impression is that most workers would judge the prior probability of hunting to be greater than that of scavenging (e.g., Haynes 1969, 1982; Frison 1978; Saunders 1980). However, much of the background information in support of this assessment is derived from cases involving mammoths. While *probably* relevant to cases involving mastodonts, such interspecific comparisons are less than compelling. In any event, we have only a general notion of the range of factors affecting the prior probabilities of H_h and H_s and only a subjective basis for evaluating and weighting those factors of which we are aware. Improvement of this situation may be a goal for future work, but it will not be the focus of this discussion. Rather, as is common practice when this type of situation is encountered in Bayesian analysis, I will treat the prior probabilities of hunting and scavenging as if they were equal. This is not a substantive assumption of equality, but rather a methodological expedient that acknowledges the uncertainty of present assessments of prior probability and therefore allows the data (i.e., $P(D|H_iX)$) to dominate the assessment of posterior probabilities.

Relating Hypotheses to Data

Even this simplified version of the problem is best handled quantitatively, since choosing among all members of the hypothesis set can be done most efficiently when the ratio (or ratios) of posterior probabilities can be evaluated numerically. This in turn requires numerical values for the probabilities conferred upon the data by competing hypotheses. In a fully quantitative Bayesian analysis, these values would usually be based on a combination of empirical and theoretical considerations, but for reasons explained below, this level of quantification is not feasible here. Nevertheless, it is critical for the Bayesian (or indeed, for any) analysis to have access to data, or attributes of the case at hand, that show differing probabilities conditional on different members of the hypothesis set. For treatment of the entire hypothesis set this requires explicit or implicit (i.e., by transitivity) specification of at least the *relative* conditional probabilities of the data on each H_i (i.e., $P(D|H_1X)$ vs. $P(D|H_2X)$ vs. ... $P(D|H_iX)$). Fortunately, this point of last retreat is accessible (at least in certain circumstances) through the procedure outlined below.

A variety of methods may be used to specify the probabilities of observed data conditional on each H_i. While any such pattern of reasoning may thus serve a Bayesian end, the assessment of these probabilities takes place outside the immediate context of Bayes' Theorem and is not at root a uniquely Bayesian activity. The present analysis will therefore shift to this more general plane. I will continue to explain arguments through reference to the Bayesian framework but will return to Bayes' Theorem only when appropriate probability statements are in hand.

The approach taken for relating data and hypotheses in this analysis is not a novel type of argument, but it is sufficiently different from previous approaches to the hunting vs. scavenging issue that some explicit comparisons are in order. Several potential approaches to relating hypotheses and data are diagrammed in Fig. 9 and discussed below. The one followed here is indicated

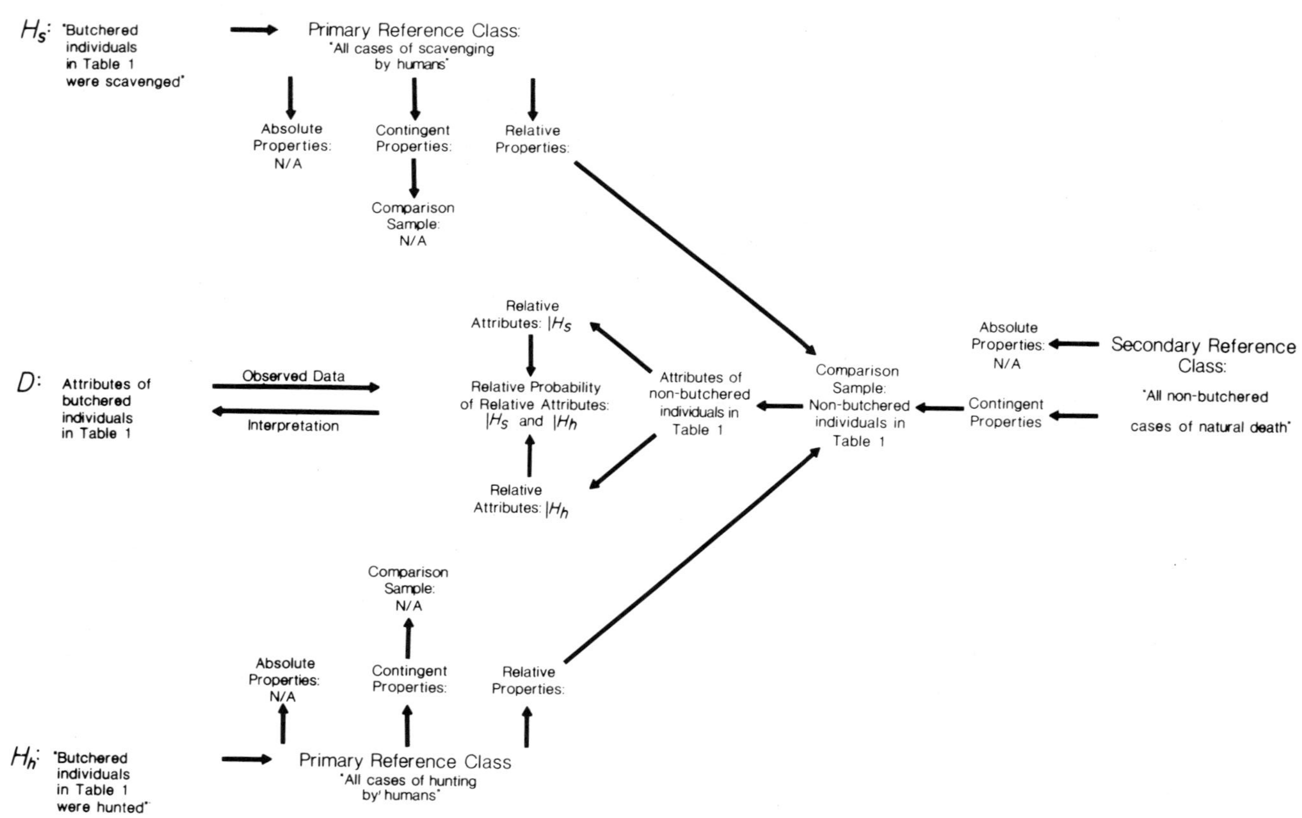

Figure 9. Simplified schematic portrayal of an indirect approach to relating hypothesis to data. H_s is the scavenging hypothesis, H_h is the hunting hypothesis, and D is the data observed for the case under study. Arrows imply direction of associative reasoning based on relationships that are variously logical (i.e., formal), empirical, theoretical (causal), or inferential. Reference classes, properties, and attributes are discussed in the text.

by a complete series of arrows between hypotheses and data and is admittedly the least direct of those shown. However, it has been adopted only after encountering difficulties with the application of alternative approaches (represented in Fig. 9 by series of arrows ending in "N/A") to the present problem. The main intent of comparing these approaches is to clarify the strategy of argumentation followed here, which because of its indirectness, might otherwise seem obscure.

Primary Reference Classes

A convenient way to portray the relationship between a hypothesis and data is to consider the hypothesis as referring to a general class of phenomena -- its "primary reference class" -- of which the case under study may be a particular instance. For a given hypothesis to be plausible, the data must correspond, in some degree, to properties of members of the primary reference class of that hypothesis. For example, H_S ("the butchered individuals in Table 1 were scavenged") refers to the primary reference class "all cases of scavenging by humans," where the qualification "by humans" is present because H_S, through its reference to butchery (carcass processing by humans), already stipulates that humans are the alleged scavengers. Recognition that this case involves utilization of mastodonts by Paleoindians could also be included in the definition of the primary reference class, but this would only add a further identical element to all competing hypotheses. Although this step is sometimes taken later in an argument, at this point it would narrow the reference class unnecessarily, making its properties more difficult to discover or generalize. I will restrict the term "properties" to the somewhat abstract sense of features shown by a member of a reference class and will use the terms "data" and "attributes" interchangeably to indicate observed features of the case at hand. For brevity, however, I will refer to data (or attributes) viewed provisionally in light of a given hypothesis, as "being," rather than "corresponding to," properties of members of the primary reference class of that hypothesis.

The most direct and unambiguous relationship between a hypothesis and data is one in which the data are absolute properties of members of the primary reference class of the hypothesis (i.e., absolute relative to the hypothesis). "Absolute" properties derive directly from the definition of the class and are possessed by "absolutely all" members of the class -- or "absolutely none." Given the hypothesis, these properties must also characterize the case under study. Absolute properties are deducible from the class definition, but in practice they may be initially encountered through induction from other cases. In terms of $P(D|H_iX)$, a hypothesis confers a probability of 1.0 (unless qualified by some component of X) upon observation of the absolute properties of members of its primary reference class and a probability of 0 upon observation of properties that are "absolutely" absent in such members.

An example of an absolute property of members of the primary reference class of H_s is that they had negligible or no chance of survival prior to being encountered by the humans who utilized them. While not all absolute properties distinguish among all competing hypotheses, this property would be especially useful, if it could be observed (i.e., identified as an attribute), because it is "absolutely" absent (i.e., its absence is an absolute property) in members of the primary reference class of H_h ("all cases of hunting by humans"). However, for an encounter that took place in the late Pleistocene, the probability of observing this particular absolute property is effectively 0 for *both* H_s and H_h, and the property is thus of no practical value in this context.

An alternative path by which hypotheses can be related to data (Fig. 9) is to consider data that represent contingent properties of a primary reference class. Whereas absolute properties are context-independent and deducible from the definition of the reference class, "contingent" properties are context-dependent. They may be deduced from knowledge of the reference class *and* sufficiently complete knowledge of context, but since the latter is often not available, they are more commonly *induced*, or derived by generalization from some comparison sample. Comparison samples are composed of cases "known" (i.e., observed or strongly inferred) to be members of their reference class. If a comparison sample were treated as a class itself, its definition would include that of its reference class, along with further qualifications or restrictions. The identities, attributes, and contexts of comparison samples are part of the background information summarized in *X*. A comparison sample is expected to be representative of (and thus informative regarding) the contingent properties of members of its reference class, and hence the conditional attributes of the case under study (i.e., attributes it would show if the given hypothesis were true). However, this cannot be assured in the way that it can for absolute properties. Rather, the validity of generalizing from attributes of the comparison sample to conditional attributes of the case under study depends on the extent to which their respective contexts are "dynamically similar." Dynamic similarity is always a matter of degree and refers to similarity in those aspects of context that are judged capable of producing, through their qualitative or quantitative variation, a significant difference in the distribution, or probability of occurrence, of a given contingent property. The assessment of dynamic similarity thus depends both on the property in question and the interpretation of what factors control that property's distribution. Dynamic similarity for a given property does not require "overall" similarity, but this latter concept is commonly treated as a sufficient condition for dynamic similarity. Even though separate contexts are rarely identical in all relevant respects, careful choice of comparison samples can make this a practical approach to relating hypotheses to data. The term "comparison sample" is clearly akin to the more conventional "control group" but is used here in preference to the latter in order to emphasize that both its relevancy and its status as an instance of the primary reference class are commonly matters of inference.

In certain instances, the empirical association between particular contingent properties and members of a reference class is so strong that these properties are possessed by "almost all" members -- or "almost none" -- of the reference class. Their distribution thus resembles that of absolute properties, even though they are not strictly deducible from the definition of the reference class. I will refer to this subset of contingent properties as "quasi-absolute," and to the complementary subset as "non-quasi-absolute." As with other categories of properties, these adjectives refer to the status of a property relative to a given hypothesis. For instance, a property may be absolute with respect to one hypothesis and contingent with respect to another, or quasi-absolute with respect to one, and nonquasi-absolute relative to another. The unconstrained nature of "almost" makes the distinction between quasi-absolute and nonquasi-absolute properties somewhat arbitrary, even when they are compared within the same context. Nevertheless, the distinction recognizes a theoretically based judgment as to what constitutes "significant" variation, and it has strong implications for the use of the property in subsequent arguments.

If quasi-absolute properties are recognized as such, and accessible, they may be treated in much the same way as absolute properties. They also frequently lead to results that are almost as clearcut. In fact, since the probabilities invoked during discussion of either absolute or quasi-absolute properties are essentially equivalent to either 0 or 1.0, much of the formalism of Bayes' Theorem becomes superfluous. In such cases, the arguments involved are for most practical purposes reducible to simple if-then statements allowing refutation by counterfactuals. I suspect that a common orientation toward treating complex problems through appeals to properties that are quasi-absolute for at least one member of a hypothesis set is part of the reason we do not usually see Bayes' Theorem as making a necessary contribution to many debates. Nevertheless, when dealing with contingent properties that are not quasi-absolute with respect to any member of the hypothesis set, and probabilities that differ appreciably from either 0 or 1.0, Bayes' Theorem offers nontrivial insight into the interaction between hypotheses, data, and background information and assumptions.

"Probability" of Properties

Within standard applications of Bayesian statistical inference, the operative sense of "probability" is usually "long-run frequency within a series of similar cases" (i.e., within a reference class). The categories of properties recognized above bear a clear relationship to "strategic breaks" in the spectrum of variation in property frequency (or distribution). A frequency-oriented concept of probability is thus sufficient for discrimination of these categories of properties and is particularly manifest in the use of comparison samples to establish the "probability" of contingent properties within particular reference classes. Nevertheless, many properties have *causal* relationships with competing hypotheses. These causal relationships are often recognized explicitly for absolute properties and at least implicitly -- through the analysis of dynamic similarity -- for contingent properties. In a world (and at a scale) that is frequently dominated by causal interactions, it is improbable that a property having no causal involvement with a hypothesis would show an absolute or quasi-absolute distribution within the primary reference class of that hypothesis. Most absolute properties are either defining properties of their reference class, or necessary conditions or consequences of the events postulated by the hypothesis with respect to which they are absolute. Contingent properties in general are more affected by interactions and tend to have a more diffuse causal involvement. They may reflect joint causes, sufficient conditions, multiple effects, or serial causes and effects (a cause of a cause or an effect of an effect). In this same sense, quasi-absolute properties tend to be more strictly determined than are nonquasi-absolute properties. If this sounds tautologous, it is only an indication of the readiness with which we already accept patterns as having a causal basis.

These causal relationships are important because they represent the "reason" behind the differences in hypothesis-dependent probability that make certain attributes inferentially useful in the first place (with respect to a certain hypothesis set). There is thus a theoretically-oriented concept of "probability" underlying even frequency-oriented applications. This is, of course, a principle common to many applications of statistics, but it is particularly important to the present problem because later in this analysis the assessment of probability will shift largely to a theoretical level. For most purposes, however, the numerical tractability of frequency observations make them the preferred basis for probability statements.

Properties in Inference and in Systematics

The "properties of properties" described above validate a generally appreciated prioritization of the approaches to relating hypotheses and data diagrammed in Fig. 9. When absolute properties are accessible for analysis, they apply without exception to members of the reference class for which they are absolute. They thus allow treatment of singular cases and, given appropriate data and property distributions in the reference classes of competing hypotheses, allow the most nearly definitive refutation of those competitors. However, for problems encountered in the fossil or archeological record, absolute properties are often inaccessible or else nearly invariant in their distribution within the reference classes of competing hypotheses. In such situations, beyond demonstrating the plausibility of a hypothesis, absolute properties do not provide definitive tests among competing hypotheses. As might be expected from the "almost"-absolute status of quasi-absolute properties, they show much of the same promise. Perhaps because they have more complex causal relationships with competing hypotheses than do absolute properties, they tend (judged subjectively) to be more numerous and more commonly accessible for analysis. They thus represent the next most convenient and effective approach (other things being equal) to relating hypotheses to data.

The principal difference that arises in dealing with nonquasi-absolute properties, especially ones that are nonquasi-absolute relative to all members of a hypothesis set, is that single (or small numbers of) instances are often inadequate for demonstrating whether or not the observed property distribution differs from the expectation associated with a given hypothesis. Small samples may still be handled in principle -- i.e., they may be judged as probable or improbable relative to an expected frequency distribution--but it often takes larger samples to produce statistically significant results. Aside from this shortcoming and the complexity it entails -- such as the need to treat the present problem as a multi-site case rather than a series of single-site cases -- nonquasi-absolute properties represent an important resource for evaluating competing hypotheses.

An interesting perspective on this problem is offered by considering an analogy between inference and the classificatory aspect of systematics, within either evolutionary biology or archeology. It is commonplace to recognize classification as involving inference, but perhaps less common to think of inference as a form of classification. In particular, the contrast between inferential strategies based on absolute and quasi-absolute properties on one hand, and nonquasi-absolute properties on the other, is somewhat analogous to the contrast between "typological" and "populational" philosophies of systematics. The latter, more familiar contrast involves alternative approaches to distinguishing species (or artifact classes) and assigning individual organisms (or artifacts) to them, while the former contrast involves the decidedly more abstract activity of distinguishing disparate patterns of evidential association with different interpretations, to which individual phenomenological cases are then assigned. While treading close to a number of controversies (e.g., an appropriate definition of "typology," or the individual vs. class nature of species) the intent of this analogy is neither to awaken them nor to disparage the use of absolute and quasi-absolute properties in an inferential context. Rather, I wish only to emphasize that while properties possessed by all or almost all members of a class are convenient, they are not essential for making rigorous decisions about class membership -- or rigorous choices between competing hypotheses.

Properties and Hierarchy

The distribution of properties has been described thus far in terms of presence or absence; they are either possessed or not possessed by a particular member of a reference class. Yet properties, like "characters" within systematics, are inherently hierarchical. "Presence" can often be further subdivided to describe different states that the property might take on in different members of a reference class. This is not an issue for absolute properties because at the level at which they are absolute, all members of the reference class have the same state, and we identify that state as the property itself. However, for quasi-absolute properties, the *exceptions* to the "almost all" generalization may either fail to possess the property in question or possess it in some different state. Nonquasi-absolute properties may also show either of these patterns. Of course, presence and absence may themselves be treated as two states. When states of a property are recognized, the relevant hierarchical level for determining the status of the property is the level of the states and not that of "the property itself." For instance, age could be treated as an absolute property of individual organisms -- as something that all individuals "have" -- but this is hardly informative. Instead, age, being implicitly divisible into multiple states, is better treated as a contingent property of individuals. The properties of butchered mastodonts on which I focus here--sex, age at death, and season of death -- are treated as contingent in this sense.

Another complication involving properties and hierarchy concerns situations in which multiple instances of a property occur within a single "case." This may arise when a case involves multiple individual organisms (as in the multi-site version of the present problem), each of which represents a single instance of a given property (e.g., age at death) or when a given property (e.g., a particular type of cutmark) is represented by multiple instances on an individual organism. Whenever multiple instances of a property are involved in a single case, it is important to distinguish between the status of an individual instance of the property and the status of the distribution of that property within the *sample* of multiple instances per case. Alternatively, it may be useful to distinguish two "different" properties -- one referring to single instances and the other referring to the distribution of instances within a sample. For example, age at death can be treated as a *contingent* property (as above) of any dead organism and thus also of individual organisms that have been either hunted or scavenged by humans. However, if a case such as the present one involves multiple dead organisms, the comparable property at the case-level (or sample-level) is the "age profile." Although the age profile can also be treated as a *contingent* property of samples of dead organisms in general, *certain types* (or states) of age profile may represent *quasi-absolute* properties of hunted or scavenged samples --

within particular contexts, if not generally (Klein 1982a, this volume). This example points up a potential ambiguity in the distinction between quasi-absolute and nonquasi-absolute status, namely, that the distinction itself (since it involves contingent properties) is context-dependent. In particular, the distribution of a property, if not single instances of that property, would be expected to be quasi-absolute within dynamically similar contexts. This suggests that the strategy of using a comparison sample to determine the distribution of a generally nonquasi-absolute property is equivalent to constraining the context narrowly enough that the distribution of the property can be treated as "locally" (i.e., for that limited context) quasi-absolute.

As noted above, the most important aspect of properties used to evaluate competing hypotheses is that they must have a different distribution within the reference classes of competing hypotheses. For a property handled in terms of presence and absence, "different distribution" means simply a different proportional (either within or between cases) or theoretically probable presence within different reference classes. Where a range of states is involved, "different distribution" means that there is a difference between classes (and thus between hypotheses) in the frequency distribution or probability distribution describing possession of the various states of a property by members of a reference class.

These distinctions between categories of property status, the property-level (property or state) at which status is determined, and the case-level (sample or single instance) at which a property (or state) occurs, form the basis for recognizing and comparing methodologically distinct approaches to relating hypotheses to data. For example, the use of cutmarks crosscutting toothmarks (Potts and Shipman 1981; Shipman 1986) to indicate that an individual organism has been scavenged by humans is an application of a property that is quasi-absolute at both the single instance- and sample-level. Because cutmarks crosscutting toothmarks are quasi-absolutely *absent* among hunted individuals, observing such marks constitutes grounds for refuting hunting (for the individuals that show them). The use of age profiles to distinguish hunted and scavenged assemblages (Klein, this volume) is based on a nonquasi-absolute property (individual age at death), but one whose distribution may be quasi-absolute over a wide range of contexts involving hunting or scavenging. In contrast, Shipman's (1986) use of data on the anatomical distribution of cutmarks on skeletal remains from Bed I, Olduvai Gorge, involves a nonquasi-absolute property (anatomical location of individual cutmarks) whose distribution is also nonquasi-absolute over most contexts involving hunting and scavenging. However, Shipman (1986) derives an expected frequency distribution for anatomical location of cutmarks from a comparison sample -- a Neolithic butchery site (Gifford et al. 1981) -- argued to be dynamically similar (i.e., similar in relevant respects) to the context in which hunting would have occurred among early hominids. The frequency distribution of anatomical location of cutmarks is thus treated as a quasi-absolute property of hunted assemblages within this restricted context. My purpose here is not to evaluate the content of these examples; rather, it is to describe their form -- which is in all cases perfectly valid -- in order to clarify the unusual aspects of the form of the arguments developed here.

The Crux of the Problem

Arguments sharing the form, and to some degree the content, of the examples listed above were among the first attempted in addressing the present problem (see text comments on cutmarks crosscutting toothmarks and on the conventional use of age profiles). Their results were, however, less than compelling. In search of alternative approaches, I have considered properties with various distributional patterns within the reference classes for hunting and scavenging. I eventually focused, however, on nonquasi-absolute properties (sex, age, and season of death) whose distribution is also recognized as nonquasi-absolute among hunted and scavenged assemblages generally (i.e., within both primary reference classes). In these respects, the arguments developed below most resemble Shipman's (1986) use of cutmark location data. However, a major difference is that, as explained in the text, I have no comparison sample that I wish to argue represents a context dynamically similar to that of the butchered mastodonts in Table 1. I therefore will not be able to shift nonquasi-absolute properties of the primary reference classes into "locally" quasi-absolute status. With respect to our limited knowledge of Paleoindian hunting and/or scavenging habits and the lack of unambiguously relevant comparison samples, the problem faced here is not so different from trying to deal with the behavior of early hominids. Few answers seem to be available as long as the analysis is restricted to the primary reference classes.

A way out of this cul-de-sac is suggested by noting that the overriding consideration in selecting properties for analysis is that they show a different distribution within the primary reference classes of competing hypotheses. If the distribution of a property within a primary reference class cannot be determined using background knowledge (and/or assumptions) dealing with that reference class alone, the same goal may be achieved by considering an *additional* reference class -- one for which an appropriate comparison sample and other background knowledge might be more readily available. If the distribution of the property can be demonstrated in this additional reference class, and if the distributions of the property in the primary reference classes are each determined *relative* to this additional distribution, then their similarity or difference relative to *each other* can be ascertained. This represents the route that will be followed through the rest of this discussion.

Secondary Reference Class

The additional reference class just proposed will be referred to as a "secondary reference class," since it enters the argument later than, and in a supporting role to, the primary reference classes. Whereas the primary reference classes were constructed "from the top, down" -- according to the composition of the hypothesis set -- the secondary reference class is constructed "from the bottom, up," according to the availability of an appropriate comparison sample. This comparison sample must be able to serve as a source of information relative to which the distributions of properties in the primary reference classes might be determined. In order to promote dynamic similarity, the comparison sample for the secondary reference class should be from the same context as the case under study. The only sample that fulfills these requirements is the sample of non-butchered individuals in Table 1. For reasons expounded in the text, I propose that this constitutes a comparison sample for the secondary reference class "all non-butchered natural deaths."

The manner in which a secondary reference class fulfills its role varies from one problem to another, but the strategy for this problem is based on the system of set-relationships between the membership of the secondary reference class and that of each primary reference class. This system is most easily described in a simplified form and then adjusted to appropriate complexity. Ignoring scavenging of moribund individuals, the primary reference class "all cases of scavenging by humans" and the secondary reference class "all non-butchered natural deaths" are complementary subsets of a further (unnamed) class "all natural deaths." Furthermore, with the same proviso, "all natural deaths" and the primary reference class "all cases of hunting by humans" are, if integrated over long enough periods of time, complementary subsets of the living population. These relationships are illustrated by the flowchart in Fig. 1, where the two primary reference classes (samples 3 and 7) and the secondary reference class (sample 8) are portrayed as abstract samples ultimately derived from the living population (sample 1). Fig. 1 also demonstrates the complication introduced by scavenging of moribund individuals: individuals scavenged by humans are represented by two complementary subsets as well, one of which is in fact the complement of "all natural deaths," within the class "all naturally moribund individuals."

The significance of these relationships is that a relatively simple *pattern* of transitions links the actual samples involved in this analysis, through the primary and secondary reference classes (or through the actual, time-integrated populations that represented them) and ultimately through the living population of this species, in a limited region and interval of time. Although the processes underlying these transitions are not necessarily simple and are certainly not all well characterized (especially the "human element"), there is a great deal known about the general *kinds* of processes involved. The argument developed below involves essentially "working backward" (though not actually in that order) along this series of transitions from the non-butchered natural death sample (sample 11) to a hypothetical scavenged sample (10) and to a hypothetical hunted sample (9). The object is to constrain, if only to a limited degree, the distribution of non-quasi-absolute properties that would be expected to characterize the butchered sample conditional on each of the two competing hypotheses, H_s and H_h.

The distributions of a property expected in successive samples in Fig. 1 will be based on analysis of the causal relationships between samples. This analysis will be an argument in part from "first principles" but equally dependent on patterns common to the natural history of most mammals. This approach amounts to proposing a causal model of the controls affecting the distributions of a property at each stage within the system of transitions diagrammed in Fig. 1. The procedure followed is conceptually related to, but not to be confused with, the analytical statistical technique known as "causal modeling" (Asher 1983).

Feasibility of a Causal Model

Generating a causal model that is general enough to apply to a wide variety of attributes and yet explicit enough to make useful statements about the potential distribution of a property in hunted and scavenged assemblages may seem prohibitively difficult. However, it has been simplified in several important ways. The first is that the derivation of both the "known" sample (the non-butchered natural deaths) and the "unknown" sample (the butchered mastodonts whose mode of procurement is in question) from the same context tends to insure that we are dealing with individuals that lived in the same climate, were equipped with the same physiological responses, were exposed to the same broad range of mortality factors, and were even to some degree subjected to similar taphonomic agents. This helps to control for the effects of many potentially confounding variables that might otherwise cause spurious similarities or differences between the "known" and "unknown" samples, independent of any similarity or difference in the causes of death represented by each. This strategy is comparable to the common practice of using "internal standards" in various analytical disciplines. For example, in x-ray diffraction studies, a powdered sample of a well characterized crystalline substance can be mixed and analyzed along with an unknown. This allows the diffraction pattern of the unknown to be measured relative to that of the known, controlling for a number of poorly constrained factors that might otherwise affect the results.

A second way in which this task is simplified is that there is no attempt to predict the exact distribution of a property in hypothetical scavenged or hunted assemblages; I only address issues of relative correspondence or disparity in property distributions between samples. The effect of this is to generate a new set of properties for the primary reference classes, referred to in Fig. 9 as "relative properties." They are "new" in the sense that their state is assessed *only* with explicit reference to an attribute distribution observed in a comparison sample of the secondary reference class. A relative property thus implies a comparison between the distribution of a (nonrelative) property in a primary reference class and the distribution of the equivalent attribute in the "internal standard." In principle, relative properties could be absolute or contingent, quasi-absolute or nonquasi-absolute. However, I am considering only contingent relative properties and treat them as nonquasi-absolute. Furthermore, although two different relative properties are analyzed, only two states are recognized for each. This greatly reduces the achievable resolution, but it also brings resolution into a range that is warranted by available information.

Third, this approach is even more profoundly "relative" than indicated by the "internal standard" analogy or the exclusive focus on relative properties. Even through the use of both of these expedients, I do not expect to be able to predict (in the sense of attributing to a given state a probability >0.5) how or to what extent one (nonrelative) distribution might differ from another. Although this may sound like giving up on the last chance for a determinate result, it is not. My goal is not to come up with an estimate of the actual probability of a disparity between any two attribute distributions -- this would be excellent, were it possible, but in most cases I do not consider it feasible using present knowledge. Rather, I only need to determine the *relative* probability of a disparity between butchered and nonbutchered samples, conditional on one hand, on interpreting butchered individuals as hunted, and conditional on the other hand, on interpreting them as scavenged. This was the minimal required input for the Bayesian analysis to have a determinate solution. The probabilities that are incorporated in Bayes' Theorem are therefore the relative theoretical probabilities of the data, conditional on each hypothesis. "The data," in turn, consist of observations organized at three hierarchical levels -- the attributes of individual mastodonts, the distributions of attributes within the samples of butchered and non-butchered individuals, and the relative attribute state comparing these two distributions. Only this third, highest level of data analysis is explicitly involved in "the probability of the data." As mentioned above, only two relative attributes are considered here (similarity/difference of attribute distributions and overlap pattern of state occupation), but these may be used on a wide variety of lower-level (nonrelative) attributes.

Scope and Content of the Causal Model

Analysis of the causal model of relative attribute distributions involves examining each transition that links samples of individuals scavenged by humans to samples of non-butchered natural deaths, and likewise each transition that links samples of individuals hunted by humans to samples of non-butchered natural deaths. Each of these transitions either preserves the distribution of a given attribute from one sample to the next, or biases (alters) that distribution, depending on whether the underlying processes operate in a nonselective or selective manner with respect to that attribute. As with the analogous process of natural selection, bias is produced by the effects of sustained, even if stochastically variable, causal interactions, integrated over the fates of many individuals and their attributes. A bias may result either from direct selection, in which the state of the attribute itself causes a selective interaction, or from indirect selection, in which the distribution of the attribute in question is altered as an effect of direct selection on a *different* attribute, correlated, and generally causally related, to the first. Although bias is produced by processes that operate in one direction only, the simultaneous operation of biasing factors along different paths within the network of relationships can result in differences between the attribute distributions of samples located on those different paths. Such differences -- the cumulative differential effects described in the text -- may also be described as biases (in one sample, relative to the other). Any significant disparity between butchered and non-butchered samples represents a bias of this type.

The outcome of any one transition depends on the context in which the interaction takes place, but also on the identity of the attribute being considered and the nature and degree of its causal involvement with the processes underlying that transition. Because of this attribute-dependency, any final statement regarding particular attribute distributions requires an attribute-specific evaluation. However, the outcomes of transitions in Fig. 1 are also determined in *part* by the nature and dynamics of the underlying processes themselves and their deployment about various "nodes" or branch-points in Fig. 1. This aspect of the model affects all attributes undergoing transition through the system.

As indicated above, one of the factors determining whether a given transition produces bias is the dynamics of the underlying processes. "Dynamics" here refers to the selective "forces" (or lack thereof) operating within a transition. In comparing processes acting within the same or different transitions, I will use a concept of dynamic similarity that is equivalent to, and even responsible for, the dynamic similarity of contexts. Processes are said to be dynamically similar if, despite differences in detail, they tend to show similar variations in outcome, in response to similar variation of input, and in particular, if they show similar selective (or nonselective) effects on the same attribute distribution.

The causal model developed here may be used for properties that are absolute or quasi-absolute relative to one or more primary reference classes, but such instances are usually simple enough that no detailed analysis is required. For nonquasi-absolute properties, use of the model is limited to attributes present within the living population (the most recent *common* source of *all* samples), though these may have accrued at any time prior (possibly *just* prior, as with age and season of death) to morbidity or death. Only such "premortem" attributes have distributions that can be linked causally between any two samples. Nonquasi-absolute attributes that accrue during transitions following death ("postmortem" attributes) bear no well constrained relationship (at least that can be handled in the general form of the model) to attributes in samples located on separate paths.

Use of the model should also be restricted to attributes that show potential for causal involvement with one or more of the processes underlying this network, since differential causal involvement is generally the basis for differing distributions in the primary reference classes of competing hypotheses. By this criterion, some heritable attributes may be included. However, the most conspicuous selective interactions, given the ecological (rather than evolutionary) timescale of this problem, are likely to involve nonheritable attributes showing considerable intrapopulational variation. It is admittedly difficult to determine degree of causal involvement on a priori grounds, but as long as multiple attributes are treated, the analysis should be robust to uncertainties of this nature.

Data (D_s *vs.* D_d)

D in Bayes' Theorem only refers explicitly to *observed* attributes (for which I am substituting observed *relative* attributes) of the "unknown" case. However, in order to discuss fully the relationship between *D* and all H_i, all of the states that might be exhibited as *D* must be considered. For the first relative attribute used to evaluate H_s and H_h, I am concerned only with whether the distribution of a given (nonrelative) attribute (e.g., season of death) within the "unknown" sample is similar to or different from the distribution of that attribute within the sample of non-butchered natural deaths. This divides all possible data configurations into two states: D_s

indicates an attribute distribution similar to that seen among non-butchered natural deaths, and D_d indicates an attribute distribution different from that seen among non-butchered natural deaths. I intend for similarity and difference of attribute distribution to be interpreted broadly, applying to either the means, ranges, or general form (e.g., modality, symmetry, kurtosis) of the distribution. Any difference in such a parameter greater than would be expected to result from random sampling error (i.e., small samples may imperfectly represent the distribution typical of their reference class) suffices to make the two distributions "different." As in any statistical problem, arbitrary thresholds of similarity must be established, but such issues can be handled separately.

Transition Analysis: Scavenged Sample to Non-Butchered Sample

If butchered individuals were procured by scavenging, then the degree to which their attribute distributions are biased relative to those of non-butchered natural deaths is determined by the interactions and cumulative effects of processes operating within the transitions that link samples 10 and 11 in Fig. 1. The common starting point from which the attribute distributions of scavenged and non-butchered samples are derived is the attribute distribution in the sample of naturally moribund individuals (sample 2). In the absence of human intervention, the 2-5 transition is nearly comprehensive, since by definition, the transition from morbidity to death allows almost no escape. Under this circumstance, it conserves attribute distributions. However, when humans scavenge moribund individuals, the 2-5 and 2-4 transitions are complementary: whatever is selected from 2 for 4 is not available for 5, and vice versa. If one of these transitions occurs in a random fashion with respect to particular attributes, such that the distributions of those attributes are conserved, then the other must also. However, if one transition shows a net bias, the other is consequently biased in the opposite direction. The magnitude of this complementary bias depends on the relative numbers of individuals involved in the 2-4, as opposed to the 2-5, transition.

This portion of the system is thus one of the places where a disparity between the attributes of individuals scavenged by humans and the attributes of non-butchered natural deaths might arise. However, there is no compelling a priori basis for *expecting* significant net bias in either transition. Further insight regarding the potential for bias at this node requires more detailed consideration of the associated processes.

One factor in the 2-4 transition is active selection, by humans, of targets for scavenging from among moribund individuals. Opposed to this, within the 2-5 transition, is the "same" process, enacted by nonhuman predator/ scavengers. [Although these include many taxa, I will focus first on those that are similar to humans in body size or scavenging abilities.] Also involved in each transition is the access to targets allowed to each set of potential scavengers (human and nonhuman) by the activities of the other. The general similarity in the mode of operation of these pairs of processes, together with the more or less similar physiological needs and capabilities of human and nonhuman (especially the large mammalian) scavengers, suggests that human and some nonhuman scavengers might have similar selective or nonselective effects on the distributions of many attributes. In other words, some of the opposed processes may show at least moderate dynamic similarity. The combination of this dynamic similarity and the link between active selection in one transition and access in the other generates damping interactions among potential biasing effects. For example, if moribund individuals with certain attributes were more easily located by Paleoindians, they would tend to enter sample 4 in disproportionate numbers, other things being equal. Suppose, for instance, that male mastodonts trumpeted more loudly than females when they were naturally entrapped or injured, making them easier than females to find from a distance. However, Paleoindians were not the only potential scavengers on the scene, sharing the same basic sensory apparatus. It is therefore highly unlikely that other things would have been equal, for the same moribund individuals would tend to be disproportionately easy for nonhuman scavengers to locate. This would decrease the residence time of such individuals in sample 2 and increase their likelihood of entering sample 5, balancing their tendency to be overrepresented in sample 4. Damping does not completely avoid bias unless dynamic similarity is "complete," but it does tend to restrict the development of bias between derived and source distributions and between two derived distributions.

Although the scavenging activity of humans and other large mammalian predator/scavengers seems likely to show considerable dynamic similarity, further information -- probably more information than is available -- would be needed to quantify the expected level of similarity. Furthermore, the activities of large mammalian predator/scavengers are only one (even if a major) component of the 2-5 transition. Since the relationship between an attribute's distributions in samples 4 and 5 (and ultimately, 10 and 11) is determined by the *overall* dynamic similarity across this node, other components of 2-5 must be examined. One such component, for which there is no counterpart in the 2-4 transition, is the simple passage of time and the nearly inevitable decline of a moribund individual. However, this process ("unprovoked" death among moribund individuals) is nonselective (one moribund individual being as likely to die, unprovoked, as another). To this extent, it may in complementary fashion *reflect* bias produced by another process, but it does not *generate* bias.

While unprovoked death of moribund individuals is not itself a biasing factor, systematic, attribute-dependent variation in the *rate* at which unprovoked death occurs may introduce bias, at least across this node. For instance, if juveniles within the moribund sample typically have a shorter "half-life" (i.e., expected time until half would have died, if unprovoked) than adults, the natural death sample would tend to be biased (relative to sample 2) in favor of juveniles. However, bias caused by accelerated natural death simply means that potential scavengers (either human or nonhuman) are more likely to encounter such an individual in a fully dead, rather than moribund, condition. The overrepresentation of such individuals in the natural death sample provides greater access to them (than was available when they were in sample 2) by both human and nonhuman scavengers operating within the 5-7 and 5-8 transitions. This in turn damps the original bias relative to both the moribund and the non-butchered natural death samples. Thus, the existence of dynamically similar processes in the 2-4 and 5-7 transitions (both, scavenging by humans) tends to counteract the most conspicuous potential bias of the 2-5 transition.

The complementary relationship between transitions 2-4 and 2-5 has an almost exact counterpart in the 5-7 and 5-8 transitions. The 5-7 selection operation involves the criteria that make individual natural deaths attractive or accessible for scavenging by humans. Once again, there are similar processes (the activities of nonhuman scavengers), with similar dynamics, controlling the 5-8 transition and in a position to balance many potential biases in the 5-7 transition. Suppose, for instance, that mastodonts dying of natural causes at a particular time of year -- such as mid-winter -- represented more attractive targets for scavenging because of the generally low availability of food at this season. While such an interaction might seem to produce bias in favor of winter deaths in scavenged assemblages, the low availability of winter food for *both* humans and nonhuman predator/scavengers would tend to minimize any such bias.

The 5-8 transition, like the 2-5 transition, includes a component that does not have a counterpart within the 5-7 transition. If a carcass is not accessible to, or not utilized by, macro-scavengers, it will nonetheless enter the 5-8 transition through the activities of micro-scavengers (e.g., insects) and decomposers (e.g., bacteria). Once again, however, this portion of the system is not known for its highly discriminating nature. Nevertheless, there is, as before, a possibility for systematic, attribute-dependent variation in *rate* of decomposition, though I expect this to involve very few attributes. If increased rate of decay were significant *relative* to the time it usually takes macro-scavengers to locate carcasses, a bias in premortem attributes might result. However, if carcasses are accessible at all, macro-scavengers usually find them rapidly enough that accelerated decay should not be an important biasing factor. I am not aware that this has been studied empirically, but it could be.

The 4-7 transition appears to offer little prospect of a change in attribute distribution. It represents the opportunity and/or choice of human scavengers to proceed with butchery of an animal that they encountered in a moribund condition but have already killed. In some contexts (e.g., where there is evidence of less technological sophistication), the ability of humans to retain control of such a carcass (against nonhuman scavengers) might be subject to considerable variation. However, this seems unlikely to be a critical issue for Paleoindians and, in any event, would result in a situation indistinguishable from a 2-5-8 transition. If the carcasses represented by sample 4 had been procured in groups of multiple individuals, it is also possible, as argued by Speth (1983), that time constraints or poor physical condition of certain carcasses would have resulted in their discard without butchery. However, the carcasses in Table 1 appear to represent isolated individuals. In such situations, physical condition and available time could probably have been assessed beforehand (i.e., before killing the animal). This would focus most of the selection associated with scavenging on the 2-4, rather than the 4-7, transition. Thus, having chosen to kill an isolated moribund animal, it seems most likely that Paleoindians would butcher it. If the 4-7 transition is indeed comprehensive, it cannot bias the distributions of premortem attributes. Since sample 7 is compounded from samples 4 and 5, the attribute distributions in samples 4 and 7 will not necessarily be identical, but the input from 4 to 7 should be nearly identical to what is seen in sample 4.

Summarizing this portion of the transition analysis, the similarity (or dissimilarity) of attribute distributions in samples 7 and 8 is controlled predominantly by processes organized about two nodes, represented by samples 2 and 5. The transitions paired across each of these nodes show obligatory complementarity for all attributes, giving these transitions a propensity toward development of opposed biases that translate directly into disparity between samples 7 and 8. However, to whatever extent the two transitions issuing from one of these nodes are dynamically similar, any tendency toward bias will be damped. Furthermore, the components of the 2-5 and 5-8 transitions that are clearly without counterpart in the 2-4 and 5-7 transitions either operate in a nonselective manner (unprovoked death or micro-scavenger activity), have biasing effects that are likely to be damped by dynamic similarity between the 2-4 and 5-7 transitions (variation in moribund half-life), or are likely to affect few attributes (variation in rate of decay). The degree of similarity of attribute distributions in samples 7 and 8 is determined by the overall dynamic similarity of the transitions paired across nodes 2 and 5 (with some contribution from dynamic similarity between 2-4 and 5-7). However, the above observations suggest that overall dynamic similarity is dominated by the degree of dynamic similarity between scavenging (of either moribund individuals or natural deaths) by humans and scavenging (of both sorts) by nonhuman macro-scavengers. Subjective assessment of the factors controlling these processes leads me to anticipate considerable dynamic similarity. If this were given, attribute distributions in samples 7 and 8 would be quite similar both to the distribution from which they were derived and to each other. However, neither the degree of dynamic similarity between human and nonhuman scavenging nor, consequently, the similarity or dissimilarity of attribute distributions can be deduced from available background knowledge. This is what forces the analysis into the "relative" mode.

The original "populations" of scavenged individuals (sample 7) and non-butchered natural deaths (sample 8) must also be compared, respectively, to the samples of each recovered for analysis today, samples 10 and 11. There may well be differences in their fates during the 7-10 and 8-11 transitions, contingent on their different early postmortem histories, but there are also strong similarities in the processes underlying the 7-10 and 8-11 transitions. Mastodonts scavenged by humans and ones not scavenged by humans represent carcasses of the same anatomy and body size, subjected to the same general climatic regime, with bones equally susceptible (or resistant) to transport or burial, and equally likely to be encountered by draglines. Whatever biasing factors *are* involved in the parallel 7-10 and 8-11 transitions should often have similar effects on distributions of a given attribute. Taphonomic bias with respect to premortem contingent attributes (e.g., age profiles biased against representation of young individuals) is by no means negligible. However, a *differential* effect, not bias per se, is the important issue. There is as yet no basis for expecting any particular differential effect, but neither can the possibility be eliminated at this time. Once again, therefore, available background knowledge is inadequate to predict an exact outcome, even at low resolution. Nevertheless, transition analysis does provide insight into the general patterns of interaction that either promote or inhibit the development of bias.

Transition Analysis: Hunted Sample to Non-Butchered Sample

If butchered individuals were procured by hunting, then the degree to which their attribute distributions are biased relative to those of non-butchered natural deaths is determined by the transitions linking samples 9 and 11 in Fig. 1. The common starting point from which the attribute distributions of these samples are derived is the attribute distribution found in the original living population (sample 1). The preservation or alteration of attribute distributions during transition 1-3 involves the interaction of all factors governing prey selection and hunting success by humans. In contrast, attribute distributions within samples of moribund individuals and natural deaths are preserved or altered relative to their condition in the living population through the combined incidence of all causes of natural death. For purposes of discussion, it will be most convenient to refer the processes underlying natural morbidity and death to the 1-2 transition, even though some natural death occurs without a prolonged period of morbidity.

One of the distinctive features of the 1-3 and 1-2 transitions is that they have a less obligatory complementarity than was seen in the transitions issuing from samples 2 and 5. This is because during intervals of time that are short relative to the lifespans of individuals, a large portion of most populations remains alive rather than contributing either to sample 3 or sample 2. For fossil assemblages of the sort treated here, which accumulate over long periods of time, the importance of this factor depends on the stability of a given attribute over the lifespan of an individual. Attributes that remain constant (e.g., sex) are handled in complementary fashion. For instance, preferential hunting of males would indeed reduce the number of males expected in a natural death sample, though the effect on the natural death sex ratio depends on the relative number of hunted and natural deaths. However, attributes that change over shorter timespans (e.g., age, season, or physical condition) show no necessary complementarity between their distributions in hunted and natural death samples. For stable attributes, where complementarity is preserved, there is the same propensity for opposed biases, but also the same potential for damping interactions, as was seen at the nodes represented by samples 2 and 5. For transient attributes, where complementarity is not preserved, there are no damping interactions. The fate of these attribute distributions thus depends on whether the processes intrinsic to each transition (1-3 and 1-2) act in dynamically similar fashion.

The processes involved in the 1-2 transition are a mixture of ones that are relatively nonselective, such as certain forms of accidental injury, and ones that are strongly selective, at least for certain attributes. The latter group of processes includes, for instance, nonhuman predation, disease, and starvation. These processes tend to show strong, direct selection on attributes such as body size, physical condition, and certain behavioral attributes, to each of which many other attributes are causally related.

Human hunting (the 1-3 transition) of single prey individuals is probably highly selective under most circumstances. Even hunting of groups of prey (as in many bison kills) may be selective (e.g., for season, or composition of herd) though in this instance, some of the selectivity may be displaced to the processing transition (3-6; Speth 1983). The unresolved issue is the nature of the selectivity shown by hunting and its relation to the overall selectivity of natural death. Human hunting could be quite nonselective with respect to attributes for which natural death is on average highly selective; it could be highly selective where natural death is nonselective; or the two sets of processes could differ in the direction of their selectivity for the same attribute. An example of this latter variety is that in some contexts hunting may select for excellent physical condition while natural death selects for poor physical condition.

The controls on the nature of selectivity in human hunting include factors such as climate, nutritional needs, and sensory abilities of both humans and prey. While these might in many contexts be responsible for departures from dynamic similarity between hunting and natural death as a whole, they may often contribute toward dynamic similarity between hunting and one component of natural death -- nonhuman predation. The implications of this depend on the fraction of natural deaths for which nonhuman predation is responsible, which, in turn, may vary systematically among mammals (Potts 1984). A model of the relationship between hunting and natural death that will cover situations in which nonhuman predation is the dominant form of natural death must consider the relationship between hunting and nonhuman predation.

The most conspicuous control on the dynamic similarity of hunting and nonhuman predation is the level of intelligence, social structure, and technology characteristic of the human side of the comparison. A suggestive, though usually only subjectively quantifiable, "measure" reflecting the degree, and to some extent the nature, of selectivity involved in a predatory encounter is the average amount of control of the situation exercised by the hunter/predator relative to the average amount of control exhibited by the prey. This might be operationalized differently in different contexts, but it is approximated in some predator/prey interaction models by the ratio of predator and prey body sizes (Roughgarden 1979). Since hunter/predators manage to survive, this "ratio" must favor them under at least some circumstances. However, human intelligence, social structure, and technology offer the potential for a more consistently favorable "ratio" than would obtain for most nonhuman predation. Such a shift in the balance of strengths and weaknesses for hunter and prey implies departure from the dynamic similarity of hunting and nonhuman predation.

Although I would not presume to predict all the dimensions of selectivity in Paleoindian hunting activity, I would expect it to be dynamically dissimilar to both nonhuman predation and natural death as a whole. The outcome is less clear, however, for some other contexts in which this model might be applied. The earliest stages of development of human hunting behavior are likely to show relatively great dynamic similarity to nonhuman predation but may still differ from natural death as a whole. Ironically, the least tractable situation may be one involving slightly more advanced stages in the development of human hunting, where the mix of selectivity and nonselectivity on different attributes might more nearly resemble the total effects of all agents of natural death. Under any circumstance, however, there is a possibility of dynamic dissimilarity based on the separation, if not total independence, of factors controlling the 1-3 and 1-2 transitions.

For the homogeneous hypothesis H_h, all instances of butchery are presumed to be associated with hunting. In the event that no scavenging by humans is occurring in conjunction with hunting, the 2-5 and 5-8 transitions must be comprehensive and unbiased. For purposes of accommodating the homogeneity of H_h, it would be justifiable -- and also most conservative relative to the claims that will be made for this series of transitions -- to assume an absence of scavenging and consequently ignore all effects of transitions 2-4, 5-7, 4-7, and 7-10 when evaluating H_h. If some scavenging by humans were occurring (but not, given H_h, represented in the available butchered sample), its potential effects could be assessed using the previous analysis of transitions related to scavenging. The outcome of this is that some additional bias could be introduced into the non-butchered natural death sample (8) beyond that already accounted for by the 1-2 transitions, but only if such a bias were strong enough to emerge as a net result of damped interactions. Although these details are not essential for analysis of H_s and H_h, they will be needed for treating inhomogeneous hypotheses in the hypothesis set.

Another transition that must be examined is 3-6. As demonstrated by Speth (1983), decisions involved in carcass processing may result in abandonment without butchery of some individuals that have been hunted. However, as noted in the discussion of transition 4-7, this is less likely to be a factor for cases involving isolated individuals. In the present case, the 3-6 transition is therefore probably comprehensive and unbiased, and no 3-8 transition is shown.

Summarizing this portion of the transition analysis for hunting, transitions 1-3 and 1-2 are complementary for some attributes, but not for others. For attributes of the former type this node shows the same general pattern of interaction as seen at nodes 2 and 5, including both the potential for damping and for opposed biases that directly augment disparity between derived distributions. Attributes for which this node is not complementary offer no potential for damping but also no requirement for opposed biases. Thus, biases relative to the source distribution (in the living population) may or may not imply bias and disparity between derived distributions (in hunted and natural death samples). For all attributes, therefore, bias between the derived distributions in samples 3 and 2 is determined by the degree of overall dynamic similarity between the 1-3 and 1-2 transitions. These transitions also appear to dominate in determining the development of disparity between samples 6 and 8. If transitions 1-3 and 1-2 *are* dynamically similar, then the attribute distributions within hunted and natural death assemblages will tend to be similar, no matter how great their bias relative to the living population. On the other hand, if transitions 1-3 and 1-2 are *not* dynamically similar, the attribute distributions of hunted and natural death assemblages will tend to differ. Subjective assessment of the processes underlying these transitions suggests considerable dynamic dissimilarity, but as in the discussion of transitions related to scavenging, the actual degree of dynamic similarity across a node cannot be determined by an analysis of this type.

Finally, comparison of transitions 6-9 and 8-11 proceeds according to the same arguments used with 7-10 and 8-11. These transitions may well involve significant bias, but only *differential* bias can affect the relative attributes at issue here. The possibility of differential bias cannot be eliminated but neither is there yet a particular reason to suspect it.

Relative Probability of the Data (D_s *vs.* D_d)

The preceding descriptions have focused on processes and patterns of interaction *within* the two series of transitions that determine how scavenged samples and hunted samples each (separately) compare to samples of non-butchered natural deaths. However, evaluating the relative magnitude of $P(D|H_sX)$ and $P(D|H_hX)$ requires comparisons *between* these two series of transitions. This will be done in two stages, the first dealing with the series of transitions linking samples 6 and 7 each to sample 8 and the second dealing with the transitions linking samples 6, 7, and 8, respectively, to 9, 10, and 11.

In the preceding transition analyses it became clear that general patterns of organization of processes may either promote or inhibit the development of disparity between any two derived attribute distributions. However, these patterns of organization do not by themselves determine the magnitude of such disparity. Disparity (either observed or expected) is determined instead by the (actual or probable) dynamic similarity of the sets of processes operating on each side of a node representing a source distribution. The *relative* probability of disparity across the nodes linking hunted and scavenged assemblages each (separately) to non-butchered natural deaths is therefore determined by the *relative* dynamic similarity of the sets of processes on each side of these nodes. In the absence of detailed knowledge of the natural history and ecology of associated humans, nonhuman predator/scavengers, and prey, relative dynamic similarity must be evaluated in a manner that shows some promise of validity across a wide range of contexts. Three indirect arguments and one direct argument are proposed for this below.

The first argument begins with the observation that more attributes are manifested by live animals than by moribund or dead ones. The premortem attributes of dead animals can all be categorized, even if somewhat loosely, as anatomical. Live, free-ranging individuals have these same anatomical attributes but in addition a full range of behavioral attributes. Moribund individuals occupy intermediate portions of this spectrum but are usually close to the "dead end." Since each attribute to some degree represents a potential basis for selection in one or more of the transition processes, there are more attributes subject to selection in a living population than in a "population" of moribund individuals or natural deaths. Not all of these attributes lead to a discernible contrast between hunting and scavenging. For instance, behavioral attributes that have no anatomical correlates may provide a basis for direct selection during hunting but are irrelevant to this analysis because only anatomical attributes can be evaluated directly in fossil material. However, most anatomical attributes, including all those considered in this type of problem (e.g., sex, age, season of death, physical condition) have behavioral correlates. During either hunting or natural death, these anatomical attributes are subject not only to direct selection, but also to indirect selection acting through their behavioral correlates. In contrast, this source of selection cannot operate during either human or nonhuman scavenging (unless the behavioral correlate is manifested commonly by moribund individuals). A greater number of attributes is thus associated with a greater number of selective forces on single attributes.

For anatomical attributes that experience direct selection or indirect selection through another anatomical attribute, the addition of indirect selection through a behavioral attribute may either enhance or inhibit biases *intrinsic* to a given process. However, in either event, the addition of selective forces probably makes the maintenance of dynamic similarity *between* transitions less likely. For anatomical attributes that experience neither direct selection nor indirect selection through another anatomical attribute, addition of indirect selection through a behavioral attribute must enhance bias intrinsic to the process in question, again decreasing the probability of dynamic similarity between transitions. The transitions deriving animals from a living population thus occur within a more complex selective regime and have lower probability of dynamic similarity than would be expected of the transitions deriving animals from the moribund and natural death "populations." Since dynamic dissimilarity across nodes has already been shown to produce disparity between derived distributions, and since a hunted assemblage and a non-butchered natural death assemblage are separated by transitions involving greater probability of dynamic dissimilarity than is the case for a scavenged assemblage and a non-butchered natural death assemblage, it follows that the probability of disparity between butchered and non-butchered assemblages is greater given hunting than given scavenging.

The second argument involves the spatiotemporal distribution of living animals as compared to moribund or dead ones in scavengeable condition. Particularly for animals of large body size, live prey are almost always more abundant than are carcasses or carcasses-to-be. Live prey also tend to be more spatiotemporally predictable, where the relevant spatiotemporal scale is determined by the food needs of the (human or nonhuman) hunter or scavenger. Utilizing live prey, whether by human or nonhuman hunting, thus allows (and to some extent requires) greater selectivity than is usually possible in human or nonhuman scavenging (cf. discussion of generalists vs. specialists in Maynard Smith 1974). If scavenging makes a sufficiently small contribution to nutritional (or other) needs, or if the scavenger can locomote cheaply enough to survey an unusually broad area (Potts 1984), this constraint may be relaxed. Nevertheless, scavenging is usually not a highly selective operation.

Since selection involves the relationship between a source distribution and a derived distribution, it may or may not (depending on the organization of processes across a node) produce disparity between two derived distributions. Nevertheless, within a "contingent world," where causal interaction is complex, higher levels of selection generally make dynamic similarity less probable. This in turn increases the probability of disparity between derived distributions (as in divergent evolution between reproductively isolated populations). The implications of this argument are thus identical to those of the first.

The third argument returns to the concept of the "ratio" of control exercized by a hunter/predator to that exercized by the prey but generalizes this concept to apply both to utilization of live prey and to scavenging. For utilization of live prey, whether by human or nonhuman predators, the interaction of capabilities of both predator and prey is a significant determinant of the nature and degree of selectivity in the predator/prey interaction. In either human or nonhuman scavenging, however, the control exercized by the prey dwindles to nothing. The nature and degree of selection in a scavenging interaction is thus determined by the *relatively* simple system of constraints represented by the scavenger's own interests and the activities of other scavengers. As a consequence, utilization of live prey involves a more complex selective regime than does scavenging. This point was also made in the first argument, but it referred there to the number and diversity of prey attributes rather than the relationship between predator and prey attributes. From this point on, though, the two arguments proceed in parallel, associating hunting with transitions involving a lower probability of dynamic similarity and a higher probability of disparity between butchered and non-butchered samples. This third argument assumes that nonhuman predation is the major agent of natural death (more likely to be the case for nonmigratory than for migratory mammals; Potts 1984). However, even if this assumption is not valid in a given case, it is likely that the degree of dynamic similarity between hunting by humans and nonhuman predation is an important determinant of the disparity between hunted and non-butchered natural death assemblages.

These first three arguments have focused on how living animals differ from dead (and moribund) ones as objects of selection and have avoided distinguishing between selective agents associated with different transitions issuing from a single node on the causal network. The fourth, more direct, argument takes this latter step. It is an extension of the third argument but has a more limited domain than the first three. While the first three are applicable to human groups regardless of their level of development, the fourth argument applies only to cases in which humans show "significant" technological development.

Repeating an observation from the discussion of transitions 1-3 and 1-2, technological development does not just make dynamic similarity between hunting and natural death less probable, it directly changes the dynamics of human hunting. While some qualification is still appropriate, dynamic dissimilarity of human hunting and nonhuman predation (and hence probably natural death as a whole) is thus highly likely. In contrast, the same change in human capabilities makes relatively little difference in a scavenging context because of the negligible influence of "prey" capabilities. Thus, "advanced" human and nonhuman scavengers, despite their different capabilities relative to each other, are in a similar situation with regard to their ability to act in a selective fashion *relative to the "prey"* -- i.e., they are likely to be dynamically similar in many respects. This implies the same evaluation of relative probability of disparity as was suggested by the previous arguments.

The four arguments above apply to relationships among attribute distributions as they would appear in butchered and non-butchered samples composed of "fresh" remains -- samples such as 6, 7, and 8. However, evaluation of the relative probability of disparate attribute distributions in fossil material requires treatment of the transitions representing preservation and later collection of such specimens. As pointed out in the discussion of individual transitions, only a differential bias between a scavenged sample and a non-butchered natural death sample, or between a hunted sample and a non-butchered natural death sample, would affect the relative attribute comprising the "observed data." Following from this, the only complication that would thwart evaluation of the relative probability of observed data, conditional on scavenging on one hand, and hunting on the other, is a *difference* in differential (or a "differential differential") bias. Allowing that there may or may not be a differential bias between butchered and non-butchered specimens, is the nature or degree of differential bias likely to be dependent on, or independent of, the mode of procurement of the butchered individuals?

This is again a question of dynamic similarity -- namely, whether taphonomic agents interacting with butchered, hunted specimens, and with butchered, scavenged specimens, operate in dynamically similar fashion. Although we have no detailed knowledge of the post-butchery taphonomic processes involved in the present case, I see no reason to expect bias dependent on the mode of procurement. In general, butchery would be expected to leave both hunted and scavenged carcasses in similar condition with respect to the kinds of factors that affect subsequent taphonomic history (e.g., extent of disarticulation). Although absence of a differential differential bias cannot be guaranteed, assuming such a bias to be negligible is at least defensible as a first approximation. Given this, the evaluation of relative probability proposed for samples 6, 7, and 8 should hold for fossil as well as "fresh" material.

In light of the foregoing arguments, $P(D_d|H_sX) < P(D_d|H_hX)$. Furthermore, since D_s and D_d are mutually exclusive and jointly exhaustive, $P(D_s|H_sX) > P(D_s|H_hX)$. These inequalities, following equation (3), imply similar inequalities for the posterior probabilities, such that $P(H_s|D_dX) < P(H_h|D_dX)$ and $P(H_s|D_sX) > P(H_h|D_sX)$. Thus, a finding of D_d supports H_h in preference to H_s, and a finding of D_s supports H_s in preference to H_h.

The arguments used to evaluate the relative probability of observed data, conditional on each hypothesis, are framed in very general terms to maximize their applicability to different contexts, including that of the present case. Nevertheless, without further information, acceptance of these arguments as applicable to the present case must be treated as one or more assumptions, on which the rest of the analysis depends. The arguments themselves, or the assumptions representing them, are included symbolically in X in Bayes'

Theorem. Their importance is recognized, for instance, in the formulation of posterior probabilities by making support for particular hypotheses conditional on *X* as well as on *Y*. Since the arguments included in *X* are eminently testable in other contexts, and to some extent have been tested by various studies in ecology and taphonomy and by concordances with common experience, it is also appropriate to refer to them as auxiliary hypotheses, as is done in the text. For convenience, I have treated the arguments included in *X* as belonging to two groups: the behavioral auxiliary hypothesis and the taphonomic auxiliary hypothesis. These are stated in attribute-specific form in the text. If new information challenges either of these auxiliary hypotheses, they can be revised and the relative probabilities of the data can be reevaluated, but they otherwise stand as the best available summary of the factors bearing on the theoretical probability of the data.

Data (D_c *vs.* D_f)

Evaluating hunting and scavenging hypotheses through observation of the similarity or dissimilarity of premortem attribute distributions may appear to favor hunting. By this, I do not mean to imply that any aspect of the methodology prevents scavenging from being the hypothesis best supported by the data. In particular, requiring attribute distributions in butchered samples to be statistically significantly different from those in non-butchered samples in order to be assessed as D_d prevents random sampling error from being interpreted as support for hunting. Nevertheless, a finding of D_d *could* be plausibly interpreted either as hunting or as significantly biased scavenging. In order to avoid premature rejection of this latter possibility, a different approach to the data is needed -- one that, if anything, favors scavenging.

This alternative approach to the data involves discrimination of a second relative attribute, or comparison, between butchered and non-butchered samples. It is based on the same lower level data--premortem attribute distributions -- as were used before to distinguish D_s and D_d. However, it focuses on a different aspect of the data: set/subset relations of ranges rather than just similarity or difference in distributions. As before, only two states will be recognized: D_c and D_f. D_c indicates data showing a correspondence of ranges such that the attribute states occupied by the butchered sample represent a subset of the attribute states occupied by the non-butchered sample. D_f indicates data failing to show this type of range correspondence; it represents instances in which the range of attribute states occupied by the butchered sample includes states significantly beyond those occupied by the non-butchered sample. Statistical recognition of D_c and D_f is discussed below.

Transition Analyses and Relative Probability of the Data (D_c *vs.* D_f)

The factors influencing D_c and D_f are aspects of the same causal model proposed for treatment of D_s and D_d. If the range of attribute states represented in the butchered sample differs at all from that of the non-butchered sample, it may be (a) because attribute states represented in the latter are not in the former, (b) because attribute states represented in the former are not in the latter, or (c) both of the foregoing. By definition of the relative attribute in question here, condition (a) and identity of attribute state ranges would each be classified as D_c, and conditions (b) and (c) would be classified as D_f. If the butchered animals were procured by scavenging, condition (a) and the component of condition (c) for which (a) is responsible are readily explained as negative scavenging bias (i.e., against certain states) on the part of humans. However, condition (b) and the component of condition (c) for which (b) is responsible usually cannot be completely explained by positive scavenging bias on the part of humans. This is due to a fundamental asymmetry of comprehensive action; put simply, you can do nothing, if you so choose, but you can't do everything. Stated in terms of Fig. 1, humans may refrain from scavenging (or possibly be unable to scavenge) individuals showing certain states of an attribute, whereupon those individuals undergo the 2-5 and 5-8 transitions -- if not through the activities of macro-scavengers, then through micro-scavengers and decomposers. However, even if humans have a strong positive scavenging bias for a certain attribute state, and even if they succeed in appropriating many of the carcasses showing that state, they usually cannot secure them all. As explained in the previous treatment of scavenging, at least a moderate degree of dynamic similarity is expected between human and nonhuman macro-scavengers, such that carcasses that are attractive and accessible to one are likely to be to the other. Even chance variation in the relative proximity of humans and nonhuman macro-scavengers to a newly available carcass would usually suffice to produce variable outcome and damp a strong positive intrinsic bias. Furthermore, whether or not dynamic similarity between human and nonhuman macro-scavengers is sufficient to guarantee this, occasional "losses" to micro-scavengers and decomposers would be expected.

The preceding arguments essentially rule out comprehensive human scavenging of individuals with certain attribute states, but 100% comprehensiveness is not exactly the issue posed by condition (b) above. Assuming scavenging as the means of procurement, condition (b) implies scavenging that is comprehensive *enough*, for individuals with certain attribute states, that their absence from the non-butchered sample is expected. Any decision as to what constitutes a statistically significant instance of conditions (b) or (c) (i.e., D_f) depends on the total and state-specific sizes of the butchered and non-butchered samples, a probability distribution for (or knowledge of state-specific) degree of scavenging comprehensiveness, and a probability distribution for (or knowledge of state-specific) magnitude of positive human scavenging bias. If all this were available, D_f could even be defined to include cases in which there is an "unexpected" paucity (rather than absence) of individuals with particular states in the non-butchered sample. However, these requirements for probability distributions or knowledge are rather imposing. Since I have no satisfactory solution to them at present, D_c and D_f will be distinguished in a subjective fashion. Although dangerous, this should at least improve with sample size. In general, the larger the number of instances of butchery for which no non-butchered instances with corresponding attributes are found, the more extreme and inherently implausible is the view of scavenging comprehensiveness and bias that must be maintained in order to retain scavenging as a plausible explanation of carcass procurement.

Interestingly, since scavenging is probably even less comprehensive in situations involving human capabilities that are less advanced, this argument is probably even stronger for early hominids than for Paleoindians.

For attributes that are constant throughout the lifespan (e.g., sex) and fossil assemblages that accumulate over long periods of time, the transitions issuing from the living population behave identically to those issuing from samples 2 and 5. For such attributes, comprehensive hunting, such that no representatives of a given attribute state appear in the non-butchered sample, is usually just as improbable as comprehensive scavenging. The actual relative probability of these two outcomes depends on the mix of hunting and scavenging biases and degrees of comprehensiveness, but since this is not generally knowable, such an attribute does not have a discernibly different distribution under competing hypotheses and would not be expected to contribute to evaluations based on D_c vs. D_f.

For transient attributes for which the non-butchered sample happens to include abundant representatives of all recognized attribute states, the situation is much the same. Such a non-butchered distribution guarantees that relative attribute state D_c will be observed, and hence, the conditional probabilities of D_c and D_f on H_s and H_h are again indistinguishable. Use of D_c vs. D_f for evaluating H_s and H_h therefore requires transient attributes for which the non-butchered sample includes less than the full range of recognizable states. To be sure, in a large enough non-butchered sample, all states should eventually be occupied, but if natural death is rare for individuals with certain states, these states may frequently be vacant in smaller samples. In such instances, however, condition (b) above, interpreted in terms of hunting, requires no intrinsically comprehensive action. It only requires a positive hunting bias relative to the distribution produced through natural death, and a natural death distribution in which certain states (found in the butchered sample) are rare *enough* that their absence from a non-butchered sample of a given size is expected. Since hunting and natural death are distinct processes, there is no requirement that the hunted sample be restricted to the same range of attribute states observed in the non-butchered natural death sample. In fact, to the extent that hunting and natural death are dynamically dissimilar, condition (b), and hence D_f, may even be highly probable.

Although the preceding arguments make some headway on determining the relative probabilities of D_c and D_f, conditional on H_s and conditional on H_h, some constraint on the relative magnitude of hunting and scavenging biases and on the relative degree of hunting and scavenging comprehensiveness is also needed. The former objective is met by most of the arguments used to evaluate the relative conditional probabilities of D_s and D_d. The second through fourth arguments of that discussion suggest not only that bias is more probable across the sample 1 node than across nodes 2 or 5, but also that biases tend to be greater in magnitude at node 1 than across nodes 2 or 5. Since degree of comprehensiveness is determined by the interaction of biases across a node, the potential comprehensiveness of human action, relative to the opposed nonhuman processes (not relative to the source population) should be greater across node 1 than across nodes 2 or 5. This statement does not determine the relative probabilities in question, but it does mean that assuming the same degree of bias and (relative) comprehensiveness for hunting and scavenging would, if anything, favor scavenging.

For any finite degree of state-specific bias and comprehensiveness in scavenging by humans, any lapse in scavenging (i.e., any individual not successfully scavenged) automatically enters the non-butchered natural death sample. Since even transient premortem attributes are fixed (by death) in the sample from which all scavenged individuals are derived, this guarantees representation of the given state within the non-butchered sample, at a frequency determined by the degree of scavenging comprehensiveness. In contrast, transient premortem attribute states are not fixed in the living population, from which both hunted individuals and natural deaths are derived. Any individual that is not hunted does not thereby "carry" its current attribute state into the non-butchered sample. Rather, it generally remains alive, exposing its transient attribute state to further change. It may then enter either the hunted or the natural death sample at a later time, with either the same, or a different, attribute state. In any event, representation of the original attribute state in the non-butchered sample is not deducible from the bias of hunting (taken, conservatively, as equal to that of scavenging). Representation of that state requires, instead, the action of a separate, natural death process. The probability that a given attribute state is occupied in both hunted and non-butchered natural death samples is thus a joint probability, whose value must be less than the probabilities of the separate events. Therefore, the probability that attribute states occupied in the butchered sample will also be occupied in the non-butchered sample is greater for scavenging than for hunting. This conclusion is equivalent to a second behavioral auxiliary hypothesis for inclusion in X. Since it is entirely compatible with the previous behavioral auxiliary and is based on the same background information, it does not represent a substantive revision of X.

Differential taphonomic biases between butchered and non-butchered samples could alter one or both patterns of state occupation. However, as long as there is no *differential* differential bias, the relative probability of particular patterns of state correspondence should not be altered. Under this condition, the arguments given here apply as well to fossil as to fresh material. Thus, accepting the same taphonomic auxiliary hypothesis as for evaluation of D_s and D_d, it follows that $P(D_c|H_sX) > P(D_c|H_hX)$. Furthermore, since D_c and D_f are mutually exclusive and jointly exhaustive, $P(D_f|H_sX) < P(D_f|H_hX)$. Following equation (3), these inequalities imply similar inequalities for the posterior probabilities, such that $P(H_s|D_cX) > P(H_h|D_cX)$ and $P(H_s|D_fX) < P(H_h|D_fX)$. Thus, a finding of D_c supports H_s in preference to H_h, and a finding of D_f supports H_h in preference to H_s.

In the derivation of equation (3), the overall probability of the data $(P(D|X))$ appeared in both numerator and denominator and thus cancelled in the equation used for choosing among competing hypotheses. $P(D|X)$ nonetheless remains important as a reflection of the differential support actually given to the preferred hypothesis (relative to competitors) by the observed data. The immediate importance of this is that $P(D_c|X)$ varies with the range of states occupied by the non-butchered sample. For cases in which the range of non-butchered states is small relative to the total range of recognizable states, the probabilities of the data (either D_c or D_f) conditional on each hypothesis are likely to differ significantly. Thus, the overall probability of both D_c and D_f

will be relatively small, and this approach to evaluating H_s and H_h will be relatively unbiased in favor of either. However, as the range of states occupied in the non-butchered sample increases, the probabilities of the data conditional on each hypothesis are likely to converge (mainly by increase in $P(D_c|H_hX)$ and decrease in $P(D_f|H_hX)$). This increases the overall probability of D_c, making this approach increasingly biased in favor of H_s. In this situation, D_c provides only a weak argument for H_s, while D_f still provides strong support for H_h.

Multi-Site Hypothesis Set

As a final comment on general methodology, it is important to return to the single-site vs. multi-site issue. The $2^n - 2$ inhomogeneous focal hypotheses whose treatment was postponed earlier may also be evaluated using the causal model described above, though this will not be undertaken in detail. For each of these inhomogeneous hypotheses, the probability of bias and the probability of occupation of corresponding attribute states in butchered and non-butchered samples will be intermediate between the respective probabilities determined for the homogeneous hypotheses. However, the conditional probability of observed data depends on the actual configuration of those data. In cases for which all (nonrelative) attribute distributions overlap broadly, but not completely, between butchered and non-butchered samples, the values of $P(D|H_iX)$ for one or more inhomogeneous H_i may well exceed the values of this expression for both H_s and H_h. In such a case, those particular inhomogeneous H_i would indeed be preferred over any other H_i (taking all prior probabilities as equal). However, in cases for which either some attribute distributions in butchered and non-butchered samples are identical, or some fail completely to overlap, all inhomogeneous H_i will have values of $P(D|H_iX)$ less than this value for one or the other of H_s and H_h. In these instances, the more probable homogeneous hypothesis is preferred relative to all others.

Postmortem Attributes as Evidence of Hunting or Scavenging

Recent work on the hunting vs. scavenging issue in other historical and geographic contexts has stressed the importance of cutmarks made by tools, crosscutting toothmarks made by nonhuman predators (Potts and Shipman 1981; Shipman 1983, 1986). This superpositional relationship is indicative of prior access (to the anatomical part on which it occurs) by nonhuman predators or scavengers, suggesting that scavenging is more likely than hunting as an explanation for the procurement episode with which the cutmark is associated. The power of this argument derives from the quasi-absolute status of cutmarks crosscutting toothmarks (CCT) relative to H_h. The probability of CCT on an animal hunted and killed by humans is treated as essentially 0, but this cannot be deduced from the definition of the primary reference class of H_h. Indeed, counterexamples are conceivable. For instance, an animal could be caught and bitten to the bone by a nonhuman predator and yet escape, otherwise unscathed, only to be killed soon afterward by human hunters and butchered in such a manner as to leave a cutmark across the prior toothmark. We correctly recognize such a turn of events as exceedingly unlikely. However, this decision emerges from a causal analysis based on knowledge of typical predation patterns and success rates among nonhuman predators, according to which the joint probability of these events is judged to be negligible. CCT are, however, a nonquasi-absolute property relative to scavenging. The probability of producing CCT, given scavenging by humans, is determined by both the probability of producing CCT when humans scavenge natural deaths produced through nonhuman predation and the relative incidence of this source of carcasses as opposed to others (not involving toothmarks). Even without well documented estimates of these values of probability and relative incidence, common experience would suggest for each a value significantly greater than 0 (though undoubtedly less than 1.0). The critical issue for hypothesis selection is that even according to these crude estimates, the probability of observing CCT, given scavenging by humans, is greater than its probability given hunting.

The low overall probability of CCT (i.e., $P(\mathrm{CCT}|X)$), or at least the negligible contribution to this value made by the term involving $P(\mathrm{CCT}|H_h X)$, makes CCT a strong argument for scavenging.

The butchered mastodonts in Table 1 have been examined for instances of cutmarks crosscutting toothmarks, but although both types of mark are found separately on certain specimens (e.g., Pleasant Lake, Russell Farm I), no instances of CCT have been found. With this negative result, H_h survives a potential refutation. However, beyond this, the outcome must be classed as indeterminate, since common experience is not sufficient to indicate the sample size for which $P(\mathrm{CCT}|H_s X)$ is great enough to expect to observe CCT. Because the overall probability of no-CCT is high, no-CCT is not a strong argument for hunting. Neither is the absence of toothmarks crosscutting cutmarks (TCC) determinate, since TCC are nonquasi-absolute with respect to both hunting and scavenging by humans. Without further information from a comparison sample, cutmark data thus offer no differential support for either hypothesis.

Other contingent properties of the primary reference classes associated with H_h and H_s might also be exploited, and a number of these involve bone or carcass modification (e.g., Binford 1984; Shipman 1986). The problem with this approach, as recognized by those who have attempted it, is in demonstrating the generality of the properties taken to typify either hunted or scavenged assemblages. This problem is minimal with attributes that are clearly quasi-absolute (such as CCT) with respect to one of the hypotheses, but very few of these have been discerned. More commonly, development of a "diagnostic" criterion requires quantitative analysis of a comparison sample, the relevance of which may always be called into question. Studies of a wide variety of comparison samples have added greatly to our understanding of the patterning that arises through cultural processing of faunal materials (e.g., Yellen 1977; Binford 1981; Gifford et al. 1981; Winterhalder and Smith 1981). However, some of the most intrinsically interesting problems -- such as the reconstruction of early hominid lifestyles -- require the results of such analyses to be generalized to contexts in which their validity cannot be easily ascertained. Consequently, substantive assumptions must be made about human behavior in the "unknown" case. For instance, Shipman assumes that if

"Oldowans" hunted, they would have followed butchery practices "comparable to those observed on more recent, butchered ... prey remains" and that if they scavenged, "carcass-processing patterns typical of hunter-gatherers would be absent" (1986, p.28). Whatever the validity of these assumptions for early hominids, I have chosen not to apply them to North American Paleoindians. We know virtually nothing of the dependency (or lack thereof) between Paleoindian carcass processing techniques and mode of carcass procurement.

Certain patterns of bone modification or of bone presence/absence or state of articulation may well represent quasi-absolute properties of hunted or scavenged assemblages. However, their postmortem character complicates causal analysis of the type proposed here. Bone and carcass modification occur independently in all paths of the causal network in Fig. 1, but not in ways whose probabilities can be constrained relative to one another without invoking additional assumptions. Although postmortem contingent properties can be handled effectively when appropriate comparison samples are available, I consider them extremely problematic in the present case.

REFERENCES

Asher, H. B. 1983. *Causal Modeling*. Second edition. Beverly Hills: Sage Publications, Inc.

Augusta, A., and Z. Burian. 1962. *A Book of Mammoths*. London: Paul Hamlyn.

Behrensmeyer, A. K., K. D. Gordon, and G. T. Yanagi. 1986. Trampling as a cause of bone surface damage and pseudo-cutmarks. *Nature* 319:768-771.

Binford, L. R. 1981. *Bones: Ancient Men and Modern Myths*. San Francisco: Academic Press.

Binford, L. R. 1984. *Faunal Remains from Klasies River Mouth*. San Francisco: Academic Press.

Bonnichsen, R. 1979. *Pleistocene Bone Technology in the Beringian Refugium*. Mercury Series, Paper 89. Ottawa: Archaeol. Surv. Canada.

Brain, C. K. 1981. *The Hunters or the Hunted? An Introduction to African Cave Taphonomy*. Chicago: University of Chicago Press.

Bryan, A. L., R. M. Casamiquela, J. M. Cruzent, R. Gruhn, and C. Ochsenius. 1978. An El Jobo mastodon kill at Taima-Taima, Venezuela. *Science* 200:1275-1277.

Burks, A. W. 1977. *Chance, Cause, Reason*. Chicago: University of Chicago Press.

Coe, M. 1978. The decomposition of elephant carcasses in the Tsavo (East) National Park, Kenya. *J. Arid Environ*. 1:71-76.

Conybeare, A., and G. Haynes. 1984. Observations on elephant mortality and bones in water holes. *Quat. Res*. 22:189-200.

Crane, H. R., and J. B. Griffin. 1959. University of Michigan radiocarbon dates IV. *Amer. J. Science Radiocarbon Suppl.* 1:173-198.

Douglas-Hamilton, I., and O. Douglas-Hamilton. 1975. *Among the Elephants*. New York: Viking Press.

Eisenberg, J. F., G. M. McKay, and M. R. Jainudeen. 1971. Reproductive behavior of the Asiatic elephant (*Elephas maximus maximus* L.). *Behavior* 38:193-225.

Farrand, W. R. 1977. Revision of the Two Rivers ("Valders") drift border and the age of fluted points in Michigan. *Anthrop. Pap., Mus. Anthropol., Univ. Mich.* 61:74-84.

Fatti, L. P., G. L. Smuts, A. M. Starfield, and A. A. Spurdle. 1980. Age determination in African elephants. *J. Mammal.* 61:547-551.

Fisher, D. C. 1981. Evidence of mastodon butchering in southeastern Michigan. *Geol. Soc. Am., Abst. Prog.* 13:452.

Fisher, D. C. 1984a. Mastodon butchery by North American Paleo-Indians. *Nature* 308:271-272.

Fisher, D. C. 1984b. Taphonomic analysis of late Pleistocene mastodon occurrences: Evidence of butchery by North American Paleo-Indians. *Paleobiology* 10:338-357.

Fisher, D. C., and P. L. Koch. 1983. Seasonal mortality of late Pleistocene mastodons: Evidence for the impact of human hunting. *Geol. Soc. Am., Abst. Prog.* 15:573.

Fitting, J. E. 1966. Part 1: The Holcombe site. The Paleo-Indian occupation of the Holcombe beach. *Anthropol. Pap., Mus. Anthropol., Univ. Mich.* 27:1-81.

Frison, G. 1978. *Prehistoric Hunters of the High Plains.* New York: Academic Press.

Garland, E. B., and J. W. Cogswell. 1985. The Powers mastodon site. *Michigan Arch.* 31:3-39.

Gifford, D. P., G. L. Isaac, and C. M. Nelson. 1981. Evidence for predation and pastoralism at Prolonged Drift: A pastoral Neolithic site in Kenya. *Azania* 15:57-107.

Graham, R. W., C. V. Haynes, D. L. Johnson, and M. Kay. 1981. Kimmswick: A Clovis-mastodon association in eastern Missouri. *Science* 213:1115-1117.

Grue, H., and B. Jensen. 1979. Review of the formation of incremental lines in tooth cementum of terrestrial mammals. *Danish Rev. Game Biol.* 11:1-48.

Gustafson, C. E., D. Gilbow, and R. D. Daugherty. 1979. The Manis mastodon site: Early man on the Olympic Peninsula. *Can. J. Archaeol.* 3:157-164.

Hannus, L. A. 1983. Of mammoths, men and ice: Making it on the margins. *Geol. Soc. Am., Abst. Prog.* 15:590.

Haynes, C. V., Jr. 1966. Elephant-hunting in North America. *Scientific Amer.* 214:104-112.

Haynes, C. V., Jr. 1969. The earliest Americans. *Science* 166:709-715.

Haynes, C. V., Jr. 1982. Were Clovis progenitors in Beringia?; pp. 383-398. In: *Paleoecology of Beringia*, ed. D. M. Hopkins, J. V. Matthews, Jr., C. E. Schweger, and S. B. Young. New York: Academic Press.

Haynes, G. 1980. Evidence of carnivore gnawing on Pleistocene and recent mammalian bones. *Paleobiology* 6:341-351.

Haynes, G. 1983. A guide for differentiating mammalian carnivore taxa responsible for gnaw damage to herbivore limb bones. *Paleobiology* 9:164-172.

Haynes, G. 1985. Age profiles in elephant and mammoth bone assemblages. *Quat. Res.* 24:333-345.

Hill, A. 1979. Disarticulation and scattering of mammalian skeletons. *Paleobiology* 5:261-274.

Hill, A., and A. K. Behrensmeyer. 1985. Natural disarticulation and bison butchery. *Am. Antiq.* 50:141-145.

Irwin-Williams, C. 1967. Associations of Early Man with horse, camel, and mastodon at Hueyatlaco, Valsequillo (Puebla, Mexico); pp. 337-347. In: *Pleistocene Extinctions: The Search for a Cause*, ed. P. S. Martin and H. E. Wright, Jr. New Haven: Yale Univ. Press.

Isaac, G. L. 1983. Bones in contention: Competing explanations for the juxtaposition of Early Pleistocene artifacts and faunal remains; pp. 3-19. In: *Animals and Archaeology*, Vol. I, ed. J. Clutton-Brock and C. Grigson. London: British Archaeological Reports.

Isaac, G. L. 1984. *The Archaeology of Human Origins: Studies of the Lower Pleistocene Lake Basin in Kenya*. Chicago: University of Chicago Press.

Iversen, G. R. 1984. *Bayesian Statistical Inference*. Beverly Hills and London: Sage Publications.

Johnson, E. 1982. Paleo-Indian bone expediency tools: Lubbock Lake and Bonfire shelter. *Can. J. Anthropol.* 2:145-157.

Kay, M. 1974. Dental annuli age determination on white-tailed deer from archaeological sites. *Plains Anthropol.* 19:224-227.

Klein, R. G. 1982a. Age (mortality) profiles as a means of distinguishing hunted species from scavenged ones in Stone Age archeological sites. *Paleobiology* 8:151-158.

Klein, R. G. 1982b. Patterns of ungulate mortality and ungulate mortality profiles from Langebaanweg (early Pliocene) and Elandsfontein (middle Pleistocene), south-western Cape Province, South Africa. *Annals of the South African Museum* 90:49-94.

Klein, R. G., and K. Cruz-Uribe. 1984. *The Analysis of Animal Bones from Archeological Sites*. Chicago: University of Chicago Press.

Klevezal, G. A., and S. E. Kleinenberg. 1969. Age determination of mammals from annual layers in teeth and bones. *Israel Prog. Sci. Trans.*, Jerusalem TT 69-55033:1-128.

Koch, P. L., and D. C. Fisher. 1986. Out of the mouths of mammoths: An isotopic signal of seasons in proboscidean tusks. *Geol. Soc. Am., Abst. Prog.* 18:660.

Kurtén, B. 1968. *Pleistocene Mammals of Europe*. London: Weidenfeld and Nicolson.

Laws, R. M. 1966. Age criteria for the African elephant, *Loxodonta africana. E. Afr. Wildl. J.* 4:1-37.

Mardia, K. V. 1972. *Statistics of Directional Data.* New York: Academic Press.

Martin, P. S., and R. G. Klein, eds. 1984. *Quaternary Extinctions: A Prehistoric Revolution*. Tuscon: Univ. of Arizona Press.

Martin, P. S., and H. E. Wright, Jr., eds. 1967. *Pleistocene Extinctions: The Search for a Cause.* New Haven: Yale Univ. Press.

Maynard Smith, J. 1974. *Models in Ecology*. Cambridge: Cambridge University Press.

Mehl, M. G., R. A. Marshall, J. A. Martin, and J. H. Williams. 1966. The Grundel mastodon. *Rep. Invest., Missouri Geol. Surv. Water. Res.* 35:5-28.

Meltzer, D. J., and J. I. Mead. 1983. The timing of late Pleistocene mammalian extinctions in North America. *Quat. Res.* 19:130-135.

Miller, S. J., and W. Dort, Jr. 1978. Early man at Owl Cave: Current investigations at the Wasden Site, eastern Snake River Plain, Idaho; pp. 129-139. In: *Early Man in America from a Circum-Pacific Perspective*, ed. A. L. Bryan. Dept. Anthropol., Univ. Alberta, Occ. Pap. 1.

Morlan, R. E. 1980. *Taphonomy and Archaeology in the Upper Pleistocene of the Northern*

Yukon Territory: A Glimpse of the Peopling of the New World. Mercury Series, Paper 94. Ottawa: Archaeol. Surv. Can.

Müller-Beck, H. 1966. Paleohunters in America: Origins and diffusion. *Science* 152:1191-1210.

Osborn, H. F. 1936. *Proboscidea*. New York: The American Museum of Natural History.

Palmer, H. A., and J. B. Stoltman. 1976. The Boaz mastodon: A possible association of man and mastodon in Wisconsin. *Mid.-Cont. J. Archaeol*. 1:163-177.

Potts, R. 1983. Home bases and early hominids. *Amer. Sci*. 72:338-347.

Potts, R. 1984. Hominid hunters? Problems in identifying the earliest hunter-gatherers; pp. 129-166. In: *Hominid Evolution and Community Ecolog*, ed. R. Foley. New York: Academic Press.

Potts, R., and P. Shipman. 1981. Cutmarks made by stone tools on bones from Olduvai Gorge, Tanzania. *Nature* 291:577-580.

Quimby, G. I. 1960. *Indian Life in the Upper Great Lakes, 11,000 BC to AD 1800*. Chicago: University of Chicago Press.

Rasmussen, L. E., M. J. Schmidt, R. Henneous, D. Groves, and G. D. Daves, Jr. 1982. Asian bull elephants: Flehman-like responses to extractable components in female elephant estrous urine. *Science* 217:159-162.

Rosenkrantz, R. D. 1977. *Inference, Method and Decision. Towards a Bayesian Philosophy of Science*. Dordrecht, Holland: D. Reidel.

Roth, V. L. 1984. How elephants grow: Heterochrony and the calibration of developmental stages in some living and fossil species. *J. Vert. Paleont*. 41:126-145.

Roughgarden, J. 1979. *Theory of Population Genetics and Evolutionary Ecology: An Introduction*. New York: MacMillan Publ. Co., Inc.

Salmon, W. C. 1975. Confirmation and relevance. *Minn. Stud. Phil. Sci*. 6:3-36.

Sanford, J. T. 1935. The Richmond mastodon. *Proc. Rochester Acad. Sci*. 7:137-156.

Saunders, J. J. 1977. Late Pleistocene vertebrates of the western Ozark Highland, Missouri. *Illinois State Mus., Rep. Invest.* No. 33, 118 pp.

Saunders, J. J. 1979. A close look at ivory. *The Living Museum* 41:56-59.

Saunders, J. J. 1980. A model for man-mammoth relationships in late Pleistocene North America. *Can. J. Anthropol*. 1:87-98.

Schour, I., and M. M. Hoffman. 1939. Studies in tooth development: I. The 16 microns calcification rhythm in the enamel and dentin from fish to man. *J. Dent. Res*. 18:91-102.

Sellards, E. H. 1938. Artifacts associated with fossil elephants. *Geol. Soc. Amer. Bull*. 49:999-1010.

Shipman, P. 1981. Applications for scanning electron microscopy to taphonomic problems. *Ann. N. Y. Acad. Sci*. 376:357-386.

Shipman, P. 1983. Early hominid lifestyle: Hunting and gathering or foraging and scavenging; pp. 31-50. In: *Animals and Archeology*, Vol. 1, ed. J. Clutton-Brock and C. Grigson. London: British Archaeological Reports.

Shipman, P. 1986. Scavenging or hunting in early hominids: Theoretical framework and tests. *Amer. Anthropol*. 88:27-43.

Shipman, P., D. C. Fisher, and J. J. Rose. 1984. Mastodon butchery: Microscopic evidence of carcass processing and bone tool use. *Paleobiology* 10:358-365.

Short, R. V. 1966. Oestrous behaviour, ovulation and the formation of the corpus luteum in the African elephant, *Loxodonta africana*. *E. Afr. Wildl. J.* 4:56-68.

Sikes, S. K. 1971. *The Natural History of the African Elephant*. London: Weidenfeld and Nicolson.

Soffer, O. 1984. *The Upper Paleolithic of the Central Russian Plain*. Ph.D. Dissertation, City University of New York, New York.

Speth, J. D. 1983. *Bison Kills and Bone Counts. Decision Making by Ancient Hunters*. Chicago: University of Chicago Press.

Speth, J. D., and K. A. Spielmann. 1983. Energy source, protein metabolism, and hunter-gather subsistence strategies. *J. Anthrop. Archaeol.* 2:1-31.

Spiess, A. 1976. Determining season of death of archaeological fauna by analysis of teeth. *Arctic* 29:53-55.

Stanford, D. 1979. The Selby and Dutton sites: Evidence for a pre-Clovis occupation of the High Plains; pp. 101-123. In: *Pre-Llano Cultures of the Americas: Paradoxes and Possibilities*, ed. R. L. Humphrey and D. Stanford. Washington, D. C.: Anthropol. Soc.

Stanford, D., R. Bonnichsen, and R. E. Morlan. 1981. The Ginsberg experiment: Modern and prehistoric evidence of a bone-flaking technology. *Science* 212:438-440.

Taylor, J. G., ed. 1969. William Turner's journeys to the caribou country with the Labrador Eskimos in 1780. *Ethnohistory* 16:141-164.

Van Valen, L. 1964. Age in two fossil horse populations. *Acta Zoologica* 45:93-106.

Voorhies, M. R. 1969. Taphonomy and population dynamics of an early Pliocene vertebrate fauna, Knox County, Nebraska. *University of Wyoming Special Contributions to Geology Paper* 1:1-69.

Vrba, E. S. 1980. The significance of bovid remains as indicators of environment and predation patterns; pp. 247-271. In: *Fossils in the Making: Vertebrate Taphonomy and Paleoecology*, ed. A. K. Behrensmeyer and A. P. Hill. Chicago: University of Chicago Press.

Watson, G. S., and E. J. Williams. 1956. On the construction of significance tests on the circle and the sphere. *Biometrika* 43:344-352.

Wheeler, S., and G. S. Watson. 1964. A distribution-free two-sample test on a circle. *Biometrika* 51:256-257.

Williams, S. 1957. The Island 35 Mastodon: Its bearing on the age of Archaic cultures in the East. *Am. Antiq.* 22:359-372.

Winterhalder, B., and E. A. Smith, eds. *Hunter-gatherer Foraging Strategies: Ethnographic and Archaeological Analyses*. Chicago: University of Chicago Press.

Yellen, J. E. 1977. Cultural patterning in faunal remains: Evidence from the !Kung Bushmen; pp. 271-331. In: Experimental Archaeology, ed. D. Ingersoll, J. E. Yellen, and W. Macdonald. New York: Columbia University Press.

Taphonomy and Hunting

Anna K. Behrensmeyer

Paleobiology Department
Smithsonian Institution
Washington, D.C.

The terms "meat-eating" and "hunting" are generally used in anthropology in the same sense as "carnivory" and "predation" are used in zoology. This book symbolizes the importance that anthropologists give to hunting as a behavioral adaptation in human evolution. Meat-eating *per se* seems to provoke less controversy in discussions about early humans than whether the meat was acquired by hunting or scavenging.

The fossil record contains far more taphonomic evidence for carnivory than for how carnivores acquired their food. I will review paleontological evidence for carnivory as it relates to questions of meat-eating by humans and their ancestors, particularly the question of hunting versus scavenging. Taphonomic evidence is of critical importance in arguments about early hominid diet and also about the timing of human immigration into North America. Fossil and recent taphonomic evidence from bone assemblages unaffected by humans provides one of the most important comparative bases for judging what evidence *is* diagnostic of human activity. Such evidence will be examined for its strengths and weaknesses with respect to the issue of early human hunting and meat-eating.

PALEONTOLOGY AND CARNIVORY

A predator is an animal that hunts, and the objects of this activity are its prey. A scavenger is an animal that utilizes a static resource, i.e., animals already dead that do not have to be pursued, stalked, or killed. Predators and scavengers are carnivores because they use other animals as food. Many mammalian carnivores both hunt and scavenge, depending on circumstance, and the distinction between these adaptive modes is less important to ecologists and ethologists than it appears to be to anthropologists. Vertebrate paleontologists have generally assumed that adaptations for meat-eating meant predation, but new information on the flexibility of modern carnivores indicates that both hunting and scavenging are possible for fossil species.

Determining carnivore behavior from fossils has been a long-standing interest in paleontology. It is usually read from the tooth and cranial structure as well as other anatomical features indicating that the species belongs in one of the carnivorous taxonomic groups. Rarely, preserved stomach contents provide even more direct evidence of meat-eating, as in *Strobodon stirtoni* from the Miocene of Nebraska, which was fossilized with a juvenile *Merycodus* in its stomach (R. Emry, pers. comm.). Strontium in bone apatite has been proposed as an indicator of carnivory since less strontium is available from meat than from plants (Parker and Toots 1980; Sillen and Kavanagh 1982). Microscopic wear patterns on teeth can also indicate meat and/or bone-eating (B. Van Valkenburgh, pers. comm.). These recent techniques have been developed because of interest in determining meat-eating adaptations in animals with unknown dietary habits, especially early hominids.

When considering hunting versus scavenging in a fossil carnivore, paleontologists interpret predatory behavior from specialized functional adaptations such as stabbing canines and from locomotor specializations for pursuit or capture prey. Conversely, bone-crushing premolars and skull-jaw structure indicate scavenging. Ancestry may also be invoked, e.g., cats are usually predators, hyenas are predator-scavengers. Interest in the evolution of the predator guild (Van Valkenburgh 1985) is encouraging paleontologists to develop ways to distinguish carnivores that were primarily hunters from those that were primarily scavengers. Van Valkenburgh is investigating tooth-wear scratches as a method to identify bone crushers and meat-slicers. Damage to carnivore teeth may also suggest how meat was acquired; broken canines are more typical of hunters (not uncommon in sabre-toothed cats) and chipped premolars in scavengers (Van Valkenburgh, pers. comm.).

Paleontologists rarely use taphonomy in interpretations of carnivory because carnivores usually are not preserved with the debris from their meat-eating activities. Given functional evidence, there is little need to resort to more indirect taphonomic methods. However, the mammalian fossil record does provide examples of peculiar bone concentrations with taphonomic features attributed to the activities of meat-eaters.

Recent owls leave their pellets at favored roosting sites (often caves). These pellets consist of distinctive associations of bones of small vertebrates, often with damage that is specific to particular owl species (Brain 1981; Dodson and Wexlar 1979). Fossil assemblages of micro-vertebrates attributed to owl carnivory are relatively common, especially in caves (Brain 1981; Levinson 1982). The predatory (as opposed to scavenging) behavior of owls is apparent from their selection of a narrow range of body sizes and, in some cases, prey species.

The cranial and dental anatomy of sabre-toothed cats strongly indicates that they were predators. However, what they preyed on and how they used their canines has been subject to debate (Ewer 1955, 1967, 1973; Emerson and Radinsky 1980). A Late Pleistocene fauna from Friesenhahn Cave, Bexar County, Texas (Evans 1961), provides taphonomic evidence for the prey of one of these cats, *Homotherium*. The assemblage, representing "several hundred" individuals, includes 441 teeth of juvenile *Mammuthus*, of which only 14 are adults. The few preserved postcranial parts are disarticulated, scattered, and bear probable tooth marks. Remains of other large mammals (bison, deer, horse, etc.) are all adult. A complete *Homotherium* skeleton and remains of juveniles were found in the same deposit, and this species is the only predator large enough to carry remains of juvenile elephants into the cave. Thus, taphonomic evidence supports the interpretation that *Homotherium* was a specialized hunter of mammoths that used Friesenhahn Cave as its den.

Similar reasoning is used in Vrba's (1976) hypothesis on hunting by sabre-toothed cats and in Brain's (1981) discussion of lion-like *Dinofelis* as a possible hunter of australopithecines in South Africa. Concentrations of particular species of animal and/or their body sizes (baboon, hyrax, and the small fossil antelope *Antidorcas*) are attributed to particular predators such as leopards (Brain 1980, 1981; Klein 1980). Patterns of bone damage are also used in support of such interpretations.

The following illustrate some common themes in the use of taphonomic evidence for interpretations of predation: (1) The context of the bone assemblage limits the number of bone-concentrating processes (i.e., in cave situations); (2) There is a concentration of remains of one species, body size group and/or age-class of the proposed prey animals; (3) The damage to bones is consistent with the size and morphology of the proposed predator; and (4) The remains of the proposed predator (especially juveniles) are present, indicating denning activity.

ANTHROPOLOGY AND HUNTING

Many anthropologists seem preoccupied with the details of the carnivorous behavior of prehistoric humans. These humans had no obvious morphological adaptations for carnivory (although there may have been subtle ones; see Trinkaus, this volume). In the absence of obvious hunting implements (i.e., before the Upper Pleistocene), hypotheses about carnivory must be based on evidence provided by taphonomic study of bone assemblages. Paleontologists are keenly aware of the difficulties of this approach, given the extent of taphonomic variability in bone assemblages not involving humans. However, the intense anthropological search for more information from bones has led to a rapid increase in knowledge of human versus nonhuman taphonomic processes and the signatures of these processes on recent and fossil assemblages (e.g., Potts 1984). Interpretations of early human behavior which anthopologists base on taphonomic evidence go far beyond what has been attempted in vertebrate paleontology.

A paleontologist might ask why it is important to establish "significant utilization of animal protein" or "successful hunting" for human ancestors. Many animals that normally do not eat meat occasionally are scavengers or even predators, therefore the anthropological concern seems to involve carnivorous behavior beyond the limits of casual opportunism. The ability to hunt implies something about the elusive quality of "being human," a behavioral and technological sophistication beyond that of other animals, including other primates, that are not morphologically adapted for hunting. Scavenging implies "creative foraging," perhaps, and tool use, but less behavioral sophistication than hunting - or at least less of what we might easily recognize as human behavior.

There are two general problems in the human evolutionary story which currently depend heavily on taphonomic analysis: (1) establishing the degree to which the carnivorous activities of hominids influenced bone assemblages preserved with stone artifacts (such as dense stone-bone concentrations at Olduvai Gorge); and (2) demonstrating evidence for human activity in bone assemblages where stone artifacts or other cultural materials are absent (such as various Wisconsin fossil sites that bear on the timing of Paleo-Indian entry into North America; see Morlan, this volume).

These two problems are controversial and have spawned numerous archeological studies and taphonomic investigations in living ecosystems. Formerly there was relatively little interchange between researchers with common taphonomic problems in North America and in Africa, but now there is a rapidly increasing international pool of information on taphonomic processes relevant to archeological site formation.

Has this increase in knowledge provided us with reliable indicators of human carnivory, hunting, or scavenging activity? Judging from current controversy and other papers in this volume, the answer appears to be "not yet," although there is a great deal of unexplored potential in the accumulated taphonomic evidence. In the following section I will review the problems and potential of such evidence as it is used in anthropological hypotheses on human behavior, especially carnivory. Taphonomic features will be considered independently of such evidence as stone tools and other cultural materials.

TAPHONOMIC EVIDENCE FOR CARNIVORY

The major types of evidence relating to carnivore activity summarized in Table 1 are used to analyze fossil bone assemblages and are applied to bones associated with stone artifacts, bone "tools," or breakage patterns that might represent human activity. Interpretation of this evidence is based on observation of cause and effect observed in recent analogues (e.g., hyena bone-collecting and den assemblages) or on inferred effects of known processes (e.g., trampling). Recent taphonomic observation and experimentation have shown many of the early inferences on processes affecting bones to be overly simplistic (the classic example being Dart's osteodontokeratic culture; Dart 1957). Establishing diagnostic criteria for specific processes, especially for human activities, has been complicated rather than clarified by increasing information because patterns thought to be caused by specific processes are now known to be caused by several processes. However, we are accumulating the basis for a more realistic interpretation of what is undeniably indirect and problematic evidence.

Spatial Concentrations of Bones

Formerly researchers had few conceptual alternatives for explaining dense patches of fossil bones other than concentration by stream action, natural traps, and human (or other) meat-eating activities. Now it is well known that many biological and physical processes concentrate bones on land surfaces, and fewer initial assumptions are made about the cause of any particular patch. Since the term "concentration" or "patch" implies more bones per unit area than in the "background" distributions, it is important to consider the overall spectrum of bone occurrence in defining what is a significant patch. This can also help to identify concentrations that might be indicative of unique processes (Isaac and Crader 1981; Potts 1984).

The different kinds of bone occurrences on a landscape, grouped according to causal processes, are summarized in Figure 1, and information on

Table 1. Taphonomic evidence used by archeologists and taphonomists to support hypotheses about carnivory.

TAPHONOMIC EVIDENCE	EXAMPLE
Dense Concentrations of Bones	- Dens, Eating Areas
Bone Modification	- Breakage, Chewing, Cutmarks
Skeletal Part Frequencies	- More Limb Elements in Dens
Species and Sizes of Animals	- Selection of Particular Species
Age Spectra in Animals	- Juveniles of Large Species
Season of Death	- Increased Mortality During Low Natural Stress Time (Fall vs. Spring)

the taphonomic features of such occurrences is given in Table 2. The lowest density of bones is represented in attritional deaths and kills dispersed throughout an ecosystem ("background"), and the highest in mass-death or natural trap situations such as caves. Studies of recent bones in Amboseli Park, Kenya (Behrensmeyer et al., 1979; Behrensmeyer and Dechant Boaz 1980; Behrensmeyer 1983), provide order-of-magnitude estimates for bones/m^2 for some of these occurrences.

Individual kill and death sites are scattered over the land surfaces of all six major habitats in Amboseli Park, reflecting the chance position of a predation event or the place where an animal died of other causes. During droughts, death sites may be relatively common. After death, skeletons tend to be consumed and/or dispersed more rapidly at kill sites. Depending on the subsequent availability of fresh carcasses to scavengers, skeletons may remain relatively complete until they are no longer a worthwhile food resource. At kill sites, the area of bone scatter is typically large (hundreds of square meters), but this depends on the types and numbers of carnivores, and on the interest of scavengers over a period of several years after the kill. Because bones of larger animals such as elephants retain food value longer they can be more widely dispersed than bones of smaller species.

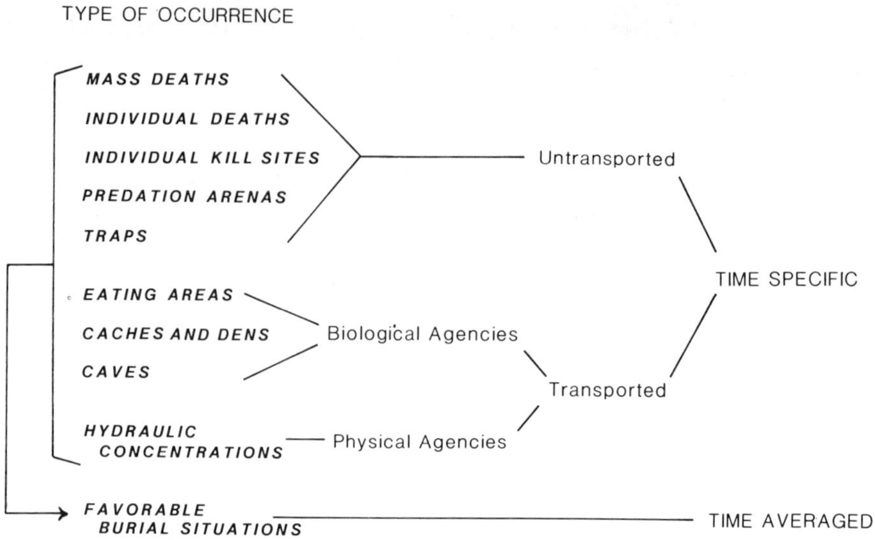

Figure 1. The major types of bone occurence, for animals larger than 15 kg, classified according to causes and how these relate to biological or physical agencies, transport and time. Time-averaging can increase the density of bones in any of these occurrences and lead to combinations of different causes of concentration.

Predators in Amboseli (and elsewhere, e.g., Schaller 1967) tend to favor particular places where there is optimal cover (for stalkers) or open ground (for pursuers) and abundant prey (e.g., near waterholes). More kill sites in such "predation arenas" lead to more carcasses through time, although surface densities do not approach those of eating areas or dens (Table 2). Similar densities of bones from non-predation deaths can occur because weakened animals congregate near critical resources such as waterholes.

Bone densities well above background can occur because carnivores transport bones to particular places. In Amboseli Park, "eating areas" with numerous bones occur under isolated *Acacia tortilis* trees in the Plains and Open Woodland habitats (Figure 2). Here the larger carnivores bring whole or partial carcasses for consumption, and such trees also are used by predatory birds. The main attraction for the mammals is probably shade, but protection is

Table 2. General taphonomic characteristics of different types of bone occurrence. The bones/m^2 figures are estimates based on current knowledge; they provide order-of-magnitude comparisons for relative densities of bones in different types of occurrences. Diversity refers to numbers of different species of animals represented. Under Major Body Parts, "Ax" refers to vertebrae and ribs (not crania) and "Lb" to limb elements; ">" means that one grouping is often more common than the other and "+" means that both groups typically occur in frequencies similar to those in an intact skeleton. Under Bone Modification, Pred. = modification by a primary predator, Scav. = scavenger modification, and Trmp. = trample modification.

SITE TYPE	DENSITY OF BONES/M^2	DIVERSITY	MAJOR BODY PARTS	BONE MODIFICATION
BACKGROUND	≤.005*	Mod.-High	Ax>Lb	Pred., Scav, Trmp.
MASS DEATH	10-100	Low	Ax+Lb	Scav., Trmp.
INDIVIDUAL DEATH	1-20*	—	Ax+Lb	Scav., Trmp.
INDIVIDUAL KILL	1-20*	—	Ax>Lb	Pred., Scav., Trmp.
PREDATION ARENA	.01*	Mod.	Ax+Lb	Pred., Scav., Trmp.
EATING AREA	.06*	Mod.	Ax<Lb	Pred. , Scav., Trmp.
CACHE	(1-5)*	Low	Lb	Scav., Pred.
DEN	75**	High	Lb>Ax	Scav., Pred.
HYDRAULIC CONCENTRATION	Var.	Var.	Lb>Ax	Above + Particle Abrasion
TIME AVERAGED	10-30***	Mod.-High	Lb>Ax	All Above

*AMBOSELI PARK

**HILL 1984

***MIOCENE ASSEMBLAGE

also a factor. Bushes beneath these trees offer cover for "lying-up" and for ambushes. Bones that accumulate in such places are often extensively trampled by ungulates also attracted to the shade. Weathering rates are slower and burial rates in the soft litter below the tree are higher than in surrounding areas. The degree of concentration of bones in the context of this "common amenity" (Isaac 1983) also depends in part on the number and spacing of such trees on a savanna.

Caching of carcass parts under water is known for hyenas (Kruuk 1972: 119). In Amboseli there are occasional low-density concentrations of relatively complete limb parts found in clumps of grass or bushes, apparently also cached by hyenas. Such occurrences have much lower densities than dens, where large numbers of transported and modified bones can accumulate in and around caves and burrows due to carnivores and bone-collecting porcupines. There are only a few published studies of bone accumulations associated with hyena dens (e.g., Sutcliffe, 1970; Hill 1983, 1984; Bunn 1982). I am not aware that hyenas accumulate bones in places where there are no subterranean burrows, i.e., on flat land surfaces resembling Pleistocene "open-air" sites such as FLK-Zinj at Olduvai Gorge (Bunn 1982, 1983).

Hydraulic processes are thought to concentrate bones by selective transport and winnowing (Hanson 1980), but this has not yet been adequately demonstrated by observation or experiment. Current action *alone* may not be capable of assembling dispersed, unmineralized bones into dense "lag" accumulations due to the relatively low density of bones as sedimentary particles (Behrensmeyer 1975). Based on my observations and experiments in recent rivers, I believe that hydraulic concentration occurs primarily when currents modify patches of bones already assembled by biological processes.

Abrasion on bones reflects the length of time they remain unburied, the intensity of particle impact, and processes such as trampling and weathering. It is a poor indicator of distance of transport (Behrensmeyer 1982), probably because much of the abrasion takes place while the bone is stablized rather than moving at speeds comparable to other clasts. Abraded bones need not indicate transport from one place to another, and this bears upon arguments concerning "derived" versus "primary" components of archeological sites such as

Figure 2. An " eating area" under an isolated *Acacia tortilis* tree in Amboseli Park, Kenya. Bones were concentrated here through transport and consumption of carcasses by various carnivores over many years.

Torralba and Ambrona (see Binford, Klein, this volume).

Accumulation of bones from any of the above processes extended through long time periods (time-averaging) can lead to high densities of bones. Under burial conditions favorable to time-averaging (e.g., soft substrate, frequent trampling, low net rates of sediment accumulation), bones from widely separated times can be preferentially buried in the same place. Estimates from the dispersed kill/death assemblage in Amboseli Park indicate that it would take 10,000 years, at a burial rate of 2-6% of the surface assemblage, to accumulate an average of 1-5 bones/m^2 in the upper part of the soil (Behrensmeyer 1983). However, in situations such as predation arenas and eating areas, where input and burial rates are significantly higher than background, tens to hundreds of years might be sufficient to create time-averaged assemblages comparable in numbers of bones/m^2 to dens. Time-averaging also means, of course, that different causes of death and bone accumulation will likely be represented in the resulting fossil assemblage.

Types of Bone Modification

Bone modification includes breakage patterns and all types of damage to bone surfaces. Breakage was believed an indicator of taphonomic activities of humans (e.g., Dart 1957; Sadek-Kooros 1966), but Myers et al. (1980) and Brain (1981) showed that breakage patterns (such as spiral fractures) typically reflect structural properties of the bones themselves rather than the process doing the breaking. Thus breakage *patterns* of individual bones are not stressed as evidence in most current hypotheses on human activity versus other processes. However, large *numbers of bone fragments*, particularly of limbs, are thought to indicate human activity (Vrba 1976; Klein 1980; Bunn 1982). Some archeologists (e.g., Bunn et al. 1980; Kroll and Isaac 1984) use conjoining bone pieces and their spatial distribution within a bone patch as evidence for local fragmentation by carnivores or hominid bone-smashing. Trampling has been added to the list of important processes in bone breakage (e.g., Brain 1967; Behrensmeyer and Dechant Boaz 1980; Bunn 1982; Fiorillo 1984). Trampling can create fractures in elephant bones (Haynes, pers. comm.) which resemble some of the breakage patterns attributed to humans (e.g., Morlan 1980, 1986).

Tooth marks on bones are one of the clearest proofs for carnivory, and with macro- and microscopic techniques they can ususally be distinguished from other marks (e.g., Potts and Shipman 1981; Shipman 1981; Shipman and Rose 1983). Scratches, grooves, and scorings of bone surfaces are common in fossils and have been attributed to trampling (e.g, Conybeare and Haynes 1984; Fiorillo 1984; Behrensmeyer et al. 1986, in press). Recent experimental work (Behrensmeyer et al. 1986) and controlled observations of natural trampling (Fiorillo 1984; Andrews and Cook 1985) confirm this interpretation. Trampling marks do not usually resemble toothmarks, but in some cases they can be confused with stone tool cutmarks (Behrensmeyer et al. 1986).

Recent carnivores inflict distinctive patterns of gnawing of bone edges and articular ends and cause rounding of fracture surfaces, as documented in North America (Binford 1981; Haynes 1980, 1982, 1983) and Africa (Hill 1975; Brain 1981; Binford 1984). These patterns reflect both the kind of carnivore and its interest in a particular carcass, which can vary markedly depending on season and longer term fluctuations in predator-scavenger/prey ratios. Although gnawing damage is recognized in a number of fossil assemblages, so far it has been used mainly as "evidence of carnivore activity" rather than for more elaborate interpretations of the intensity of carcass utilization or behavior of hunters versus scavengers.

The subset of surface marks identifiable as cutmarks are the definitive signature of hominid tool use, but not necessarily of the purpose of the tool-user. Shipman (1983) has interpreted marks on nonmeat bearing limb shafts as indicative of removal of tendons and skin by a scavenging hominid. In a number of fresh kills by lions and hyenas observed in Amboseli Park, limbs were disarticulated at the scapular or pelvic joint and pulled from that end so the skin turned inside out (Figure 3), peeling off down to the point where it is attached to tendons in the lower limbs. A hominid wanting to remove the limb would have to cut the skin at this point, leaving "disarticulation" marks on the *shaft*s of radii-ulnae, tibiae, or metapodials. Such a pattern of carcass dismemberment, if common in modern ecosystems, would provide an alternative hypothesis about the cause of cutmarks on limb-shafts.

Given large enough samples and limited overprinting by other taphonomic processes, it should be possible to identify a particular carnivore from its characteristic bone modification, especially if this includes distinctive marks such as cutmarks. The pattern of such marks should also provide evidence for "time of access" to the carcass, whether before or after removal of the bulk of the meat (Binford 1981; Potts 1983, 1984; Bunn and Kroll in press). If we admit the possibility of scavenging whole or nearly whole carcasses, however, marks left during butchery of a scavenged animal probably would not differ from those on carcasses obtained directly by hunting (Potts 1982, 1983, 1984).

Skeletal Part Frequencies

If all the bones in a skeleton could respond independently and randomly to taphonomic processes, there would be a very large number of possible

Figure 3. Wildebeeste carcass one hour after abandonment by the lions that killed it, showing the disarticulation of the left front limb. The limb has been pulled and the skin has turned inside out along the lower leg. A hominid or other animal wishing to remove the leg would have to sever the skin around the lower radius-ulna or metacarpal.

combinations and relative abundances in bone assemblages. Such independent behavior is impossible because of the ways bones are tied, their strength, and their attached muscle (i.e., meat-bearing versus nonmeat-bearing). These constraints limit the possible patterns of relative abundance, as discussed originally by Brain (1967, 1969).

However, the relative frequencies of skeletal parts can be sensitive to different taphonomic processes such as carnivory, transport, trampling, weathering, and burial. Relative frequencies (or skeletal part ratios) have been used extensively as evidence for carnivore behavior, including the presumed activities of early hominids (e.g., Binford 1981, 1984; Bunn 1982; Klein 1978; Bunn and Kroll 1986). A major problem with such interpretations is that there can be many different causes of similar skeletal part frequencies. This is supported by documentation of skeletal part abundances from denning activities in carnivores (Brain 1981; Bunn 1982; Bunn and Kroll 1986) and background bone scatters from attritional deaths (Hill 1975; Gifford 1977; Behrensmeyer

1983). Skeletal part sorting also occurs during current transport and is a function of bone size, shape and density (Voorhies 1969; Behrensmeyer 1975; Hanson 1980). The physical strength of bones (Brain 1967, 1969) has considerable effect on their abundances and may be the dominant signal in assemblages subjected to multiple taphonomic processes.

Another problem with skeletal part data is that skeletal part abundances cannot be validly compared because assemblages of broken bones are not treated uniformly by different workers, e.g., when limb shaft pieces are counted or not counted (Bunn and Kroll 1986). There are attempts to standardize recording and presentation of skeletal part data (e.g., Gifford and Crader 1977; Klein and Cruz Uribe 1984), and many workers are now more careful to report what goes into their decisions regarding numbers of different skeletal parts.

Binford (1981, 1984), Bunn (1982), Potts (1984), and Bunn and Kroll (1986) have focused on skeletal part groupings rather than individual bones, using ratios of major body parts classed as cranial, axial and limb, and ratios of limb elements classed as meat-bearing or nonmeat-bearing. The concentration in fossil assemblages of such lower limb parts as metapodials and podials has been proposed as an indicator of "marginal scavenger behavior" in hominids (Binford 1981, 1984, 1985). Based on observations in Amboseli Park and simple deduction from known carnivore behavior (e.g., Kruuk 1972), there are at least three possible causes of lower limb part concentration (Figure 4). The middle taphonomic pathway might indicate a "minimal scavenger," but the upper and the lower options indicate hunter or scavenger activity followed by further scavenging. In the lower pathway, the relatively greater burial potential of lower limb parts would further enhance their abundance in the preserved record.

Successive taphonomic "overprints" can lead to similar skeletal part abundances because survival depends on morphological properties of these parts (e.g, strength of lower limb parts) more than the taphonomic processes affecting them. This also suggests tests of which hypothesis best fits a given assemblage. If only lower limbs were scavenged for marrow or sinew or both, then many should be broken (and/or cutmarked), and there should be few

Figure 4. Three "taphonomic pathways" that would lead to a concentration of lower limb parts (metapodials, podials, phalanges) at a bone accumulation site. In the upper two pathways, bones are accumulated by predators or scavengers, while in the lower pathway they occur at the kill site.

fragments from other bones in the assemblage. If whole limbs were brought to a site and then eaten and scavenged, there should be fragments of the meatier elements, few of the ends of such bones, and probably more complete lower limb elements. If ease of burial helped to increase the number of lower limb parts in a fossil site, then this might be indicated by lower vertical positions in the excavation profile and by less advanced stages of weathering relative to other body parts.

Species and Body Sizes of Prey

Assemblages of bones representing highly selected samples of prey animals provide compelling evidence for hunting. Determining patterns of prey selection, however, is possible only when a bone sample is large enough to show that the relative abundance of particular species or body sizes is anomalous with respect to "normal" death assemblages. Thus, concentrated remains of baboons, klipspringers, or hyraxes in caves (Brain 1980, 1981) indicate carnivore activity because such concentrations are rarely formed by other causes of mortality in recent ecosystems. Recurring patterns of particular species or body sizes at different sites, such as in owl pellets, is even stronger evidence for the persistent activity of a selective carnivore.

The tendency of a carnivore to select prey of a particular size and/or species implies predation since scavengers must be more opportunistic, taking whatever is available (Vrba 1976; Isaac and Crader 1981). However, this view of scavenging should not be applied too strictly to the past. It is apparent among lions, hyenas, cheetahs, leopards and jackals that "early scavenging" (= prey theft) of freshly killed carcasses is a common occurrence (Kruuk 1972). A scavenger specializing in taking freshly killed prey from a particular predator might create bone assemblages very similar to those of a specialized predator. It is possible that early hominids obtained meat from cheetahs and jackals, if not from the more intimidating predators (Potts 1984). Moreover, during stress periods such as droughts, particular species might be more available for scavenging, and more of their remains would end up in bone assemblages formed at these times.

Documentation of an "anomalous" abundance of species or body sizes in a fossil assemblage depends on reconstruction of relative species abundance from fragmentary remains. Minimum Numbers of Individuals (MNI) and Minimum Animal Units (MAU) have been widely discussed as measures of relative abundance (Binford 1981, 1984; Bunn 1982; Klein and Cruz-Uribe 1984). The methodological problems encountered with such indices may not affect strong patterns in the assemblage but can obscure weaker ones. Also, size-related taphonomic processes preferentially destroy bones of smaller animals, thereby reducing their abundance relative to the original body size distribution at a site (Behrensmeyer et al. 1979).

In her work on South African cave faunas, Vrba (1975, 1976, 1980) proposed criteria for distinguishing scavenging from hunting by comparing body weights and ages of the fossil bovids. Vrba assumed that bones were brought to caves by biological agencies - mainly by carnivores. She reasoned that "primary assemblages" accumulated by predators transporting whole carcasses should have a lower variance in body weight than "secondary assemblages" composed of parts of carcasses brought by scavengers. Based on relatively low weight variance, she interpreted most of the cave assemblages as primary, with larger body-sized prey indicating larger predators such as sabre-toothed cats. One site (SE - Sterkfontein West Pit) has a wide variance in body weight and was classed as a secondary or scavenger accumulation. Since it contained stone artifacts, Vrba speculated that the scavengers were hominids (Vrba 1976).

Later Vrba (1980:269) was concerned about the certainty of correlating a body weight "peak" with a predator accumulation, since this "only becomes definitive when contrasted with the weight distribution of the underlying live population" (Vrba 1980:269). Such comparative data cannot be determined for her cave sites, leaving open the possibility of a significant scavenger component in the supposedly "primary" predator assemblages.

In Brain's (1981) comprehensive treatment of all the faunal remains from the cave sites, he concludes that the bone accumulations resulted from numerous agents, including scavengers. Large numbers of smaller prey animals, particularly primates, led him to propose that a specialized hunter, perhaps the cat *Dinofelis*, was responsible for part of the bone accumulations. Although Brain recognizes the contributions of scavengers, the patterns of species and size selection indicate to him that predators were the major bone-accumulating agents.

Early East African archeological assemblages (Bunn 1982; Potts 1982, 1984; Bunn and Kroll 1986) typically have a wide diversity of species and body sizes, indicating a lack of strong selectivity. At Olduvai most of the fauna is from archeological sites, and comparisons with nonarcheological samples of the vertebrate community are very limited. Faunal samples at Koobi Fora are available from a number of archeological sites and nonarcheological assemblages in different paleoenvironments (Behrensmeyer 1975, 1985; Bunn 1982), but detailed comparative analysis of the proportion of different body-size classes is not yet available.

It has been suggested that abundant remains of relatively small, adult ungulates (e.g., bovids <50 kg) in archeological bone concentrations indicate human hunting since small carcasses are supposedly completely consumed by nonhuman predators, leaving little to scavenge (Isaac and Crader 1981; Bunn and Kroll 1986). In Amboseli Park, the proportion of a carcass left for scavenging depends on the size ratio of predator to prey, number of predators, and the hunger of the predators. A number of defleshed but intact, marrow-bearing, complete and partial skeletons of small bovids such as Thomson's gazelle occurred in surface bone surveys (Figure 5). Whether there are small carcasses to scavenge depends on diverse circumstances of a particular vertebrate community, and it cannot be assumed that these always will be rare. The smaller prey of small predators such as jackal and cheetah would have been likely targets for early scavenging of whole carcasses by hominids.

Age Spectra

Due to the softness (and high protein content) of their bones, juveniles are often more completely destroyed by carnivory and other taphonomic processes than adults of the same body weight (Pienar 1969; Vrba 1976). This results in strong taphonomic biases against the preservation of juveniles. Their percentages at the accumulation site are thus conservative estimates of original numbers and skeletal part representation. Vrba (1975, 1976, 1980) and Brain (1981) use the high frequencies of juveniles in the Sterkfontein Valley sites, along with body weight and species data, to support primary predator hypotheses. Increased percentages of juveniles of larger species provide further evidence of size selectivity by predators.

The numbers of juveniles recorded in early sites in East Africa (Bunn 1982; Potts 1982, 1984) are not unusual in comparison with nonarcheological assemblages or with the recent assemblage in Amboseli Park. So far there is not enough published data on juveniles in nonarcheological assemblages for quantative comparisons.

Klein (1976, 1978, 1980, 1982, this volume) and Klein and Cruz-Uribe

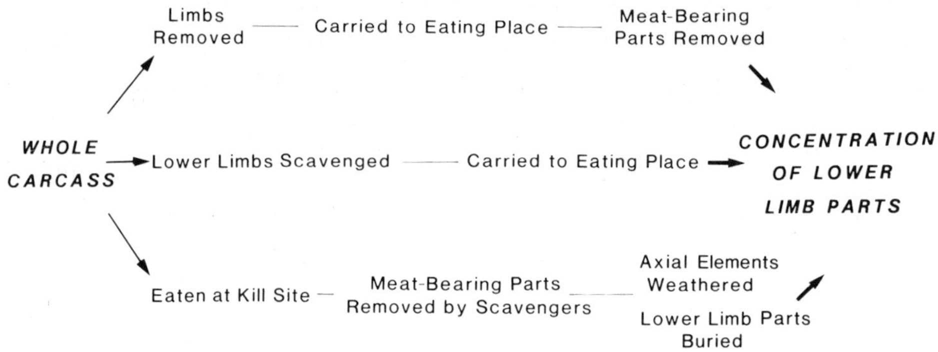

Figure 5. Intact but mostly defleshed carcass of Thomson's gazelle after it was killed and partially eaten by jackals. The limbs are undamaged and a source of marrow for scavengers equipped to break them open.

(1983, 1984) used age categories as evidence of carnivore selectivity in later archeological sites. Based on the high frequency of juvenile remains, primarily teeth, Klein makes a case for predation by Late Pleistocene-Holocene hominids on juvenile buffalo and other large bovid species at Klasies River Mouth and Nelson Bay caves in South Africa. Many of these teeth were from very young individuals, i.e., lacking significant wear on their deciduous teeth (Klein 1978:211), and this supports the predation hypothesis since Klein believes that young individuals were seldom available for scavenging. Binford (1984:213) proposed that the "schlepp effect" would cause the same pattern due to scavenging of heads of smaller (i.e., juvenile) individuals. Since there is little published information on the frequency of juvenile postcranials at these sites, and since such parts may be poorly represented due to preferential destruction relative to teeth, such evidence does not provide strong support for either the hunting or scavenging hypothesis. It seems likely, however, that fewer juveniles than adults would be available for scavenging unless there was frequent theft of juvenile carcasses from smaller predators.

Season of Death

Most ungulates have a seasonal period of birth, and for those with high infant-juvenile mortality, this also implies a concentrated period of death. Little is known about seasonal mortality of African ungulates during "normal" years, but increased mortality during periods of unusual drought is well studied. Ungulates in temperate latitudes are subject to predictable winter stress, and this is the basis for a new hunting hypothsis formulated by Fisher (this volume).

Due to their size, the largest ungulates are relatively free of pressure from nonhuman predation, so with the possible exception of juveniles, their pattern of mortality should correspond to periods of environmental stress. If this is true, as Fisher maintains for his Michigan mastodonts, then any deviation from it suggests a different cause of death, such as predation, disease, or accident. Contextual evidence should help to distinguish among these alternatives, and the strongest support for human predation is a recurring pattern of taphonomic features indicating butchery *plus* an anomalous season of death in a number of different sites.

This evidence would be more difficult to apply to faunal assemblages from tropical latitudes. However, if season of death could be established for animals at an archeological site, (e.g., by analysis of growth lines on teeth that lack closed roots), there might be added evidence for interpreting patterns of carnivore (including hominid) activity. This could, for example, indicate seasonal occupation of the site for meat-eating and bone processing.

SUMMARY AND CONCLUSION

Each type of taphonomic evidence from fossil bone assemblages contributes to interpretations of carnivore activity. From the standpoint of anthropological hypotheses about the carnivorous behavior of early humans, the strengths and weaknesses of such evidence are as follows:

1. Dense concentrations of bones indicate bone collectors but may be

caused by predators, scavengers, or both. With the exception of porcupines, such collectors are carnivores with a wide range of body sizes and functional equipment for carcass transport and utilization. Natural traps, mass deaths, hydraulic processes, and time-averaging can also contribute to dense bone concentrations. Geology and taphonomy may help distinguish these from collector assemblages, but many fossil concentrations represent the combined effects of different processes that are very difficult to disentangle.

2. Bone modification patterns are often diagnostic of carnivore activity, and the degree of modification or destruction indicates the intensity of utilization by one or more bone modifying agents. Repeated patterns of modification in large samples, such as the placement of cutmarks on bones from FLK-Zinj at Olduvai Gorge (Bunn and Kroll 1986), are compelling evidence for the carcass-processing activities of a specific carnivore, i.e., hominids. Such patterns can support hypotheses for time of access to edible parts of a carcass. Evidence of early access may mean predation, but cannot exclude "early" scavenging of meat-bearing carcasses or carcass parts.

3. Ideally, skeletal part frequencies can indicate taphonomic processes such as selective transport of limb elements versus transport of whole carcasses. In practice, however, there are multiple taphonomic overprints that affect skeletal parts. Methods of counting and reporting numbers of bones in fragmented assemblages are also problematic. These factors may result in skeletal part frequencies reflecting the indestructibility (and identifiability) of specific bones more strongly than the effects of any particular taphonomic process. In favorable circumstances, skeletal part frequencies may indicate bone collecting and transporting, but the question of whether the original carcasses were hunted or scavenged cannot be resolved by this evidence alone.

4. Unusual frequencies of particular species or body sizes can be evidence for selective predation by carnivores. However, the strength of this evidence depends on whether there is a basis of comparison for species and body size abundances in the community from which the selected sample was taken. High frequencies of particular taxa might be due simply to large populations during the period of bone accumulation. The transporting abilities of either predators or scavengers can produce a "schlepp effect," another cause of assemblages

with unusual body size distributions. Specialized scavenging from particular predators also might create distinctive size modes. Thus, an assemblage with a strong pattern of body size selection suggests a number of alternatives to selective predation.

5. Age spectra with unusually large numbers of juveniles can be evidence for predation, particularly in a relatively narrow range of species or body sizes in large fossil assemblages. Scavenging could duplicate this pattern if (a) it occurred during unusually high juvenile mortality or (b) if there were early scavenging or "prey theft" from predators that selected juveniles. The assumption that juvenile carcasses are rarely available for scavenging needs further testing in modern ecosystems.

6. New evidence for selective predation by humans is based on documenting a season of death which deviates from a "normal" time of stress. This is only likely to be convincing for animals such as elephants that normally die during stress periods rather than being preyed upon throughout the year.

The taphonomic evidence given above demonstrates the difficulties in distinguishing predation from scavenging. With such limitations, at best we might be able to estimate the relative importance of these two behaviors by documenting percentages of taphonomic characters attributable to hunting, scavenging, and/or indeterminate causes (Isaac and Crader 1981). There are, however, other possible ways to use taphonomic evidence to demonstrate hunting versus scavenging in the fossil record:

1. Research unique *combinations* of evidence which are more indicative of specific meat-acquiring behavior than is any single type of evidence, e.g., isolate associations of skeletal part frequencies *and* cutmark patterns indicative of early or late access to carcasses of particular species, body sizes or age categories.

2. Propose multiple hypotheses, based on knowledge of recent predators and scavengers (including humans) predicting expected carcass-processing patterns and bone debris associated with hunting versus scavenging. The possible variations in butchery following hunting, competitive early scavenging, noncompetitive early scavenging, and late scavenging could be proposed and

tested using recent analogue studies and experimentation.

3. Shift emphasis in research on archeological assemblages from whether carcasses were hunted or scavenged to how they were processed, a problem which can be studied with careful analysis of fossil material and modern analogues. Perhaps the question of hunting versus scavenging is not the most productive question to ask of our data, given the limitations of most fossil samples. Instead, questions such as "What are the recurring modes of carcass utilization by carnivores in the Plio-Pleistocene?" will more likely lead to a significant increase in understanding of ancient hominid behavior.

A number of researchers have begun to make the transition from the oversimplified "hunting versus scavenging" controversy to more productive approaches outlined above. As we learn how to deal effectively with the strengths and weaknesses of the taphonomic evidence, it will tell us more about early human carnivory. Results may differ, however, from what appears most important about meat-eating from the perspective of modern humans.

ACKNOWLEDGEMENTS

Ideas for this paper were drawn from discussions with Glynn Ll. Isaac over our years of collaboration. His inspiration will continue to influence my desire to provide a taphonomic perspective on problems of early human dietary adaptations as well as many other issues. Allison Brooks, John Damuth, Bob Emry, Fred Grady, Kathleen Gordon, Rick Potts, Kathy Stewart, Martha Tappen, Blaire Van Valkenburgh, and John Yellen provided references and discussions essential to the development of this paper, though, of course, I accept full responsibility for the perspective I have presented on taphonomic evidence for hunting versus scavenging.

REFERENCES

Andrews, P., and J. Cook. 1985. Natural modifications to bones in a temperate setting. *Man* 20: 675-691.

Behrensmeyer, A. K. 1975. The taphonomy and paleoecology of Plio-Pleistocene vertebrate assemblages east of Lake Rudolf, Kenya. *Bulletin of the Museum of Comparative Zoology*. 146:473-578.

Behrensmeyer, A. K. 1982. Time resolution in fluvial vertebrate assemblages. *Paleobiology*. 8: 211-227.

Behrensmeyer, A. K. 1983. Patterns of natural bone distribution on recent land surfaces: Implications for archaeological site formation. In: *Animals and Archeology*, Vol. 1, *Hunters and Their Prey*, ed. J. Clutton-Brock and C. Grigson, pp. 93-106. *British Archeological Report* #163.

Behrensmeyer, A. K. 1985. Taphonomy and the paleoecologic reconstruction of hominid habitats in the Koobi Fora Formation. In: *L'environment des Hominides au Plio-Pleistocene*, ed. Y. Coppens, pp. 309-324. Paris: Masson. (Foundation Singer-Polignac).

Behrensmeyer, A. K., and D. E. Dechant Boaz. 1980. The recent bones of Amboseli National Park, Kenya, in relation to East African paleoecology. In: *Fossils in the Making*, ed. A. K. Behrensmeyer and A. Hill, pp. 72-93. Chicago: University of Chicago Press.

Behrensmeyer, A. K., K. D. Gordon and G. Yanagi. 1986. Trampling as a cause of bone surface damage and pseudo-cutmarks. *Nature* 319:768-771.

Behrensmeyer, A. K., D. Western and D. E. Dechant. 1979. New perspectives in vertebrate paleoecology from a recent bone assemblage. *Paleobiology* 5:12-21.

Behrensmeyer, A. K., G. T. Yanagi and K. D. Gordon. In press. Non-human bone modification in Miocene fossils from Pakistan. In: *Bone Modification*, ed. R. Bonnichsen and M. Sorg. Orono: Center for the Study of Early Man, University of Maine at Orono.

Binford, L. R. 1981. *Bones: Ancient Men and Modern Myths*. New York: Academic Press.

Binford, L. R. 1984. *Faunal Remains from Klasies River Mouth*. Orlando: Academic Press.

Binford, L. R. 1985. Human ancestors: Changing views of their behavior. *Journal Anthropological Archaeology*. 4:292-327.

Brain, C. K. 1967. Hottentot food remains and their bearing on the interpretation of fossil bone assemblages. *Scientific Papers of the Namib Desert Research Station* 32: 1-11.

Brain, C. K. 1969. The contribution of Namib Desert Hottentots to an understanding of australopithecine bone accumulations. *Scientific Papers of the Namib Desert Research Station* 39:13-22.

Brain, C. K. 1980. Some criteria for the recognition of bone-collecting agencies in African caves. In: *Fossils in the Making*, ed. A. K. Behrensmeyer and A. Hill, pp. 108-130. Chicago: University of Chicago Press.

Brain, C. K. 1981. *The Hunters or the Hunted? An Introduction to African Cave Taphonomy*. Chicago: University of Chicago Press.

Bunn, H. T. 1982. Meat-eating and Human Evolution: Studies on the Diet and Subsistence Patterns of Plio-Pleistocene Hominids in East Africa. Ph.D. Dissertation, University of California, Berkeley, California.

Bunn, H. T. 1983. Evidence on the diet and subsistence patterns of Plio-Pleistocene hominids at Koobi Fora, Kenya and Olduvai Gorge, Tanzania. In: *Animals and Archeology*, Vol. 1, *Hunters and Their Prey*, ed. J. Clutton-Brock and C. Grigson, pp. 21-30. *British Archeological Report* #163.

Bunn, H. T., J. W. K. Harris, G. Isaac, Z. Kaufulu, E. Kroll, K. Schick, N. Toth, and A. K. Behrensmeyer. 1980. FxJj50: An early Pleistocene site in northern Kenya. *World Archaeology* 12:109-136.

Bunn, H. T., and E. M. Kroll. 1986. Systematic butchery by Plio/Pleistocene hominids at Olduvai Gorge, Tanzania. *Current Anthropology* 27:431-452.

Conybeare, A., and G. Haynes. 1984. Observations on elephant mortality and bones in water holes. *Quaternary Research* 22:189-200.

Dart, R. A. 1957. The osteodontokeratic culture of Australopithecus prometheus. *Transvaal Museum Memoirs*, no. 10.

Dodson, P., and D. Wexlar. 1979. Taphonomic investigations of owl pellets. *Paleobiology* 5: 275-284.

Emerson, S. B., and L. Radinsky. 1980. Functional analysis of sabertooth cranial morphology. *Paleobiology* 6:295-312.

Evans, G. L. 1961. The Friesenhahn Cave. *Bulletin of the Texas Memorial Museum*. No. 2, Pt. 1:7-22. Austin: Museum of the University of Texas.

Ewer, R. F. 1955. The fossil carnivores of the Transvaal caves: Machairodontinae. *Proceedings of the Zoological Society of London*. 124:587-615.

Ewer, R. F. 1967. The fossil hyaenids of Africa - A reappraisal. In: *Background to Evolution in Africa*, ed. W. W. Bishop and J. D. Clark. Chicago: University of Chicago Press.

Ewer, R. F. 1973. *The Carnivores. The World Naturalist*. London: Weidenfeld and Nicolson.

Fiorillo, A. R. 1984. An introduction to the identification of trample marks. *Current Research* 1:47-48.

Gifford, D. P. 1977. Observations of Modern Human Settlements as an Aid to Archaeological Interpretations. Ph.D. Dissertation, University of California, Berkeley, California.

Gifford, D. P., and D. C. Crader. 1977. A computer coding system for archaeological faunal remains. *American Antiquity* 42:225-238.

Haynes, G. 1980. Evidence of carnivore gnawing on Pleistocene and Recent mammalian bones. *Paleobiology* 6:341-351.

Haynes, G. 1982. Utilization and skeletal disturbances of North American prey carcasses. *Arctic* 35:266-281.

Haynes, G. 1983. A guide for differentiating mammalian carnivore taxa responsible for gnaw damage to herbivore limb bones. *Paleobiology* 9:164-172.

Hill, A. P. 1975. Taphonomy of Contemporary and Late Cenozoic East African Vertebrates. Ph.D. Dissertation, Univerity of London.

Hill, A. P. 1980. Early postmortem damage to the remains of some contemporary East African mammals. In: *Fossils in the Making*, ed. A. K. Behrensmeyer and A. Hill, pp. 131-154. Chicago: University of Chicago Press.

Hill, A. P. 1983. Hyaenas and early hominids. In: *Animals and Archeology*, Vol. 1, *Hunters and Their Prey*, eds. J. Clutton-Brock and C. Grigson, pp. 87-93. *British Archeological Report* #163.

Hill, A. P. 1984. Hyaenas and hominids: Taphonomy and hypothesis testing. In: *Hominid Evolution and Community Ecology*, ed. R. Foley, pp. 111-124. New York: Academic Press.

Isaac, G. L. 1983. Bones in contention: Competing explanations for the juxtaposition of early Pleistocene artifacts and faunal remains. In: *Animals and Archeology*, Vol. 1 *Hunters and Their Prey*, ed. J. Clutton-Brock and C. Grigson, pp. 3-20. *British Archeological Report* #163.

Isaac, G. Ll., and D. C. Crader. 1981. To what extent were early hominids carnivorous? An archaeological perspective. In: *Omnivorous Primates: Gathering and Hunting in Human Evolution*, ed. R. Harding and G. Telecki, pp. 37-103. New York: Columbia University Press.

Klein, R. G. 1976. The mammalian fauna of the Klasies River Mouth Sites, southern Cape Province, South Africa. *South African Archaeological Bulletin*. 31:75-98.

Klein, R. G. 1978. Stone Age predation on large African bovids. *Journal of Archaeological Science*. 5:195-217.

Klein, R. G. 1980. The interpretation of mammalian faunas from Stone-Age archeological sites, with special reference to sites in the Southern Cape Province, South Africa. In: *Fossils in the Making*, ed. A. K. Behrensmeyer and A. Hill, pp. 223-272. Chicago: University of Chicago Press.

Klein, R. G. 1982. Age (mortality) profiles as a means of distinguishing hunted species from scavenged ones in Stone Age archeological sites. *Paleobiology* 8:151-158.

Klein, R. G., and K. Cruz-Uribe. 1983. The computation of ungulate age (mortality) profiles from dental crown heights. *Paleobiology* 9:70-78.

Klein, R. G., and K. Cruz-Uribe. 1984. *The Analysis of Animal Bones from Archeological Sites*. Chicago: University of Chicago Press.

Kroll, E. M., and G. Ll. Isaac. 1984. Configurations of artifacts and bones at early Pleistocene sites in East Africa. In: *Intrasite Spatial Analysis in Archaeology*, ed. H. J. Hietala, pp. 4-31. Cambridge: Cambridge University Press.

Kruuk, H. 1972. *The Spotted Hyena*. Chicago: University of Chicago Press.

Levinson, M. 1982. Taphonomy of microvertebrates - from owl pellets to cave breccia. *Annals of the Transvaal Museum* 33:115-121.

Morlan, R. E. 1980. Taphonomy and archaeology in the Upper Pleistocene of northern Yukon Territory: A glimpse of the peopling of the New World. *Archaeological Survey of Canada Paper #* 94, Ottawa: National Museum of Man.

Morlan, R. E. 1986. Pleistocene archaeology in Old Crow Basin: A critical reappraisal. In: *New Evidence for the Pleistocene Peopling of the Americas*, ed. A. L. Bryan, pp. 27-48. Orono: Center for the Study of Early Man, University of Maine at Orono.

Myers, T. P., M. R. Voorhies and R. G. Corner. 1980. Spiral fractures and bone pseudotools at paleontological sites. *American Antiquity*. 45:483-490.

Parker, R. B., and H. Toots. 1980. Trace elements in bones as paleobiological indicators. In: *Fossils in the Making*, ed. A. K. Behrensmeyer and A. Hill, pp. 197-207. Chicago: University of Chicago Press.

Pienar, U. de V. 1969. Predator-prey relationships amongst the larger mammals of the Kruger National Park. *Koedoe* 12:108-176.

Potts, R. B. 1982. Lower Pleistocene Site Formation and Hominid Activities at Olduvai Gorge, Tanzania. Ph.D. Dissertation, Harvard University, Cambridge, Massachusetts.

Potts, R. B. 1983. Foraging for faunal resources by early hominids at Olduvai Gorge, Tanzania. In: *Animals and Archeology*, Vol. 1, *Hunters and Their Prey*, ed. J. Clutton-Brock and C. Grigson, pp. 51-62. *British Archeological Report* #163.

Potts, R. B. 1984. Hominid hunters? Problems of identifying the earliest hunter/gatherers. In: *Hominid Evolution and Community Ecology*, ed. R. Foley, pp. 129-166. London: Academic Press.

Potts, R. B., and Shipman, P. 1981. Cutmarks made by stone tools on bones from Olduvai Gorge, Tanzania. *Nature* 291:577-580.

Sadek-Kooros, H. 1966. Jaguar Cave: An Early Man Site in the Beaverhead Mountains of Idaho. Ph.D. Dissertation, Harvard University, Cambridge, Massachusetts.

Schaller, G. B. 1967. *The Deer and the Tiger*. Chicago: University of Chicago Press.

Shipman, P. 1981. Application of scanning electron microscopy to taphonomic problems. *Annals of the New York Academy of Sciences* 276:357-385.

Shipman, P. 1983. Early hominid lifestyle: Hunting and gathering or foraging and scavenging? In: *Animals and Archaeology,* Vol. 1, *Hunters and Their Prey*, ed. J. Clutton-Brock and C. Grigson, pp. 31-50. *British Archeological Report* #163.

Shipman, P., and J. Rose. 1983. Early hominid hunting, butchering, and carcass-processing behaviors: Approaches to the fossil record. *Journal of Anthropological Archaeology* 2: 57-98.

Sillen, A., and M. Kavanagh. 1982. Strontium and paleodietary research: A review. *Yearbook of Physical Anthropology* 25:67-90.

Sutcliffe, A. J. 1970. Spotted hyaena: Crusher, gnawer, digester and collector of bones. *Nature* 292:57-64.

Van Valkenburgh, B. 1985. Locomotor diversity within past and present guilds of large predatory mammals. *Paleobiology* 11:406-428.

Voorhies, M. 1969. Taphonomy and population dynamics of an early Pliocene vertebrate fauna, Knox County, Nebraska. *Contributions to Geology Special Paper* No. 1. Laramie: University of Wyoming Press.

Vrba, E. S. 1975. Some evidence of chronology and palaeoecology of Sterkfontein, Swartkrans and Kromdraai from the fossil Bovidae. *Nature* 254:301-304.

Vrba, E. S. 1976. The Fossil Bovidae of Sterkfontein, Swartkrans and Kromdraai. *Transvaal Museum Memoirs* 21:1-166. Pretoria: Transvaal Museum.

Vrba, E. S. 1980. The significance of bovid remains as indicators of environment and predation patterns. In: *Fossils in the Making*, ed. A. K. Behrensmeyer and A. Hill, pp. 247-271. Chicago: University of Chicago Press.

Contributors

Anna K. Behrensmeyer, Paleobiology Department, NHB-E207 MRC 121, Smithsonian Institution, Washington, D.C. 20650

Lewis R. Binford, Department of Anthropology, University of New Mexico, Albuquerque, NM 87131

Daniel C. Fisher, Museum of Paleontology, University of Michigan, Ann Arbor, MI 48109

George C. Frison, Department of Anthropology, University of Wyoming, Box 3431, University Station, Laramie, WY 82071

Richard G. Klein, Department of Anthropology, University of Chicago, 1126 E. 59th Street, Chicago, IL 60637

Richard E. Morlan, Archaeological Survey of Canada, Canadian Museum of Civilization, Ottawa, ONT KIA OM8

Matthew H. Nitecki, Department of Geology, Field Museum of Natural History, Roosevelt Road at Lake Shore Drive, Chicago, IL 60605

Lawrence Guy Straus, Department of Anthropology, University of New Mexico, Albuquerque, NM 87131

Erik Trinkaus, Department of Anthropology, University of New Mexico, Albuquerque, NM 87131 and Unité Associé #376 du C.N.R.S., Laboratoire d'Anthropologie, Université de Bordeaux I, 33405 Talence, France

Lawrence C. Todd, Department of Anthropology, University of Denver, Denver, CO 80208

Index